The Active Modeler:
Mathematical Modeling
with Microsoft® Excel

Erich Neuwirth

Deane Arganbright

THOMSON

BROOKS/COLE

Australia • Canada • Mexico • Singapore
Spain • United Kingdom • United States

THOMSON
BROOKS/COLE

Publisher/Executive Editor: Curt Hinrichs
Assistant Editor: Ann Day
Editorial Assistant: Katherine Brayton
Technology Project Manager: Earl Perry
Marketing Manager: Joe Rogove
Marketing Assistant: Jessica Perry
Advertising Project Manager: Bryan Vann
Project Manager, Editorial Production:
Andy Marinkovich
Art Director: Rob Hugel

Print/Media Buyer: Emma Claydon
Permissions Editor: Sarah Harkrader
Production Service: Shepherd, Inc.
Text Designer: John Edeen
Copy Editor: Jeanne Patterson
Cover Designer: Denise Davidson
Cover Image: © Photodisc
Cover Printer: Webcom
Compositor: Shepherd, Inc.
Printer: Webcom

Printed in Canada

1 2 3 4 5 6 7 07 06 05 04 03

For more information about our products, contact us at:
Thomson Learning Academic Resource Center
1-800-423-0563
For permission to use material from this text, contact us by:
Phone: 1-800-730-2214
Fax: 1-800-730-2215
Web: http://www.thomsonrights.com

Library of Congress Control Number: 2003114962

ISBN 0-534-42085-0

Brooks/Cole—Thomson Learning
10 Davis Drive
Belmont, CA 94002-3098
USA

Asia
Thomson Learning
5 Shenton Way #01-01
UIC Building
Singapore 068808

Australia/New Zealand
Thomson Learning
102 Dodds Street
Southbank, Victoria 3006
Australia

Canada
Nelson
1120 Birchmount Road
Toronto, Ontario M1K 5G4
Canada

Europe/Middle East/Africa
Thomson Learning
High Holborn House
50/51 Bedford Row
London WC1R 4LR
United Kingdom

To Gabi and Susan

Brief Contents

Contents

Prologue

Why is there a bicycle on the cover of this book? There are two reasons: one obvious and one not so obvious. The obvious one is that in Section 6.1 we describe a project that involves computing the speed of a bicycle from wheel diameters, gear ratios, and other inputs. But a bicycle is also a useful paradigm for how this book approaches the modeling process. A bicycle is a vehicle with which you can reach a reasonable speed, but you have to supply all of the energy yourself. Cars are faster, but they require external energy sources. Moving at bicycle speed you still are able to view and assimilate the places that you pass by. You can see the finer details of the landscape around you, whereas in a car you tend to disregard a lot of the things you drive by. Most people have a reasonable mental model of how a bicycle works, and they understand it fully. In contrast, almost no one really grasps the finer details of engines, transmissions, and clutches—not to mention computer-driven fuel injection systems and carbon dioxide sensors. The workings of a bicycle are obvious; the workings of a car are mostly hidden under the hood. When a bicycle does not work you can see why it is not working, and most likely you can fix it yourself. With a car, you will probably need a mechanic.

Spreadsheets are similar to bicycles, while most other numerical programs are like cars. With spreadsheets, you can solve problems of reasonable complexity and size. You have to set up the model yourself, and while you are creating it you come to understand all the details of the modeling process. It is obvious to you that when a formula in a cell uses another cell as input, its contents will change when you change the input cell. This is the "engine" behind a spreadsheet. You can see all of the numbers that make up your model; nothing is hidden from your view under the hood. If your model does not work properly, you just trace your formulas visually and you can usually find where the errors are.

One might say that a bicycle is a very human means of transportation—it takes you where you want to go with a minimum of fuss, while engaging you in the process. A spreadsheet is a very human way of performing extensive computations with a powerful tool. Our purpose in writing this book is to introduce you to this tool so that you may join us in exploring the hills and valleys and country lanes of the world of mathematical modeling.

Preface

This book presents a new approach to the field of mathematical modeling. It uses the most widely employed spreadsheet, *Microsoft Excel,* not only as a tool to carry out the numerical computation of models, but also as "conceptual notepaper" for designing the models. A fundamental strength of using a spreadsheet in mathematics lies in its ability to create models in the way we might do by hand, but with a further interactive, visual component. This method of creating models "as we go" is the fundamental paradigm of our book.

The basic philosophy of the book can be characterized as guided discovery learning, using spreadsheets as conceptual tools to develop important mathematical and modeling ideas through a wide range of models. We start with a problem and then build a model of it in a way that makes it easy to see the relation between the original problem and the model on the computer.

Why We Wrote the Book

The software used in mathematics is usually highly specialized and employed only by mathematicians—a major obstacle for potential users of mathematical models who are not mathematicians themselves. We show that "serious" mathematics can be done with a tool that is available on practically every desktop computer, namely a spreadsheet program.

Spreadsheets have a slightly different way of representing mathematical structures that is much more visual and action oriented than mathematical programming languages. Our experiences with students indicate that this approach to modeling is helpful in making mathematical modeling seem more "natural" for non-mathematicians.

Courses in mathematical modeling generally use computer programs to implement models and study their numerical behavior only after mathematical formulation of the problem. In contrast, we employ a more intuitive approach, making use of the computer program a part of the modeling process itself.

Why Spreadsheets?

Modeling is a very important use of mathematics, not only because it produces applicable results for problems in applied fields, but also because many mathematical concepts are best presented and developed by starting with a modeling problem.

Spreadsheets are excellent tools for developing mathematical models because they allow us to implement and develop modeling concepts and techniques in an intuitive manner.

Spreadsheets possess many attributes that make them particularly effective for use in teaching mathematical modeling. First, because spreadsheets are employed so widely, most readers will already be familiar with their general use, and they will not find the concepts of our approach and the related spreadsheet tools difficult to learn. Moreover, spreadsheets are ubiquitous in today's world. They are readily available and are the primary mathematical and modeling tools of the workplace. Thus, skills learned in creating mathematical models on spreadsheets have immediate practical value.

Our main focus is not to provide our readers with well-established models in their final version, but to consider modeling as a process, and believe that building a model by oneself instead of using someone else's is a key activity in mathematics. The purpose of our book is to help our readers to develop skills in model building.

A key aspect of our modeling philosophy is direct manipulation. Spreadsheets are different from most other programs used in mathematics in that they are not language based. Our models are created not by describing the relations in a programming language, but by arranging objects on a spreadsheet grid and indicating the relations by "pointing and clicking." We have designed our arrow notation to faithfully reproduce this constructive aspect of our modeling philosophy.

Innovative Features

- We have created our models using *Microsoft Excel*. One tool that we use is the slider control, or scroll bar, which is used extensively to enhance our models. When a model contains a well-designed graph, the scroll bar allows us to include animation. In essence, this tool allows us to replace "discrete" variation of a parameter's value accomplished by typing a new value into its cell with a "continuous" variation of a parameter through moving the slider on the scroll bar.
- We also illustrate new and creative ways to enhance models through the inclusion of such spreadsheet features as the Data Table, Solver[C1], and Goal Seek commands.
- The principal models of each section are provided in two formats on an accompanying CD in both PC and Macintosh versions. The basic format presents a model just as it is produced from the fundamental construction developed in the book. The amplified format provides an expanded version of a model employing more advanced formatting and presentation styles designed for productive user interaction.
- The CD contains about 40 minutes of video lessons explaining the use of the principal *Excel* tools and techniques employed in designing our models.
- More than 150 exercises are included throughout the book. Some call upon the reader to investigate the model described in the section by experimenting with different values of the parameters, while others ask them to extend the model or to enhance its formatting and graphs. Many sections also contain exercises that ask readers to develop separate models that are in the same

spirit of the primary example of the section. More difficult exercises have been marked with an asterisk [*], and some have been designed specifically for readers who possess knowledge of calculus.

- Construction summaries at the end of each section provide a fast motion view of building models.

Audience

The projects in the book are designed to make the fundamental concepts and techniques of mathematical modeling accessible to a wide audience—at different levels—and from many fields of study. Instructors will find a wide range of models from which to choose, and by selecting appropriate topics and adjusting the pace of a course, an instructor can readily adapt the material to be taught to classes at virtually any undergraduate level, from first-year through fourth-year.

The book can be used in:

- A modeling course that does not emphasize calculus;
- Modeling courses that does employ calculus, differential equations, or linear algebra as a supplementary text;
- An applied mathematics course directed at students in the managerial, life and social sciences;
- A mathematics course for in-service teachers in disciplines involving mathematics or problem solving. The spreadsheet approach affords practicing teachers with an innovative way to refresh and expand their mathematical talents, while providing them with ideas and applications to be integrated into their classroom teaching;
- A self study setting for people interested in mathematical modeling as a personal skill. The video lessons should help one to learn the basic techniques even without an instructor.

Organization of the Book

The book begins with a two-chapter study of dynamical models that build upon concepts such as financial applications and growth processes that will be known to most readers, and concludes with new ways to look at such topics as iterated functions and chaos. Throughout these introductory chapters we introduce new spreadsheet concepts and build spreadsheet modeling skills. All of the necessary basic techniques are illustrated in the video lessons on the accompanying CD.

The topics in the remaining chapters are largely independent of one another, and focus on examples from diverse areas of applications in addition to mathematics. We have strived to select interesting models that are accessible to a general audience and illustrative of the ideas of spreadsheet modeling. We incorporate a broad range of interactive and animated graphical techniques into the examples, providing creative ways to visualize the underlying mathematics. Finally, in an extensive Appendix, we include details about implementations in *Excel* and the use of *Excel's* powerful tools and commands.

Each section of the book describes the construction of a primary model. We develop a description for the spreadsheet implementation of that model in a step-by-step fashion, while simultaneously developing the underlying mathematical concepts. The steps in the construction of each spreadsheet model are described through our arrow diagrams. At the end of each section we present a compact summary of the steps used in a model's construction. Occasionally we insert additional brief discussions of advanced mathematics for further exploration. Most sections finish with exercises stimulating students to practice their skills and even investigate further problems.

To Teachers

Many of the topics in this book will be familiar to you, although the approach may be new. The pace of the introductory sections is designed to allow even spreadsheet novices to get up to speed. Experienced spreadsheets users may be surprised by the versatility and mathematical usefulness of a tool they already know well.

In designing the book we have purposely avoided the use of calculus, with one exception, and have minimized the use of algebra in developing our models. However, spreadsheet models utilizing the book's approach with topics from algebra, calculus, and more advanced mathematics also can be incorporated into courses when the students have adequate backgrounds from these fields.

Moreover, one thing that you will find is that students frequently will create completely unexpected approaches to solving a problem. This is one of the side benefits of using a spreadsheet. It encourages creativity among students as they use an innovative approach for doing mathematics.

To Students

While using the book, not only will you see ways of doing mathematical modeling, but you will also be acquiring spreadsheet skills that are highly valued in their own right beyond the course you are presently taking. Throughout this book we emphasize that spreadsheets are true mathematical tools. In each section you will encounter examples that introduce you to some valuable mathematical ideas. and provide you with the background and motivation to enable you to develop models on your own. We encourage you to be inventive and to try out new ideas and approaches that come to you as you explore and investigate the examples that we have provided, and to find enjoyment in pursuing the mathematics of modeling.

To describe our models we have created a new way of presenting spreadsheet descriptions designed to overcome one of the inherent shortcomings of spreadsheets: while we can implement mathematical expression in a spreadsheet in a very natural manner, it can be difficult to describe the resulting creation in a suggestive way.

Our approach presents a new way to describe a spreadsheet model's construction that closely imitates the way in which we carry out the construction. Thus, if we were describing the computation of taxes, we might first write down values for income and tax rate and then describe the computation of the tax by saying that we multiply "income" by "rate" as we gesture to the respective values. This is more intuitive and less confusing than referring to the physical location of the written values or using an algebraic formula.

To describe the "gesturing" approach that is employed in this book we utilize a new and suggestive notation. To illustrate this with our current example, consider the diagram presented below. The dots represent the sources of the components of a formula, while the arrows indicate where they are used in the formula. Thus, the display shown below for computing the tax is interpreted as "multiply the income by the tax rate."

Income	● 5000
Tax Rate	●	0.15
Tax	↓ * ↓

This display, or "arrow diagram", contains all of the information describing the computation to be performed, so it is fully equivalent to the algebraic description. We point out that we have specially created these arrows for the book's descriptions; they are not a built-in component of a current spreadsheet.

In working with a spreadsheet, all of our constructions are done in a similar fashion. That is, rather than typing in the cell locations of a formula directly, we begin a formula and then use either the mouse or the keyboard to point to the cell locations that we reference as we create the formula. This way we keep our focus on the model itself, rather than on the physical cell locations of its parameters and formulas.

To assist in the development of both modeling concepts and spreadsheet skills, this book presents a series of carefully designed models that describe our approach. As we progress through these examples and applications we will gradually build up a repertoire of modeling techniques and useful spreadsheet tools. In the process we not only build and examine many models that are directly applicable, but we also encounter models that are designed to help us to understand mathematical techniques and concepts.

Acknowledgements

We wish to express our gratitude to those who provided valuable suggestions based on an early draft of the book: Donald Cathcart of Salisbury University, Alan Davies of the University of Hertfordshire, Janice Epstein of Texas A & M University, Jenny Sendova from the Bulgarian Academy of Sciences, Bart Stewart of the U. S. Military Academy, and David Tucker of Midwestern State University. Professor Tucker also provided us with well-appreciated editing ideas. In addition, we wish to thank the Mathematical Association of America for the opportunity to present mini-courses in spreadsheet modeling together with Robert S. Smith of Miami University, the great "mathematical spreadsheets public relations officer," who has supplied us with his beneficial wisdom. Finally, the three people whose encouragement and willingness to engage in discussions and provide constructive criticism have been of great value to us: Wally Feurzeig, Paul Horwitz, and Viera Proulx.

Introductory Models

CHAPTER OUTLINE

1.1 Fibonacci Numbers

Fibonacci numbers appeared in early population models developed by the Italian mathematician Leonardo Fibonacci who lived in Pisa during the late 12th and early 13th centuries. In this introductory section, we supplement our standard description of the model by also providing details of its *Microsoft Excel* implementation together with views of the *Excel* screen displays that are generated.

A nice way of introducing these numbers is to examine the idealized growth of a population of rabbits. We start with one pair of rabbits, a male and a female. We assume that these rabbits will produce another pair of offspring (a male and a female) but only after a given period of time has elapsed during which they will reach adulthood. Later on these offspring in turn will also have offspring of their own, again after the same period of time has passed during which they reach adulthood. We want to know how many pairs of rabbits we will have after one, two, three, . . . periods if during each period each pair of adult rabbits gives birth to a pair of offspring and no rabbits die.

We start our calculations by setting up a spreadsheet table. Our standard display, which is not dependent upon the cell locations that are used, is shown in Figure 1.1, while a typical *Excel* screen display is shown in Figure 1.2.

period	rabbit pairs
1	1
2	1

Figure 1.1

	A	B
1	period	rabbit pairs
2	1	1
3	2	1
4		

Figure 1.2

We start with a pair of newborn rabbits in Period 1. They are still around and reach adulthood in Period 2, but at that time they have not yet produced offspring. Consequently, we still have only one pair of rabbits in the second period. Since our original pair will have become adults in Period 2, they produce a pair of offspring who will be present in Period 3. Thus, in Period 3 we have two pairs of rabbits, the original pair and their first pair of offspring. As a result, we can extend the table and enter another row in the *Excel* spreadsheet, as shown in Figure 1.3 and Figure 1.4.

period	rabbit pairs
1	1
2	1
3	2

Figure 1.3

	A	B
1	period	rabbit pairs
2	1	1
3	2	1
4	3	2

Figure 1.4

Instead of typing the number indicating the period each time, we can make use of the fact that each of the period numbers is just one more than the preceding period number. We can indicate this with the diagram in Figure 1.5 in which the counter for the second period is calculated by adding one to the preceding cell.

period	rabbit pairs
● 1	1
▼ + 1	1

Figure 1.5

We can almost "literally" translate this diagram into a formula. Practically all spreadsheet programs allow us to create a formula in a cell by using the cursor to point at the other cells containing the "input values" and typing the mathematical operators (like +) to create the formula. The screen display showing the formula that is generated in *Excel* is shown in Figure 1.6.

	A	B
1	period	rabbit pairs
2	1	1
3	=A2+1	1

Figure 1.6

Although it is possible to type in the formula as =A2+1, our approach is to create the formula through a "gesturing" method that is independent of cell locations. This approach allows us to focus on a natural way of interacting with our model rather than "following a recipe."

Thus, in Cell A3 we start the formula by typing in =. We then click on the cell that contains the number of the previous period (in this case, Cell A2). The screen display at this point appears in Figure 1.7 showing the partial formula of =A2 with the cell that is referenced highlighted by a dashed border. We finish the formula by typing +1 to obtain the formula =A2+1, which is displayed in Figure 1.8, and then press the Enter key. This generates the number 2 in Cell A3.

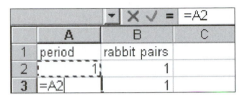

Figure 1.7   Figure 1.8

Next, we create the sequential numbers for the subsequent periods by copying the cell contents of the cell just above. In practice we can do this by clicking on the cell to be copied. There we see a "fill handle" in the lower right corner (see Figure 1.9). We now use the mouse to place the cursor, which has been a wide white plus sign, on the fill handle. At this point when the cursor turns into a thin black plus sign, we hold down the left mouse button and drag it down as far as we want to copy the expression (see Figure 1.10).

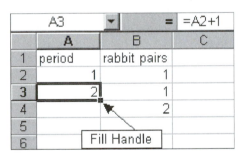

Figure 1.9   Figure 1.10

When we release the mouse, the formula will be copied into the cells, giving the display shown in Figure 1.11. If we display the formulas, as illustrated in Figure 1.12, we see that the cell reference in our formula has been treated as a relative reference in the process, or as "the cell above," to reproduce the formula repeatedly as "the cell above +1."

	A	B
1	period	rabbit pairs
2	1	1
3	2	1
4	3	2
5	4	
6	5	

Figure 1.11

	A	B
1	period	rabbit pairs
2	1	1
3	=A2+1	1
4	=A3+1	2
5	=A4+1	
6	=A5+1	

Figure 1.12

We indicate this copying process by using shading as shown in Figure 1.13. The darker cells show what is copied, while the lighter cells indicate the region into which the formula is copied.

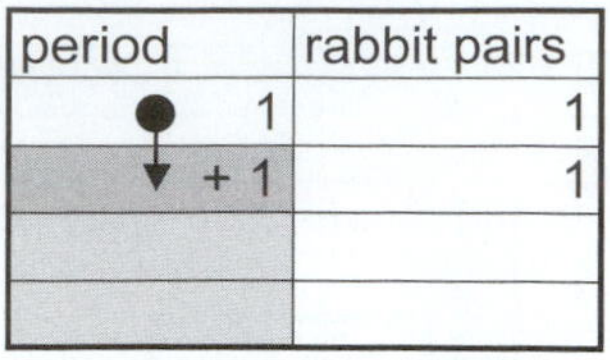

period	rabbit pairs
1	1
+ 1	1

Figure 1.13

Returning to our rabbit population model, how many pairs of rabbits will we have in the fourth period? At least two pairs, because we assume that none of the rabbits will die during the time that we study the system. So in any period we continue to have all of the rabbit pairs from the previous period. But additionally we will have the rabbit offspring that were born in the previous period. How many of these do we get? We know that all mature pairs of rabbits will produce exactly one pair of offspring in each period. So we have to find out how many of the rabbit pairs living in Period 3 will be old enough to have offspring during Period 4. We can determine that by observing that all of the rabbits living in Period 2 will be adults in Period 3. So the number of pairs of rabbits born during Period 4 is the same as the number of pairs of rabbits living in Period 2. Therefore the number of rabbits in Period 4 equals the number of rabbits in Period 2 plus the number of rabbits in Period 3. We can indicate this in a table as shown in Figure 1.14.

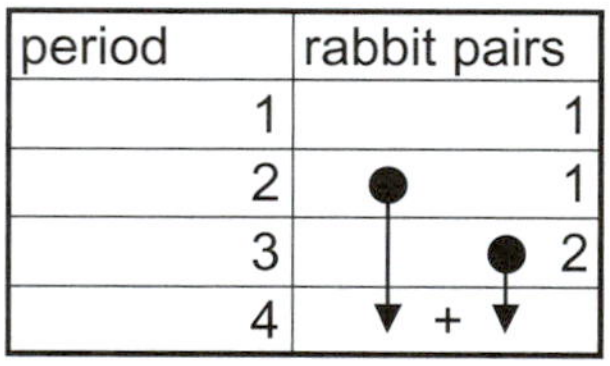

period	rabbit pairs
1	1
2	1
3	2
4	+

Figure 1.14

The computer screen of course will not show this arrow diagram representation of the formula. Instead it will show the resulting values, as shown in Figure 1.15.

period	rabbit pairs
1	1
2	1
3	2
4	3

Figure 1.15

We will use our "gesturing" approach to accomplish the previous step. Thus, in Figure 1.16 we begin by clicking in Cell B5 where we enter the = symbol to indicate that a formula is being entered and then gesture to Cell B3 by clicking in it. We next type + and click on Cell B4 and finish the process by pressing the Enter key (see Figure 1.17). Thus we have a formula that is interpreted as "two cells above" plus "the cell above."

		=	=B3
	A	B	C
1	period	rabbit pairs	
2	1	1	
3	2	1	
4	3	2	
5	4	=B3	

Figure 1.16

		=	=B3+B4
	A	B	C
1	period	rabbit pairs	
2	1	1	
3	2	1	
4	3	2	
5	4	=B3+B4	

Figure 1.17

In reviewing how we get the numbers in the last row of our current table we see that the period number is calculated by adding one to the number above. Just copying this formula down will create a column containing the sequence of numbers 1, 2, 3 . . . . So to get the first column in the way we need it, we simply have to copy the formula down far enough.

The number of rabbit pairs in Period 4 is calculated by adding the number of pairs in the two previous periods. But that is also the general rule for how to calculate the number of rabbit pairs in any period: all the ones from the previous period will still be there, and among them all those who were living during the period before the previous period are old enough to produce offspring. So the new total of pairs of rabbits is the sum of the numbers of pairs of the two preceding periods. The convenient consequence of this fact is that we just have to copy down the formula giving the number of rabbits in Period 4 to get the number of rabbits for all following periods. The table can be expanded as far as needed by copying the formulas as indicated in Figure 1.18 using our copying scheme.

period	rabbit pairs
1	1
2	1
3	2
4	3
5	5
6	8
7	13
8	21

Figure 1.18

To carry out this copying process in *Excel,* we first highlight the block consisting of the two indicated cells in Row 5. To do this we click in the center of Cell A5, hold down the left mouse button, and drag it into Cell B5 (Figure 1.19). We then place the cursor on the fill handle in the lower right corner of that block, hold down on the left mouse button, and drag down as far as we care to (Figure 1.20).

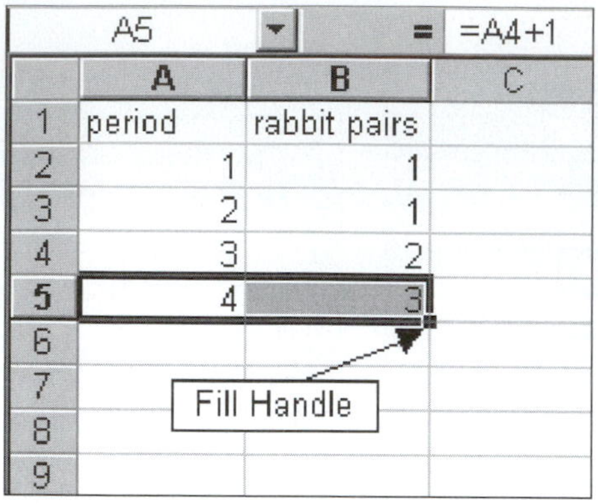

	A5	▼	=	=A4+1
	A	**B**	**C**	
1	period	rabbit pairs		
2	1	1		
3	2	1		
4	3	2		
5	4	3		
6				
7				
8				
9				

Fill Handle

Figure 1.19

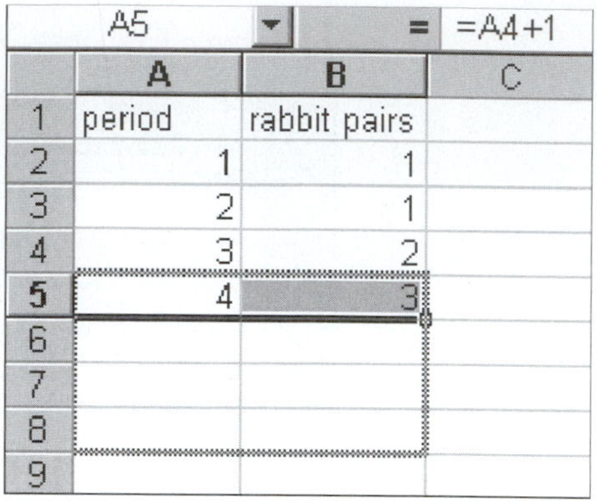

	A5	▼	=	=A4+1
	A	**B**	**C**	
1	period	rabbit pairs		
2	1	1		
3	2	1		
4	3	2		
5	4	3		
6				
7				
8				
9				

Figure 1.20

The resulting output and underlying formulas are shown in Figure 1.21 and Figure 1.22, respectively. We show how to display the *Excel* formulas in the Appendix. Notice that in copying the formulas, each of the cell references has been treated as a relative reference.

	A	B
1	period	rabbit pairs
2	1	1
3	2	1
4	3	2
5	4	3
6	5	5
7	6	8
8	7	13

Figure 1.21

	A	B
1	period	rabbit pairs
2	1	1
3	=A2+1	1
4	=A3+1	2
5	=A4+1	=B3+B4
6	=A5+1	=B4+B5
7	=A6+1	=B5+B6
8	=A7+1	=B6+B7

Figure 1.22

Notice when we copied the formula for Period 4 (see Figure 1.23) what happened (see Figure 1.24).

Copying (or *filling down,* as it is also called in some spreadsheet programs) normally preserves the form of the arrow diagrams representing the formulas. Therefore, the input cells for the copy of the formula and the input cells of the original formula differ. Instead of the absolute position of the input cells, the relative position of the input cells with respect to the formula cell(s) is preserved.

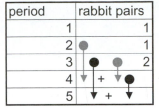

period	rabbit pairs
1	1
2	1
3	2
4	
5	

Figure 1.23

period	rabbit pairs
1	1
2	1
3	2
4	
5	

Figure 1.24

These numbers of rabbit pairs in the second column, starting with 1, 1, 2, 3, 5, 8, . . . , are called the *Fibonacci numbers*. They arise in a surprising variety of settings including the spirals in the arrangements of the seeds on the heads of sunflowers, the segments on pinecones, and the branches on plants.

Now let us look at the Fibonacci numbers in a classical notation. Usually they are defined by the following recurrence formula:

$$a_1 = 1$$
$$a_2 = 1$$
$$a_n = a_{n-1} + a_{n-2}$$

We can use our spreadsheet layout to convey the same idea, as shown in Figure 1.25.

period	rabbit pairs
1	1
2	1
n-2	a_{n-2}
n-1	a_{n-1}
n	a_n

Figure 1.25

The classical formula shows that to get any number beyond the first two we add the two immediately preceding numbers of the Fibonacci sequence. Our arrow diagram notation conveys the same idea without the need to introduce a formal subscript notation.

Note: Instead of entering the number of pairs for Period 3 directly, we could have started to use formulas for this period already. This would have reduced the number of keystrokes necessary to set up the table, but it probably would have made it more difficult to see the essence of the whole problem.

Construction Summary: Fibonacci Numbers

1. Enter 1 as the initial value of the period counter.
2. Increment the counter by adding 1 to the previous value, copy.
3. Enter 1 as the number of pairs of rabbits in the first two periods.
4. Add the two cells above, copy.

	period	rabbit pairs
1	1	1
2	2	1
	3	2
	4	3
	5	5
	6	8
	7	13
	8	21
	9	34

Exercises

1. Supplement our Fibonacci number model with an additional column that computes the ratios of successive Fibonacci numbers. Observe that the ratios appear to converge to a constant value. See if you recognize it. It appears in later sections of the book.

2. Examine the biological model (Figure 1.26) described on pages 8–11 of Edelstein-Keshet (1988). An annual plant produces k seeds in the autumn of a year. Of these seeds, a given proportion (p) will survive over the winter. A proportion (α) of the surviving seeds germinate the following spring to produce new plants, and another proportion (β) of the remaining seeds that survive a second winter germinate the next spring. No seeds survive beyond that period. Create a spreadsheet model to determine the number of plants that will be present in future years. Find a recurrence relation for the number of these plants. For which values of the parameters will the plants die out? The diagram in Figure 1.26 was created through an *Excel* bubble chart to illustrate the situation.

3. In this section's Fibonacci model, the rabbits live forever. Modify the model so that in their fifth period of life the rabbits cease to produce offspring and in the next period they die.

4. Generalize the model in the previous exercise so that the number of years that the rabbits live is a parameter. The concept of parameter is discussed in more detail in the next section.

5. Modify the basic Fibonacci model to investigate the recurrence relation $a_1 = 1$, $a_2 = 1$, $a_n = ua_{n-1} + va_{n-2}$ for $n > 2$, where u and v are non-negative real numbers. Discover any patterns that arise. For which values of u and v does the sequence decrease to 0?

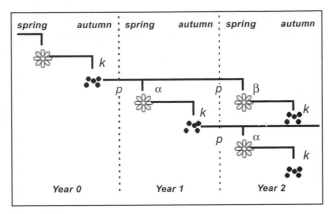

Figure 1.26
Adapted from *Mathematical Models in Biology* by Leah Edelstein-Keshet, McGraw-Hill, 1988, p. 9. Used by permission of the McGraw-Hill Companies.

6. [*] Modify our Fibonacci growth model so that the number of periodic offspring and the survival percentage of the rabbits vary at fixed rates from period to period. See page 13 of Hoppensteadt and Peskin (1992) for a discussion of Euler's renewal equations.

7. Those with advanced spreadsheet graphing skills are invited to use *Excel* to create the chart of Exercise 2 or to design a similar chart using the xy-type chart. You may want to consult later material on creating graphs and the Appendix.

8. Create an xy-chart to illustrate the Fibonacci model. One possible chart is illustrated in Figure 1.27.

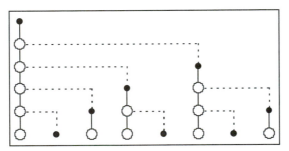

Figure 1.27

1.2 Growth Models

This section deals with simple growth models. One of the most basic growth models involves a population of a species of animals living in a well-defined region. When we start with a certain number of animals, we want to know how many animals we will have after one period, two periods, and so on.

We will express the number of newborn animals in a period as being in proportion to the number of animals already living in our region. For the sake of simplicity, we could assume that in each period our animal population will grow by 5 percent.

Starting with 1000 animals we can set up the table shown in Figure 1.28.

period	animals
1	1000

Figure 1.28

The number of newborn animals is calculated according to the diagram in Figure 1.29, which yields the table in Figure 1.30.

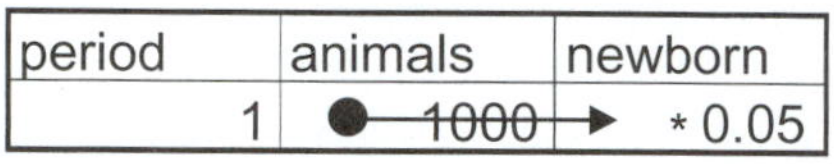

period	animals	newborn
1	●──1000──→	* 0.05

Figure 1.29

period	animals	newborn
1	1000	50

Figure 1.30

In the next line of our table, we need the number of the next period. As in our last example, we create the consecutive numbers by just adding 1 to the number directly above, as shown in Figure 1.31.

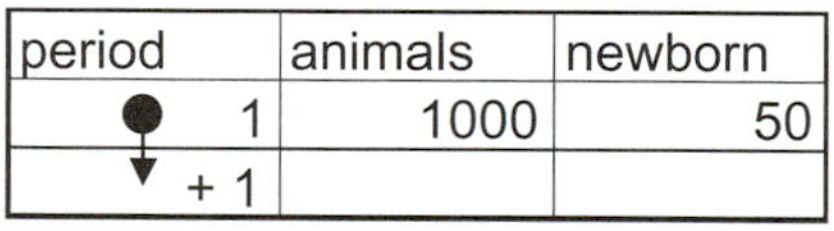

period	animals	newborn
● 1	1000	50
↓ + 1		

Figure 1.31

The number of animals at the beginning of Period 2 is just the sum of the number of animals at the beginning of Period 1 and the number of newborn animals in Period 1 (see Figure 1.32).

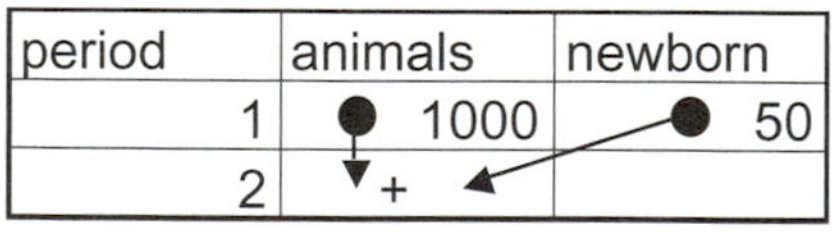

period	animals	newborn
1	● 1000	● 50
2	↓ + ←	

Figure 1.32

To calculate the number of newborn animals in Period 2, we can copy down the formula giving the number of newborn animals in Period 1 to multiply the number of existing animals with the "birth rate" (see Figure 1.33).

period	animals	newborn
1	1000	50
2	1050	

Figure 1.33

All other rows in this table can be obtained by copying down the row for Period 2, since our formulas for Row 2 were designed so as to calculate all the new values either from the values for the previous period or from the same period (for the newborn animals). We can represent this by the diagram in Figure 1.34.

period	animals	newborn
1	1000	50
2	1050	52.5

Figure 1.34

The beginning of the table as it appears on our screen is shown in Figure 1.35.

period	animals	newborn
1	1000	50
2	1050	52.5
3	1102.5	55.125
4	1157.625	57.88125
5	1215.506	60.77531

Figure 1.35

For simplicity's sake, we ignore the fact that fractional parts of animals are impossible in reality.

This mathematical model describes not only simple biological growth effects but also compound interest calculations. Instead of studying the number of animals, we study the amount of money in a savings account. The interest per period corresponds to the newborn animals, and the interest rate corresponds to the birth rate in the biological model.

The formulas do not change since the interest amount is found by multiplying the interest rate and the current amount, and the new amount is determined by adding the old amount and the interest amount. What we need to do to change our biological model into a financial model is to adjust the headings (see Figure 1.36).

period	amount	interest
1	1000	50
2	1050	52.5
3	1102.5	55.125
4	1157.625	57.88125
5	1215.506	60.77531

Figure 1.36

This model has the interest rate (5% or 0.05 in our case) written explicitly into the formulas. How can we calculate the effect of different interest rates on the amounts? We have to change the formulas in the third column of our table. We could enter a new formula (for example, for an interest rate of 7% or 0.07) in the row for Period 1 and then copy this formula down again for all the following periods. But this is somewhat cumbersome. It would be much more convenient to get the interest rate into each of these cells from a single cell so that all of the interest values calculated change automatically when we change the interest rate in this single cell. We have to be careful to achieve this goal.

Look at the table in Figure 1.37 where we set aside a separate cell for the parameter interest rate.

interest rate		0.05
period	amount	interest
1	1000	*
2		

Figure 1.37

When we copy the formula for the interest in Period 1 into the next row, the reference to the interest rate shifts away from 0.05 as shown in Figure 1.38.

interest rate		0.05
period	amount	interest
1	1000	50
2	1050	*

Figure 1.38

In other words, copying this formula down the column produces either an incorrect value or an error value! So what we need now is a way to "nail down" or "glue" the reference in the original formula to the cell with the interest rate.

We need a formula that may be represented as shown in Figure 1.39, where we "pin down" the reference to the interest rate.

interest rate		0.05
period	amount	interest
1	1000	*
2		

Figure 1.39

After copying, the formula that we obtain in the cell below is represented as shown in Figure 1.40.

interest rate		0.05
period	amount	interest
1	1000	50
2	1050	*

Figure 1.40

Therefore, when copying the formula, the absolute (instead of the relative) position of this input cell is preserved. All spreadsheet programs provide a mechanism to make a reference to another cell absolute when setting up a formula with the pointing method; this is usually done by pressing one of the function keys directly after the cursor has been moved to the input cell when setting up the formula. In *Excel*, we can create an absolute reference by pressing the F4 key. At this time *Excel* provides $ signs to the cell reference, for example $C$1, to indicate an absolute reference. This topic is discussed in the Appendix.

Setting up our formula for the interest amount in such a way (using the facility of making absolute references), we can again copy down the formulas from Row 2 into as many rows below as we are interested in. Changing the single value in the cell for the interest rate now will automatically recalculate the model and give all the values of the new interest rate.

The interest rate in our model would usually be considered a *model constant,* or *parameter,* whereas the period, the amount, and the interest amount are considered *model variables.* Generally speaking, in many spreadsheet models, the model parameters are represented as absolute references in formulas and the model variables are represented as relative references. Later on in this book, we will see that it is also possible to name constants.

In classical mathematical notation, this model is expressed by the following equations:

$$a_{n+1} = a_n + d_n$$
$$d_n = \alpha \cdot a_n$$

where

a_n is the amount at the beginning of period n,

d_n is the change of the amount during period n, and

α is the interest rate (per period).

Algebra can be used to show that the value of a_n is given by $a_n = (1 + \alpha)^n a_0$. Because of this, the resulting form of increase is often called either *exponential* or *geometric growth*.

In using this model, note that the interest rate must correspond to the compounding period. Thus, if interest is compounded once a year at a 4% annual rate, then the periods represent years with 0.04 being the periodic rate. However, if 4% annual interest is compounded four times a year, then the periods represent quarters (that is, 3-month periods) and the periodic rate is $0.04/4 = 0.01$.

A little-known but very useful spreadsheet tool that we can employ at this time is the data table. The screen display in Figure 1.41 was generated using this feature. After using the techniques presented in this section to compute a savings account's balance and annual interest for 10 years, in Column E we then create a list of interest rates that are of interest to us. The Data Table command then causes *Excel* to substitute these rates one at a time into Cell B1, recompute the model, and generate the resulting 10-year balance in Column F. This value is set by the formula that is entered as the prototype into Cell F3, that is, =B13. A discussion of how to use the Data Table command is presented in the Appendix.

	A	B	C	D	E	F
1	Rate	0.06			Summary	
2	Year	Balance	Interest		Rate	Balance
3	0	1000.00	60.00			1790.85
4	1	1060.00	63.60		0.01	1104.62
5	2	1123.60	67.42		0.02	1218.99
6	3	1191.02	71.46		0.03	1343.92
7	4	1262.48	75.75		0.04	1480.24
8	5	1338.23	80.29		0.05	1628.89
9	6	1418.52	85.11		0.06	1790.85
10	7	1503.63	90.22		0.07	1967.15
11	8	1593.85	95.63		0.08	2158.92
12	9	1689.48	101.37		0.09	2367.36
13	10	1790.85	107.45		0.10	2593.74

Figure 1.41

Finally, it is possible to extend the number of periods of the model, enhance the format of the output, and create an animated graph of the results. The formatted model of this section on the accompanying CD contains such a graph. A description of the process used to create these graphs is described in the Appendix using the model of the next section. Illustrative output similar to that on the CD is shown in Figure 1.42. The scroll bar is linked to the growth rate. Moving the slider on the scroll bar adjusts the rate and, thereby, the resulting curve in a smooth, almost continuous fashion.

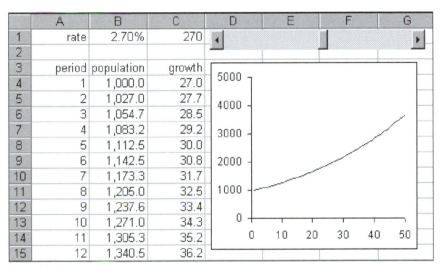

Figure 1.42

Construction Summary: Exponential Growth

1. Enter the periodic interest rate.
2. Create column as period counter.
3. Enter the onetime deposit.
4. Compute interest by multiplying the amount by the rate.
5. Compute the new amount by adding the previous interest to the previous amount.
6. Copy as indicated.

interest rate			1	0.05
period		amount		interest
2	1	3	1,000.00	4	50.00
	2	5	1,050.00		52.50
	3		1,102.50		55.13
	4		1,157.63		57.88
	5		1,215.51		60.78
	6		1,276.28		63.81

Exercises

1. Often one describes growth models by starting the period count at 0 and making the original deposit at the end of that period. Also, providing additional columns as we do in Figure 1.43 can give us new insights into a model. Design a model whose output appears in the figure. Notice that we have formatted the output.

rate	0.05		
period	start balance	interest	end balance
0			1,000.00
1	1,000.00	50.00	1,050.00
2	1,050.00	52.50	1,102.50
3	1,102.50	55.13	1,157.63
4	1,157.63	57.88	1,215.51

Figure 1.43

2. Modify your model in Exercise 1 to allow for interest compounded n times per year.

3. Create a model to compute the growth of a regular sequence of annual payments, A, into a savings account that earns interest compounded annually at the rate of r per year. Compare your results with those obtained using *Excel*'s built-in financial functions. Generate a summary of the results in a Data Table.

4. Modify Exercise 3 to allow for n periods per year in which a regular payment is made and interest is computed.

5. An individual takes out a loan of P at the annual interest rate of α. Design a spreadsheet model to show the interest and the outstanding balance of a loan each year if the individual makes an annual payment of the amount A. Use your model to determine the size of the annual payment A that is needed to pay off the loan in n years. Use either trial-and-error, built-in functions, or the solver and goal seek commands with your model.

6. A savings account pays 5% interest compounded annually. How much should an individual deposit in the account as a onetime deposit so that 20 annual payments can be withdrawn, leaving the balance at 0 after 20 years? Create a spreadsheet model to investigate this question.

7. The owner of a winning lottery ticket can either accept $n = 20$ annual payments of size $p = \$100,000$ or an immediate single payment, P. Determine the size of the single payment if the lottery organization can invest the money at the rate of $\alpha = 0.05$ per year for 20 years. Then modify your model so that n, p, α, and P are parameters.

8. Create additional models for various financial computations that arise in business, including loans, annuities, and such concepts as present and future values. Allow for different periods of compounding interest.

9. Design effective graphs that incorporate scroll bars for some of the exercises in this section.

10. Individuals who have acquired advanced skills in spreadsheet graphics are encouraged to design alternative graphs to show the effects of geometric growth using current data for countries of interest. Link the growth rate to a scroll bar. Three possibilities are illustrated in Figure 1.44 for Papua New Guinea using an initial population of 5 million, and an annual growth rate of 2.4%. In the first, a point is plotted for the current year in addition to the standard curve. In the second, the spreadsheet's random number generator is used to produce one dot for each million people. As the population grows, the available space fills up. The third draws the relative size of land per person as the years increase. The use of a scroll bar to vary the growth rate makes these into particularly useful visualization tools.

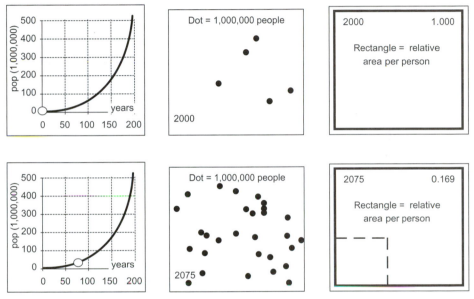

Figure 1.44

11. The same approach used in modeling growth problems can be applied to modeling a wide range of other situations in which a quantity varies over time. Although the following problems are typically found in differential equations books (see Zill (2001)), we can obtain very good approximations to their solutions following the concepts presented in this section. Create spreadsheet models for each of the following settings. Where possible, design the models so that any given values are entered via parameters. Readers with skills in differential equations are encouraged to compare their results with analytic solutions to the differential equations. They also may examine texts in this field to obtain other modeling problems. Incorporate scroll bars effectively.

   a. An object whose temperature is 100° F is placed in a room whose temperature is 20° F. Newton's Law of Cooling states that the object will cool at a rate proportional to the difference in the temperature of the object and that of the surrounding medium. Using time steps of 0.1 minutes, approximate the temperature of the object over a period of time if it is known that its temperature has fallen to 40° F after 5 minutes. Hint: One can determine the constant of proportionality either by trial and error or by using the spreadsheet's solver or goal seek tools.

b. A 300-gallon tank filled with water initially contains 50 pounds of dissolved salt. Salt water containing 2 pounds of salt per gallon is pumped into the tank at the constant rate of 3 gallons per minute, and the well-stirred solution is pumped out at the same rate. Find the amount of salt in the tank as time increases. How much salt will be contained in the tank in the long run? How long will it take for the amount of salt in the tank to reach 500 pounds? (See Zill (2001), p. 100.)

c. Bacteria increase at a rate proportional to the amount present. If initially the amount of bacteria is 100 grams and after 1 hour the level has reached 150 grams, find the amount of bacteria over the next several hours. Use time steps of size 0.1 hours. (See Zill (2001), p. 97.)

d. The amount of carbon 14 in an object decays at a rate proportional to the age of the object. If half of the carbon 14 is gone after 56 centuries, approximate the age of an object that has lost a given percent of its carbon. (See Zill (2001))

e. Tank A and Tank B each contain 50 gallons of water (see Figure 1.45). Initially Tank A contains 25 pounds of salt and there is no salt in Tank B. Pure water is pumped into Tank A at the rate of 3 gallons per minute. A well-stirred mixture from Tank A is pumped into Tank B at the rate of 4 gallons per minute. The mixture in Tank B is pumped into Tank A at the rate of 1 gallon per minute and out of the system at the rate of 3 gallons per minute. Find the amount of salt in each tank over a period of time. Create a graph showing the amounts in each of the tanks. (See Zill (2001), p. 122.)

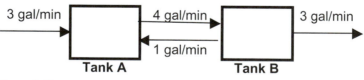

Figure 1.45

1.3 Extended Growth Models

The fundamental growth model of the previous section produces a population that increases without limit. Since such population growth cannot be sustained in nature, we modify our assumption that the growth rate remains constant over time. We replace it by the assumption that there is a limit to the size of a population that can be supported and that the growth rate will be reduced continuously as that limit is approached.

In our previous example, we had one variable (the number of animals or the amount in an account) whose value increased over successive periods. The change from one period to the next was directly proportional to the last value of this variable. Sometimes this variable is referred to as the *state variable*. So the change of the variable is the result of multiplying the state by the birth rate (or the interest rate).

If we write down the corresponding equations in an informal way we get the following equation:

$$\text{change} = \text{current state} \times \text{growth rate}$$

or equivalently:

$$\frac{\text{change}}{\text{current state}} = \text{growth rate}.$$

The last equation expresses the most important fact about our simple model: the growth rate is assumed to be constant. In many cases this is an oversimplification. Therefore we will try to adjust our model to deal with a nonconstant growth rate.

One of the most fundamental models based on a nonconstant growth rate additionally assumes that there is an upper limit for the state variable. For example, we can think of an environment that can supply food and space only for a certain number of animals. We also could imagine a bacteria culture that only allows for a certain number of bacteria. In this kind of situation, the capacity limit not only influences the behavior of the system when the limit is reached, but also tends to reduce the growth rate when the state variable approaches the limiting level. Using this capacity, we can define the saturation of the model:

$$\text{saturation} = \frac{\text{current state}}{\text{capacity}}.$$

The saturation can assume values between 0 and 1, and 1 is reached once the state variable (for example, the number of animals) has reached the capacity limit.

As opposed to the constant growth model, this model assumes that the growth rate decreases as the saturation increases. The "initial" growth rate essentially describes the growth when the system state is far below the capacity limit. When the system reaches a saturation level of $\frac{1}{2}$, the "real" growth rate is only half of the initial value; when the saturation level is $\frac{3}{4}$, the growth rate is only $\frac{1}{4}$ of the initial growth rate; and so on.

Thus, we have

$$\text{growth rate} = \frac{\text{change}}{\text{current state}} = \text{initial growth rate} \times (1 - \text{saturation})$$

and

$$\text{saturation} = \frac{\text{current state}}{\text{capacity}}.$$

To set up a spreadsheet model for our system, we need numerical values for the initial growth rate and the capacity. Additionally, we need a starting value for the state (that is, the number of animals). So let us set up the table shown in Figure 1.46.

initial growth rate		0.05
capacity		5000
period	animals	
1	1000	

Figure 1.46

Before we can calculate the change in the state variable we need the saturation, which is calculated as shown in Figure 1.47.

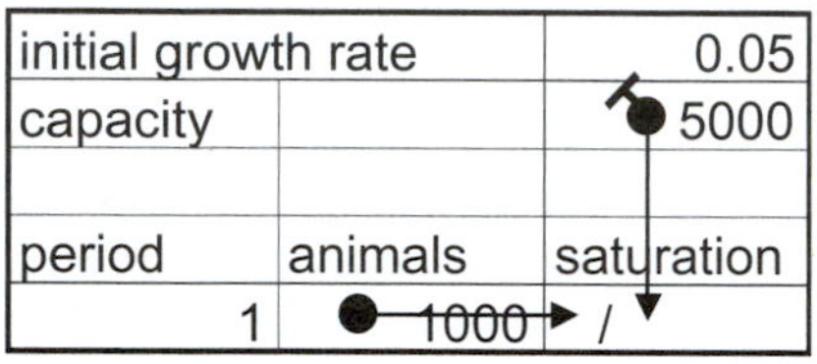

initial growth rate		0.05
capacity		5000
period	animals	saturation
1	1000	/

Figure 1.47

Now we can calculate the change, as shown in Figure 1.48.

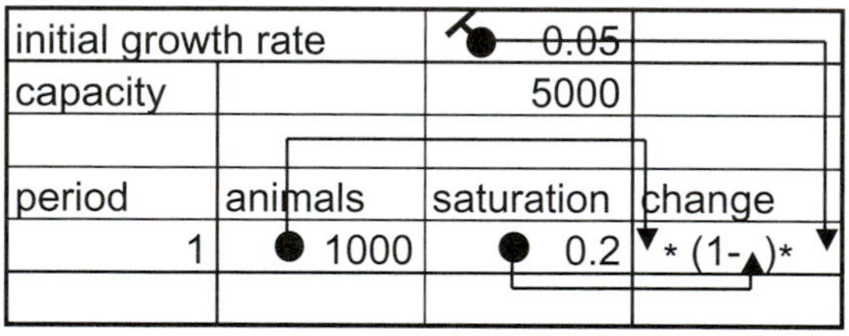

initial growth rate		0.05	
capacity		5000	
period	animals	saturation	change
1	1000	0.2	* (1-▲)*

Figure 1.48

For Period 2, we have to calculate the period number and the new number of animals (see Figure 1.49).

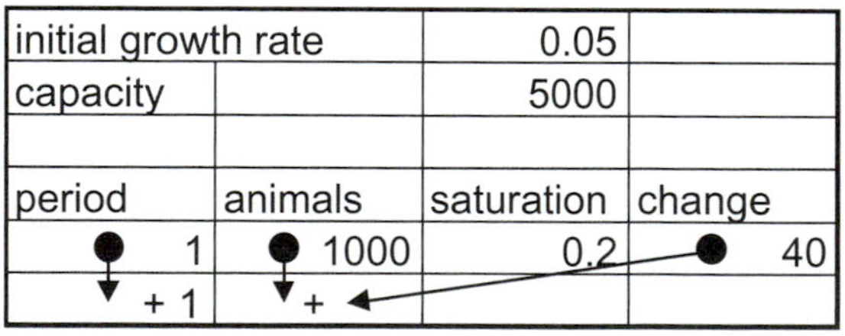

initial growth rate		0.05	
capacity		5000	
period	animals	saturation	change
1	1000	0.2	40
+ 1	+		

Figure 1.49

Saturation and change are calculated from the current number of animals. Therefore we copy down the formulas from the line for Period 1, as shown in Figure 1.50.

initial growth rate		0.05	
capacity		5000	
period	animals	saturation	change
1	1000	0.2	40
2	1040		

Figure 1.50

The line for Period 2 now contains the full description of how the system changes from a current state to the next state, and therefore we can copy it down as far as we need (see Figure 1.51).

initial growth rate		0.05	
capacity		5000	
period	animals	saturation	change
1	1000	0.2	40
2	1040	0.208	41.184
3	1081.184	0.216237	42.36961
4	1123.554	0.224711	43.55395
5	1167.108	0.233422	44.73398
6	1211.842	0.242368	45.90648
7	1257.748	0.25155	47.0681
8	1304.816	0.260963	48.21535

Figure 1.51

We can run the system over 20, 40, or even 100 periods and the amount of work will be independent of the length of the model run.

It is also interesting to see the development of our system graphically. The usual way is to chart the time on the *x*-axis and the number of animals on the *y*-axis. In most spreadsheet programs, this is done by highlighting the columns of the data to be charted and selecting a menu item indicating the type of graph to be drawn. For our example, the relevant data are in the first and the second columns, and we need an xy-graph. The marking of the values to be charted can be done visually. If we wanted to graph our system for eight periods, we would have to mark the shaded region in our sheet shown in Figure 1.52.

initial growth rate		0.05	
capacity		5000	
period	animals	saturation	change
1	1000	0.2	40
2	1040	0.208	41.184
3	1081.184	0.216237	42.36961
4	1123.554	0.224711	43.55395
5	1167.108	0.233422	44.73398
6	1211.842	0.242368	45.90648
7	1257.748	0.25155	47.0681
8	1304.816	0.260963	48.21535
99	4861.512	0.972302	6.732605
100	4868.245	0.973649	6.414169

Figure 1.52

For our model with the values given in Figure 1.52, we should get a graph similar to the one in Figure 1.53. The creation of such a graph is described in the Appendix.

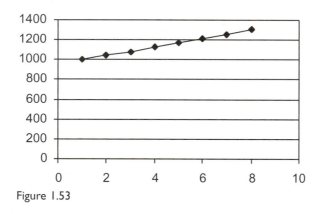

Figure 1.53

Once our graph is set up, we can easily change the values for the initial growth rate and the capacity to see how the numerical values and the graph describing the system behavior change.

To learn more about the curve of a growth phenomenon, it is important to study not just a few points on the curve but many points. So the next step in our model is to extend it to 100 points by copying down the last line sufficiently far. Graphing this larger model with the current parameter values gives the picture in Figure 1.54. A substantially embellished version of this model and its graphs is discussed in the Appendix and provided on the accompanying CD. The S shape of this curve is characteristic for this kind of model.

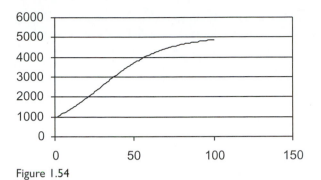

Figure 1.54

In classical mathematical notation, our system could be written in the following way:

$$s_n = \frac{a_n}{C}$$

$$d_n = a_n(1 - s_n)\alpha$$

$$a_{n+1} = a_n + d_n$$

where

a_n is the number of animals at Time n,

C is the capacity of the system,

s_n is the saturation level of the system at Time n,

α is the initial growth rate, and

d_n is the change in the number of animals during period n.

We can simplify this model algebraically by substituting the equation for s_n into the equation for d_n, giving

$$d_n = a_n\left(1 - \frac{a_n}{C}\right)\alpha.$$

By defining

$$D = \frac{a}{C}, \qquad \frac{\alpha}{C} \quad ?$$

we can rewrite the equation

$$d_n = Da_n (C - a_n).$$

Then $a_{n+1} = a_n(1 + D(c - a_n))$

This equation is called the *logistic equation*, and the resulting increase is called *logistic growth*, since the solution of the corresponding differential equation is the logistic function. The difficulty with this equation is that it is not easy to see how the model was developed originally once it is transformed into this form. In the equation, α and C are "natural" parameters for our model, while D is not "natural" any more. It only makes the model easier to deal with algebraically. Quite often discussions of growth models start with these equations for describing growth with capacity constraints, but the danger in doing so is that the connection between the growth rate and the saturation level is overlooked. Describing the model with the saturation made explicit has the advantage of making it more understandable and the disadvantage of having more formulas to deal with. Since spreadsheet programs make it easier to deal with a number of formulas, this disadvantage becomes less important.

Knowing the equation also can give us some more insight into details of the process. Using simple differential calculus, we easily see that

$$d_n = a_n \left(1 - \frac{a_n}{C} \right) \alpha$$

has its maximal value for

$$a_n = C/2$$

so the growth curve has its steepest ascent when it passes through half of the saturation level. Approaching this level from the left, the curve gets steeper and steeper; and going beyond that level and to the right, the curve becomes flatter and flatter; and that is the mathematical explanation for the S shape of this curve.

The examples given so far illustrate an important concept. Quite often a *dynamic system,* that is, a system changing over time, can be described by an equation calculating the change for one period from the current state of the system. Such an equation is called a *difference equation* because it connects the difference of the system states with the system states.

Construction Summary: Logistic Growth

1. Enter parameter values.
2. Use column as period counter.
3. Enter the initial population size.
4. Compute the saturation level by dividing population by capacity.
5. Compute the current growth rate by multiplying the initial rate by (1 − saturation), and multiply that by the population to compute the change.
6. Add the change to the previous population.
7. Copy as indicated.

initial growth rate	1	0.05	
capacity		5000	
period	animals	saturation	change
2 1	3 1000	4 0.200	5 40.0
2	6 1040.00	0.208	41.2
3	1081.18	0.216	42.4
4	1123.55	0.225	43.6
5	1167.11	0.233	44.7
6	1211.84	0.242	45.9
7	1257.75	0.252	47.1
8	1304.82	0.261	48.2

Exercises

1. Modify the logistic growth model to incorporate a time delay in the growth factor. For example, from Period $n = 5$ onward, use the rate based on the population that existed five periods earlier. Include an xy-graph showing the growth. Notice the effect that this causes, and investigate how the long-term result is influenced by the base growth rate. Use normal logistic growth for the first five periods. Examine Renshaw (1991) for additional sources of similar applications.

2. [*] Repeat Exercise 1 using the number, n, of periods of delay as a parameter of the model.

3. The logistic growth model is an excellent model in which to incorporate scroll bars to produce animation in investigating logistic growth. Graphics and format for this example are discussed in the Appendix, and a layout is shown in Figure 1.55. Design some additional useful scroll bars on your own.

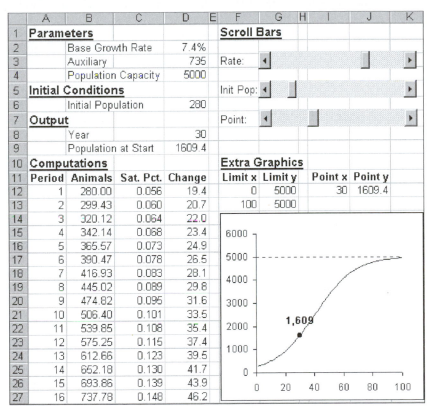

Figure 1.55

1.4 Comparing Different Growth Models

In this section, we create a spreadsheet model and graph that enable us to compare the two different growth models studied in the previous sections. To accomplish this, we have to set up both models in the same spreadsheet table. (Figure 1.56)

growth rate		0.05		initial growth rate		0.05
capacity		5000		capacity		5000
period	model1	change1		model2	saturation	change2
1	1000			1000		

Figure 1.56

Model 1 will have a constant growth rate, Model 2 will have a growth rate depending upon saturation. So far the table contains only the starting value. Let us add the formulas for Period 1 as shown in Figure 1.57 and Figure 1.58.

growth rate		0.05		initial growth rate		0.05
capacity		5000		capacity		5000
period	model1	change1		model2	saturation	change2
1	1000 ▶ *			1000 ▶ /		

Figure 1.57

growth rate		0.05		initial growth rate		0.05
capacity		5000		capacity		5000
period	model1	change1		model2	saturation	change2
1	1000	50		1000	0.2	*(1-)*

Figure 1.58

For Period 2 we have to set up the formulas shown in Figure 1.59.

growth rate		0.05		initial growth rate		0.05
capacity		5000		capacity		5000
period	model1	change1		model2	saturation	change2
1	1000	50		1000	0.2	40
+ 1	+			+		

Figure 1.59

The formulas for the changes and for the saturation in Period 2 can be copied down from the row for Period 1 (see Figure 1.60).

growth rate		0.05		initial growth rate		0.05
capacity		5000		capacity		5000
period	model1	change1		model2	saturation	change2
1	1000	50		1000	0.2	40
1	1050			1040		

Figure 1.60

Now we have to copy down the formulas for Period 2 as far as we need (Figure 1.61).

growth rate		0.05		initial growth rate		0.05
capacity		5000		capacity		5000
period	model1	change1		model2	saturation	change2
1	1000	50		1000	0.2	40
2	1050	52.5		1040	0.208	41.184
3	1102.5	55.125		1081.184	0.216237	42.36961
4	1157.625	57.88125		1123.554	0.224711	43.55395
5	1215.506	60.77531		1167.108	0.233422	44.73398
6	1276.282	63.81408		1211.842	0.242368	45.90648
7	1340.096	67.00478		1257.748	0.25155	47.0681
8	1407.1	70.35502		1304.816	0.260963	48.21535

Figure 1.61

In the next step, we will create a graph for Model 1, as it was done in the previous example. We have to mark the values for the x-axis and for the y-axis (see Figure 1.62).

growth rate		0.05		initial growth rate		0.05
capacity		5000		capacity		5000
period	model1	change1		model2	saturation	change2
1	1000	50		1000	0.2	40
2	1050	52.5		1040	0.208	41.184
3	1102.5	55.125		1081.184	0.216237	42.36961
4	1157.625	57.88125		1123.554	0.224711	43.55395
5	1215.506	60.77531		1167.108	0.233422	44.73398
6	1276.282	63.81408		1211.842	0.242368	45.90648
7	1340.096	67.00478		1257.748	0.25155	47.0681
8	1407.1	70.35502		1304.816	0.260963	48.21535

Figure 1.62

Then we can create the xy-graph for the constant growth rate model (Figure 1.63).

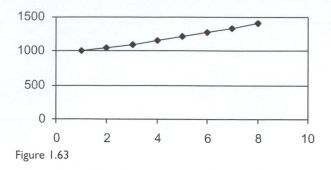

Figure 1.63

Now we want to add a second curve, corresponding to Model 2, to our graph. Since we use the same *x*-axis values, this can be done easily—we only have to mark the data values for the *y*-axis for our second data series and drag them into the existing chart (Figure 1.64), as described in the Appendix.

initial growth rate		0.05		initial growth rate		0.05
capacity		5000		capacity		5000
period	model1	change1		model2	saturation	change2
1	1000	50		1000	0.2	40
2	1050	52.5		1040	0.208	41.184
3	1102.5	55.125		1081.184	0.216237	42.36961
4	1157.625	57.88125		1123.554	0.224711	43.55395
5	1215.506	60.77531		1167.108	0.233422	44.73398
6	1276.282	63.81408		1211.842	0.242368	45.90648
7	1340.096	67.00478		1257.748	0.25155	47.0681
8	1407.1	70.35502		1304.816	0.260963	48.21535

Figure 1.64

After having done this successfully, we should see the graph shown in Figure 1.65.

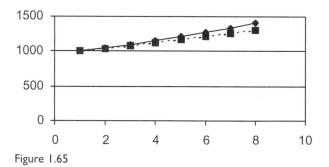

Figure 1.65

Since the growth rate for Model 1 and the initial growth rate for Model 2 are identical, the actual growth rate for Model 2 is always less than the actual growth rate for Model 1, since it is multiplied by a number less than 1.

We can now experiment with different combinations of values for the growth rates and capacity and starting values. Since Model 2 is the one having more parameters with which to experiment, we simply adjust Model 1 so as to use the same starting value as Model 2 and also to have Model 1's growth rate equal to Model 2's initial growth rate. We do this by adding formulas to our spreadsheet table as shown in Figure 1.66.

initial growth rate		←		initial growth rate		●	0.05
capacity		5000		capacity			5000
period	model1	change1		model2	saturation	change2	
1	←1000	0		● 1000	0.2	40	

Figure 1.66

To study how the system behavior depends upon the parameter values, it is quite helpful to have more than just a few data points, so let us start with 200 points. To enlarge the model to this size, we only have to copy down the last row far enough in our table. We also have to extend the range of the data columns in the menus setting up the graph. Then we get the graph shown in Figure 1.67.

In this graph, the unrestricted growth model dominates the restricted growth model, so we cannot see the S shape of the second curve and we cannot see the similarities between these curves over a larger range. However, if we set the maximum scale on the y-axis to 10,000, we see the graph of Figure 1.68.

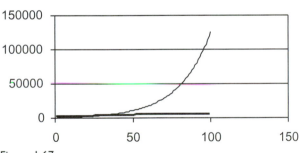

Figure 1.67

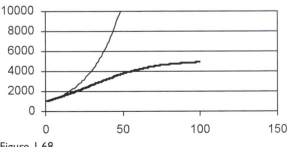

Figure 1.68

Next, let us reset the default setting of the graph's y-scale. Suppose that we change the capacity to a much higher value (for example, 130,000). If we do this, we see that at least for a noticeable period in the beginning there is not a large difference between the saturation independent model and the saturation dependent model. We still do not see the S shape for the restricted growth curve. Setting the saturation level to 30,000 produces a graph that makes the S shape visible again.

Another suggestion for studying the model is to lower the initial growth rate (for example to 0.005) and keep the high capacity value, which makes the two curves still more similar. But this is easy to understand: as long as the current values for both systems are almost equal and the saturation level is almost 0, the change for both models is also almost equal and so the new values again will be almost equal. It is quite interesting to change the two parameters of our model and study the differences in behavior for different parameter value combinations. Therefore, you should "play" with both parameters and study the effects.

Construction Summary: Comparing Growth Models

1. Enter parameter values.
2. Use column as period counter.
3. Enter initial population.
4. Compute change by multiplying population by growth rate.
5. Compute saturation by dividing population by capacity.
6. Compute the change as product of (1 – saturation), initial growth rate, and current population.
7. Add the change to the previous population.
8. Copy as indicated.

growth rate		1	0.05	initial growth rate			1	0.05
capacity		1	5000				1	5000
period		model1	change1	model2		satr.	change2	
[2] 1 [3]		1000 [4]	50	[3] 1000	[5]	0.20	[6] 40	
2	[7]	1050.0	52.5	[7] 1040.0		0.21	41.18	
3		1102.5	55.1	1081.2		0.22	42.37	
4		1157.6	57.9	1123.6		0.22	43.55	
5		1215.5	60.8	1167.1		0.23	44.73	
6		1276.3	63.8	1211.8		0.24	45.91	
7		1340.1	67.0	1257.7		0.25	47.07	
8		1407.1	70.4	1304.8		0.26	48.22	
9		1477.5	73.9	1353.0		0.27	49.34	

Exercises

1. Modify the logistic growth model in the following way:
   Instead of the equation

   $$\text{growth rate} = \frac{\text{change}}{\text{current state}} = \text{initial growth rate} \times (1 - \text{saturation}),$$

   use the equation

   $$\text{growth rate} = \frac{\text{change}}{\text{current state}} = \text{initial growth rate} \times (1 - \text{saturation})^k$$

   and compare two of these models with the same capacity and the same initial growth rate for different values of k ($k = 1$ is the original logistic growth model).

2. Design interactive graphs and use scroll bars to enhance the models of this section.

3. Create a spreadsheet model to compare the effects of differing growth rates of two or more nations using the exponential growth model. Create a graph that compares the projected populations. Incorporate scroll bars to vary the growth rates. A useful place to find current data is the International page on the U.S. Census Bureau site, http://www.census.gov.

Dynamical Models and Difference Equations

CHAPTER OUTLINE

2.1   Simple Epidemics

In addition to creating models for determining the change in the size of an entire population, we can also design models that show how the sizes of various subgroups of a population change due to various influences. We examine how a disease can spread through a population by using the well-known Kermack-McKendrick, or SIR, model that was first published in 1927 (see Edelstein-Keshet (1988), p. 242). Individuals in a population are classified into three states relative to an infectious disease: susceptibles, infectives, and removals (or healed). Our model illustrates how the disease spreads throughout the population.

In this model we suppose that all individuals in a population of a fixed size can be classified into three categories. Category S, the *susceptibles*, consists of those who are susceptible to the disease but have not yet contracted it. Category I, the *infectives*, is made up of those who have been infected by the disease and are still capable of spreading it to the susceptibles. Category R, the *removals*, is composed of those who have had the disease but are no longer infectious or susceptible because they have acquired an immunity.

The model studies the number of individuals in these three categories over time. Since we assume a fixed-size population, we already know that the sum of the number of individuals in all three categories is constant. So all of the changes in our system occur because the individuals are moved from one category to another one. How can individuals "change their category"?

Of course, removals remain removals once they are members of this category.

Our model assumes that once infected one cannot contract the disease again, so infectives either can stay infectives or they can become removals. Furthermore, the model assumes that in each period there is a fixed percentage of infectives that become removals. We might name this percentage the "healing rate."

Susceptibles either can become infectives or they can remain susceptibles. They become infectives by making contact with an already infected individual (an infective). The model, as a simplification, assumes that each contact of an infective with a susceptible individual turns the susceptible one into an infected (and infective) individual. Therefore, we have to calculate the number of susceptibles making a contact with infectives to get the number of newly infected individuals. For these calculations we use a "contact probability," that is, the chance of any individual who makes a contact with any other individual in a given period.

Let us consider two individuals, A and B. If the contact probability is 0.02 (or 2%), this means that there is a 2% chance that in a given period these two people will have contact; therefore, there is a 98% chance that these two people will not meet during that period. We further assume that the contacts in our population are random and independent of each other, meaning that considering another individual C, the probability of A contacting C in a given period is the same as for contacting B. Independence implies that for A the probability of meeting B and C in the same period is obtained by multiplying the probabilities of meeting B and of meeting C. So in our case the probability of A meeting both B and C in a given period is $0.02^2 = 0.0004$. Likewise, the probability of A meeting neither B nor C in a given period is the product of the probabilities of A not meeting B and of A not meeting C. Therefore, its value is $0.98^2 = 0.9604$. As a result of this, the probability of A meeting at least one of B or C becomes $1 - 0.98^2 = 0.0396$. For any given susceptible individual to remain noninfected, we need to know the probability that this individual does not have a contact with any infective individuals. Therefore, for any given susceptible individual, the probability of not becoming infected in any given period is given by

probability of not getting infected = $(1 - \text{contact probability})^{\text{number of infective individuals}}$

and, of course,

probability of getting infected = 1 − probability of not getting infected

$$= 1 - (1 - \text{contact probability})^{\text{number of infective individuals}}.$$

The number of susceptible individuals becoming infected in any period therefore is calculated the following way:

susceptibles becoming infected

$$= \text{susceptibles} \times (1 - (1 - \text{contact probability})^{\text{number of infective individuals}})$$

Therefore, we can model the flow of the individuals through the three possible states with the symbolic diagram shown in Figure 2.1.

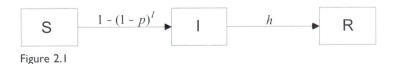

Figure 2.1

In Figure 2.1, the boxes indicate the individuals in a given state, and the arrows indicate the individuals changing from one state to another one during a given period. The expressions above the arrows indicate the percentages of the individuals changing from one state to the other state. So in our example in any period the number of infected individuals becoming removals is $h \times I$ and the number of susceptible individuals becoming infectives is $S \times (1 - (1 - p)^I)$. In these equations, p is the contact probability for any two individuals and h is the healing rate.

So let us start setting up the table. We have to supply values for the healing rate and for the contact probability, and we also have to supply initial values for the number of susceptibles and the number of infectives. Assuming that there are no removals when we start our model, we can calculate the number of susceptibles. We will assume some reasonable values. Our value for the healing rate is 1, implying that the infection never lasts longer than one period and any infected individual will be a removal in the next period (see Figure 2.2).

Population	●	1000		
contact				
probability		0.002		
healing				
rate		1		
period	susceptibles	infectives	removals	
0	▼ - ◄	● 3	0	

Figure 2.2

After having set up the initial state of our system, we have to calculate the number for susceptibles, infectives, and removals for the next period. So let us extend our table and calculate the number of susceptibles becoming infected during the first period (Figure 2.3).

Population	1000				
contact					
probability	⬦● 0.002				
healing					
rate	1				
				new	new
period	susceptibles	infectives	removals	infectives	removals
0	● 997	● 3	0	▲* (1-(1-▼)^▲)	

Figure 2.3

Since we want to create "reusable" formulas, we use an absolute reference for the contact probability. The formula for the new removals is straightforward to set up, again using an absolute reference for the healing rate (Figure 2.4).

Population	1000				
contact					
probability	0.002				
healing					
rate	1			new	new
				infectives	removals
period	susceptibles	infectives	removals	new infectives	new removals
0	997	3	0	5.97	*

Figure 2.4

We note that the numbers in our table are not integers. Of course, this cannot happen in reality, but the model is still a reasonable approximation of the process that we want to study.

In the next step, we will set up the table to calculate the number of individuals in the three categories for Period 1. So we set up some more formulas. We start by incrementing the period number by 1. Using our "state flow" diagram, we see that the following holds:

$$\text{susceptibles}_{\text{next period}} = \text{susceptibles}_{\text{current period}} - \text{new infectives}$$

so we build the table shown in Figure 2.5.

Population	1000				
contact					
probability	0.002				
healing					
rate	1			new	new
				infectives	removals
period	susceptibles	infectives	removals	new infectives	new removals
0	997	3	0	5.97	3

Figure 2.5

For the number of infectives in the next period we have the following formula:

$$\text{infectives}_{\text{next period}} = \text{infectives}_{\text{current period}} + \text{new infectives} - \text{new removals}$$

(See Figure 2.6).

Population	1000			new	new
contact probability	0.002				
healing rate	1				
				new infectives	new removals
period	susceptibles	infectives	removals	infectives	removals
0	997	3	0	5.97	3
1	991.03				

Figure 2.6

The number of removals in the next period is calculated the following way:

$$\text{removals}_{\text{next period}} = \text{removals}_{\text{current period}} + \text{new removals}$$

(see Figure 2.7).

Population	1000			new	new
contact probability	0.002				
healing rate	1				
				new infectives	new removals
period	susceptibles	infectives	removals	infectives	removals
0	997	3	0	5.97	3
1	991.03	5.97			

Figure 2.7

Calculating the new infectives and the new removals for Period 1 is uncomplicated. We can just use the formulas that we created for Period 0 since we designed those formulas to be reusable. So we copy these formulas as shown in Figure 2.8.

Population	1000				
contact					
probability	0.002				
healing					
rate	1				
				new	new
period	susceptibles	infectives	removals	infectives	removals
0	997	3	0	5.97	3
1	991.03	5.97	3		

Figure 2.8

Our row containing the formulas for Period 1 is general in the sense that it simply contains the "recipe" for how to calculate any period from the previous one. So, to have our model running for, say, 200 periods, we just copy down this row far enough as indicated in Figure 2.9.

Population	1000				
contact					
probability	0.002				
healing					
rate	1				
				new	new
period	susceptibles	infectives	removals	infectives	removals
0	997	3	0	5.97	3
1	991.03	5.97	3	11.77	5.970044
2	979.26	11.77	8.970044	22.81	11.77433
3	956.44	22.81	20.74437	42.70	22.81331
4	913.74	42.70	43.55769	74.87	42.7004
5	838.88	74.87	86.25808	116.76	74.86679
6	722.11	116.76	161.1249	150.52	116.7645
7	571.59	150.52	277.8894	148.72	150.5243

Figure 2.9

Graphing the numbers of susceptibles, infectives, and removals on the y-scale against the period number on the x-scale, and doing this in a joint graph, as we have seen in the last section, produces a graph similar to the one in Figure 2.10. Notice that not everyone will become infected.

Extending this graph any farther by showing more periods does not help us to gain any additional insight. The system stabilizes and remains in the same state for any longer time.

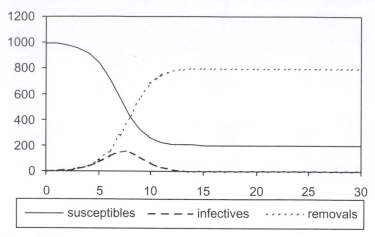

Figure 2.10

Since all the "model equations" refer to the "model parameters" (namely contact probability and healing rate), we can easily change these values and study the effect upon the behavior of the model. For example, changing the contact rate from 0.002 to 0.0013 changes the tables and the graph, and we get the picture shown in Figure 2.11, wherein fewer are ultimately infected.

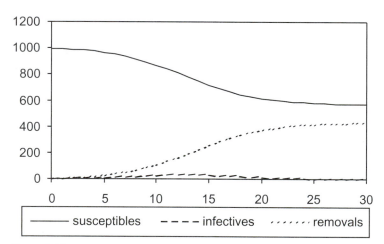

Figure 2.11

Varying the parameters of our model (contact probability and healing rate) and also the initial conditions (population size and initial number of infectives) allows us to study the behavior of our system under different circumstances. Generally speaking, we can say that the system is quite sensitive to variations in the contact probability. It turns out that in most cases the system reaches a "steady state" where it does not change any more. In this state there are always no more infectives. Since the infectives are the only ones whose presence can change the state of the system, any system with no infectives clearly has to be stable. It is quite interesting to investigate the number of people ultimately remaining uninfected for different values of the contact probability and the healing rate.

Our models will be enhanced significantly by incorporating scroll bars that allow us to investigate the effects of small changes in the values of initial conditions and parameters. Figure 2.12 provides a view of a model that allows us to investigate the effects of varying the contact probability. The scroll bar is linked to an auxiliary cell, D3, which takes on integer values. The contact probability is then computed in Cell B2 by dividing this value by 100,000.

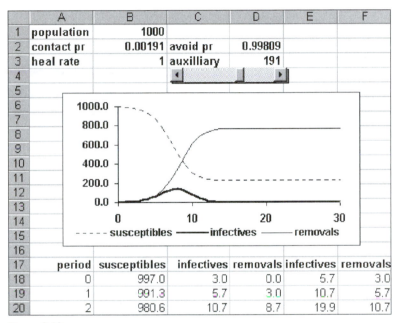

	A	B	C	D	E	F
1	population	1000				
2	contact pr	0.00191	avoid pr	0.99809		
3	heal rate	1	auxilliary	191		
4						
17	period	susceptibles	infectives	removals	infectives	removals
18	0	997.0	3.0	0.0	5.7	3.0
19	1	991.3	5.7	3.0	10.7	5.7
20	2	980.6	10.7	8.7	19.9	10.7

Figure 2.12

We also can write the equation of our model in standard mathematical notation, so let us use the following notation:

p is the contact probability,

h is the healing rate,

S_n is the number of susceptibles in Period n,

I_n is the number of infectives in Period n,

R_n is the number of removals in Period n,

and S_0, I_0, and R_0 are the initial values in Period 0 (fulfilling $S_0 + I_0 + R_0 = N$ with N being the fixed and constant population size). Using our schematic representation, our model may be written as

$$S_{n+1} = S_n - S_n \left(1 - (1 - p)^{I_n}\right)$$

$$I_{n+1} = I_n + \left(1 - (1 - p)^{I_n}\right)S_n - I_n h$$

$$R_{n+1} = R_n + I_n h$$

This can be simplified algebraically to

$$S_{n+1} = S_n (1 - p)^{I_n}$$

$$I_{n+1} = I_n (1 - h) + \left(1 - (1 - p)^{I_n}\right)S_n$$

$$R_{n+1} = R_n + I_n h$$

Construction Summary: Simple Epidemics

1. Enter parameter values.
2. Create column as period counter.
3. Enter initial infectives, removals.
4. Compute initial susceptibles: subtract infectives from population.
5. Compute new infectives: multiply susceptibles by an appropriate value.
6. Compute new removals: multiply infectives by healing rate.
7. Compute current susceptibles from previous level: subtract new infectives.
8. Compute infectives from previous level: add new infectives, subtract new removals.
9. Compute removals from previous level: add new removals.
10. Copy as indicated.

Pop	1000				
cntct					
prob	0.002	1			
heal					
rate	1				
				new	new
period	suscept	infects	remov	infect	remov
2 0 4	997 3	3 3	0 5	6.0 6	3
1 7	991.0 8	6.0 9	3.0	11.8	6.0
2	979.3	11.8	9.0	22.8	11.8
3	956.4	22.8	20.7	42.7	22.8
4	913.7	42.7	43.6	74.9	42.7
5	838.9	74.9	86.3	116.8	74.9
6	722.1	116.8	161.1	150.5	116.8
7	571.6	150.5	277.9	148.7	150.5
8	422.9	148.7	428.4	108.9	148.7
9	314.0	108.9	577.1	61.5	108.9
10	252.5	61.5	686.0	29.2	61.5

Exercises

1. Investigate the model of this section by varying the healing rate. In particular, look at the effects of the healing rates 0.9, 0.5, 0.1, and 0.0. Also vary the values of the other parameters.

2. Our epidemic model assumes that the population size remains constant. However, suppose that there is a continual influx of new residents (either by birth or immigration) into the community at a rate proportional to the population. This presents a continuing supply of new susceptibles. Create a model in which an initial population grows at the rate of 0.008 per day, with a contact probability of 0.0015 and a healing rate of 1.0. Include an xy-chart showing each component of the population. Notice that this can produce periodic recurrences, or "waves," of a disease to occur. See Edelstein-Keshet (1998) for a discussion of this topic.

3. In a certain country there are two political parties, Liberal and Conservative. It has been observed that between elections a fixed percentage, p_a, of Liberals switch to the Conservative party, and a fixed percentage, p_b, of Conservatives switch to the Liberal party. The other voters remain in the same party. Create a spreadsheet model for this situation (see Figure 2.13). What can you conclude from its output?

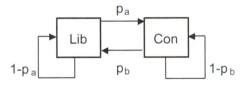

Figure 2.13

4. Modify the previous exercise to allow for three or more political parties.

5. Incorporate scroll bars into your epidemic model to assist in investigating the model by varying the values of the initial conditions and parameters.

2.2 Harvesting Models

So far we have examined several population growth models. In this section we look at another variation of those concepts to create models of the changes that occur in the population of a species, such as deer, that is regularly harvested. We will examine two specific models. In the first, a constant number of the species is harvested during a year or another convenient time period. In the second model, the number of the species that is harvested in a year is proportional to the number of the species that is available.

In both examples we extend the ideas presented in our earlier logistic growth model. In particular, we assume that a given geographical area can support a known limited number of the species. This is called the *carrying capacity* of the area. We include as parameters of our model an initial population and a growth rate that would result if there were no constraints on growth. However, our model takes into account the fact that the growth rate will be constrained by the carrying capacity.

I. Constant Harvesting Policy

The construction of this model closely parallels the logistic growth model, so we will shorten our initial discussion. We assume that there is a population limit (carrying capacity) of 1,000, an initial population of 200, and a base growth rate of 60 percent and that 50 animals are harvested each year. After creating a period counter in the first column (Figure 2.14), we enter the initial population as the population at the start of Year 1. The next column computes that year's growth rate just as in the logistic model, except that we use one column instead of the previous two. Here, the natural growth rate is multiplied by 1 – saturation, where the saturation is the ratio of the current population to the carrying capacity.

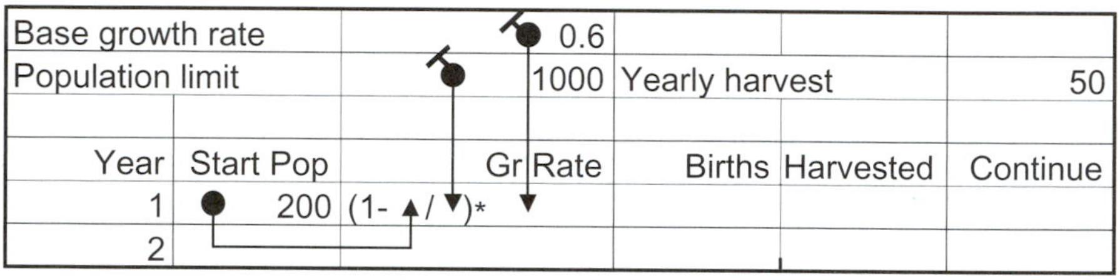

Base growth rate			0.6			
Population limit			1000	Yearly harvest		50
Year	Start Pop		Gr Rate	Births	Harvested	Continue
1	200	(1- ▲/ ▼)*	▼			
2						

Figure 2.14

Next we compute the number of births in the first year as the product of the growth rate and the population. We next include a column for the amount harvested during the year (Figure 2.15). Since this model assumes that there will be a constant number harvested each year, this is an absolute reference. Finally we include a column to compute the number of individuals continuing from the present year to the next by subtracting those harvested from the number at the start of the year. Note that this assumes that the harvesting is done after the new births for the year occur, a common policy in countries that have controlled hunting seasons.

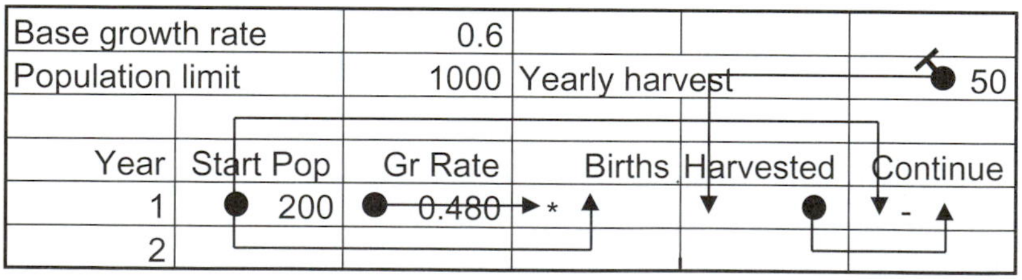

Base growth rate		0.6			
Population limit		1000	Yearly harvest		50
Year	Start Pop	Gr Rate	Births	Harvested	Continue
1	200	0.480	* ▲	▼	▼ - ▲
2					

Figure 2.15

Finally, we compute the population at the start of the next year by adding the previous year's births to the number of the species that continues (Figure 2.16). We then copy the formulas in each column to complete the model for at least 20 years.

Base growth rate		0.6			
Population limit		1000	Yearly harvest		50
	Year	Start Pop	Gr Rate	Births	Harvested	Continue
	1	200	0.480	● 96.0	50.0	● 150.0
	2	▲ + ◄				

Figure 2.16

Part of the output is shown in Figure 2.17. If we examine this on our spreadsheet, we will see that the population has reached equilibrium and that by Year 20 the number of the species being harvested is balanced by the number of new births. At this time we can take advantage of the spreadsheet's capabilities and experiment with different harvesting rates, growth rates, starting populations, and carrying capacities.

Base growth rate		0.6			
Population limit		1000	Yearly harvest		50
Year	Start Pop	Gr Rate	Births	Harvested	Continue
1	200.0	0.480	96.0	50.0	150.0
2	246.0	0.452	111.3	50.0	196.0
3	307.3	0.416	127.7	50.0	257.3
19	908.1	0.055	50.1	50.0	858.1
20	908.2	0.055	50.0	50.0	858.2

Figure 2.17

The ability to gain insights into the effects of changes in the model can be effectively visualized by including graphs, such as the column charts shown in Figure 2.18 and Figure 2.19 (see Appendix for details). Figure 2.18 uses the assumptions in our model. Figure 2.19 has the annual harvest amount increased to 95. Such a situation can shed light on the fragility of some policies. Here we annually harvest so many of the species that it takes a long time until their numbers increase. However, our models can never be so precise as this. One consequence of this is if the annual number harvested is increased only slightly, say to 97 per year, the species quickly becomes extinct in the area, as is shown in our model.

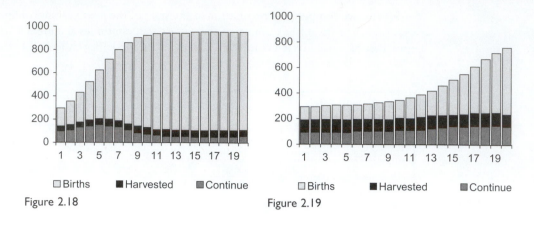

Figure 2.18   Figure 2.19

2. Introducing Mathematical Notation

Throughout this book you will see that many mathematical models can be set up, solved, and investigated in a substantial manner without using an excess amount of formal notation from algebra. However, we also need to develop skills in this area. Using our spreadsheet experience can be a help in doing this. Consequently, let us now reexamine the previous situation using more traditional notation.

First, we recall that if we assume that a population grows at the unrestricted annual growth rate of $r = 0.6$, then the numerical growth from year A_n to year A_{n+1} is given by

$$A_{n+1} - A_n = rA_n.$$

However, instead of assuming that the annual growth multiplier is a constant, because of the limiting carrying capacity, L, we assume that it is a function, $f(A)$, of the population, A. In particular, we assume that this function has the property that $f(A)$, will be essentially r when the population A is small relative to L, and is essentially 0 as A approaches the carrying capacity, L. The simplest function with this property is linear. This is the same idea that we used in our logistic growth model, but this time we present it in a different format that will later help us in our graphical analysis.

The linear equation can be obtained by observing that $f(0) = r$ and $f(L) = 0$. Then the graph is a line that passes through the points $(0, r)$, and $(L, 0)$. Consequently it has slope $-r/L$. From this it follows that the function's equation is obtained as

$$f(A) - f(0) = -(r/L)(A - 0), \text{ or } f(A) = r - (r/L)A = r(1 - A/L).$$

Hence, when $A = 0$, $f(A) = r$ and when $A = L$, $f(A) = 0$. We note that A/L is exactly the saturation component and $(1 - A/L)$ the saturation multiplier that we introduced in the logistic model.

Using the expression derived for $f(A)$, it follows that the numerical increase in population between Year n and Year $n + 1$ is then

$$f(A_n)A_n = rA_n - (r/L)A_n^2.$$

For our constant harvest model, where b is the number of animals harvested per year, we can then get a recurrence relation for the number of individuals at the start of Year n as

$$A_{n+1} = A_n + rA_n - (r/L)A_n^2 - b = (1 + r)A_n - (r/L)A_n^2 - b.$$

We use this expression in investigating a new component of our model. Also, another graph that is convenient to use to analyze the model is the cobweb diagram that is shown in the graph in Figure 2.20. The idea is that population equilibrium occurs when $A_{n+1} = A_n$, or where the graphs of the functions $g_1(x) = x$ and $g_2(x) = (1 + r)x - (r/L)x^2 - b$ intersect. Thus, we graph those two functions. Next, we observe that if we start with a population of A_0 on the horizontal axis, then the next population, $A_1 = f(A_0)$, is the y-coordinate on the g_2 curve. Since the x- and y-values are the same on the line, we can now find A_1 on the x-axis by moving horizontally until the line $y = x$ is intercepted, and then repeat the process. Thus, the cobweb diagram illustrates increasing growth and convergence to a steady state.

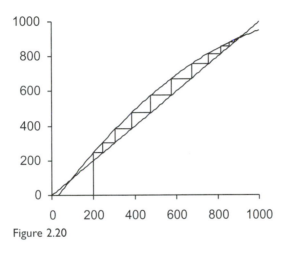

Figure 2.20

We create the cobweb graph as outlined in Figure 2.21. Further details are provided in the Appendix. First we create a graph for the two functions $y = g_1(x)$ and $y = g_2(x)$ by constructing the first three columns and using them for the graph. Next

Rate	0.6					
Limit	1000	Harvest	50			
	line & curve				cobweb	
x	$g_1(x)$	$g_2(x)$		k	x	y
0	0	-50		0	200.0	0.0
10	10	-34.06		1	200.0	246.0
20	20	-18.24		0	246.0	246.0
30	30	-2.54		1	246.0	307.3

Figure 2.21

we use the scheme presented in the figure in the right three columns to create the cob-web. We then highlight the last two columns at the x- and y-coordinates for another graph and drag that block into the graph that has just been created (see the sections on creating graphs, Sections 5.1-5.2, and the Appendix).

The outline of the process for creating the coordinates for the graphs of $y = g_1(x)$ and $y = g_2(x)$ is shown in Figure 2.22.

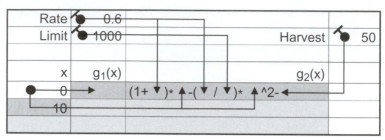

Figure 2.22

Next we outline a scheme for creating the cobweb (see Figure 2.23). The column labeled k uses the spreadsheet's MOD function to alternately generate the numbers 0 and 1. These are used in forming the x- and y-coordinates for the cobweb. If $k = 1$, then the previous x-coordinate is reproduced to generate a vertical segment. If $k = 0$, then the previous y-coordinate becomes the next x-coordinate to generate a horizontal segment.

	k		x	y
	0			0
MOD(1+ ▼,2)		IF(▶ =1, ▼, ▼)		

Figure 2.23

Similarly with the y-coordinates, if $k = 1$, the new y-coordinate is the population of the next year, generated through the basic recurrence relation. If $k = 0$, the y-coordinate of the previous row is reproduced (see Figure 2.24).

Rate	0.6										
Limit	1000	Harvest	50								
x	$g_1(x)$	$g_2(x)$		k	x						y
0	0	-50		0	200.0						0
10	10	-34.06		1	200.0	IF(▼=1,(1+▼)* ▲ -(▼/ ▼)* ▲^2-▼ , ▼)					
20	20	-18.24		0	246.0						
30	30	-2.54		1	246.0						

Figure 2.24

This graph also helps us to see that this harvesting scheme can be very sensitive to the number of the species that are selected each year. The graphs shown in Figure 2.25 and Figure 2.26 come from our model with a base growth rate of $r = 0.8$, a carrying capacity of 1,000, and an initial population of 200. In Figure 2.25, $b = 120$ animals are harvested each year and the population grows to reach 816. In Figure 2.26, everything is the same except that $b = 130$ animals are harvested each year and the population quickly dies out.

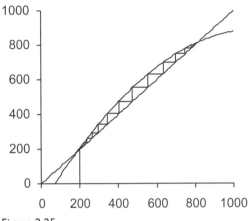

Figure 2.25

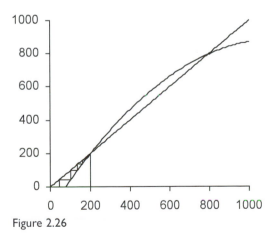

Figure 2.26

From the preceding charts and what we have examined so far, what do you predict will happen if we keep $r = 0.8$ and $b = 130$ and the initial number is increased to 500? Will the population die out or increase? After you have made your prediction, test your conclusion using the model.

3. Proportional Harvesting Policy

To create a model for a proportional harvesting policy requires only a simple change. Instead of using a constant annual harvesting amount, we enter an annual percentage harvesting rate. Here we use a rate of 0.1, so that 10% of the population will be harvested each year. We simply enter the indicated formula for the first year's harvest as the product of the starting population and the percentage harvest rate, where the latter is an absolute reference. We can then copy this formula down its column to complete our modifications (see Figure 2.27).

Base growth rate		0.6				
Population limit		1000	Percent harvested			0.1
Year	Start Pop	Gr Rate		Births	Harvested	Continue
1	200.0	0.480		96.0	*	180.0
2	276.0	0.434		119.9	27.6	248.4
3	368.3	0.379		139.6	36.8	331.5

Figure 2.27

The output is shown in Figure 2.28. Again, it is instructive to investigate the effect of changes in parameters, especially the harvesting percentage.

Base growth rate		0.6				
Population limit		1000	Percent harvested			0.1
Year	Start Pop	Gr Rate		Births	Harvested	Continue
1	200.0	0.480		96.0	20.0	180.0
2	276.0	0.434		119.9	27.6	248.4
3	368.3	0.379		139.6	36.8	331.5
19	833.3	0.100		83.3	83.3	750.0
20	833.3	0.100		83.3	83.3	750.0

Figure 2.28

We can also create a column graph similar to the earlier ones (see Figure 2.29). The creation of the cobweb diagram is more involved because we must find a new function g_2. This is left as an exercise for those who are interested.

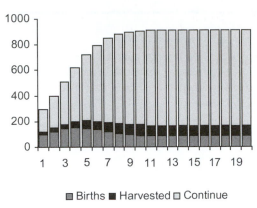

Figure 2.29

1. Enter parameter values.
2. Enter amount of yearly harvest.
3. Generate a counter column.
4. Enter the initial population.
5. Compute the current growth rate as = (1 − pop/limit) × base rate.
6. Compute births as rate × pop.
7. Reproduce fixed harvest amount.
8. Compute continuing population as population − harvest.
9. Compute the next population by adding births to the continuing population.
10. Copy as indicated.

Base growth rate	1	0.6			2	
Population limit		1000	Yearly harvest		2	50
Year	Start Pop	Gr Rate	Births	Harv		Cont
3	1	4	200	5	0.480	6	96	7	50	8	150
2	9	246	0.452	111	50	196
3	307	0.416	128	50	257
4	385	0.369	142	50	335
5	477	0.314	150	50	427
6	577	0.254	146	50	527
7	673	0.196	132	50	623
8	755	0.147	111	50	705

Exercises

1. Compute the steady-state populations that are reached under the following constant harvesting strategies. In each, assume that the sustainable population is 1,000.
   a. $r = 0.5$, $P_0 = 120$, $b = 50$ b. $r = 0.5$, $P_0 = 100$, $b = 50$
   c. $r = 0.8$, $P_0 = 120$, $b = 60$ d. $r = 1.2$, $P_0 = 120$, $b = 120$

2. Compute the steady-state population and annual harvest that are reached under the following proportional harvesting strategies. In each, assume that the sustainable population is 1,000.
   a. $r = 0.6$, $P_0 = 200$, $c = 40\%$ b. $r = 1.0$, $P_0 = 200$, $c = 60\%$
   c. $r = 0.8$, $P_0 = 200$, $c = 50\%$ d. $r = 0.5$, $P_0 = 1000$, $c = 50\%$

3. In the proportional harvesting model, use the spreadsheet to find the harvesting percentage that produces the maximal annual harvest.

4. Develop the function $y = g(x)$ needed to create the cobweb graph for the proportional harvesting model, and implement it in a model.

5. Design scroll bars for the models of this section that allow you to vary the values of the models' parameters and initial values.

6. In the discipline of economics, supply and demand curves express relationships between price, P, and quantity, Q. As price rises, the quantity demanded tends to fall, and conversely. This is expressed via a demand function. Similarly, as price rises, the quantity supplied tends to rise, and conversely. This is expressed through a supply function. The diagrams in Figure 2.30 and Figure 2.31 (see Morrison (1991), p. 117) illustrate the economic concepts involved. If there is oversupply, then prices are cut (1) because of lower demand; then, as demand falls, production is cut (2); next, after production has been cut, increased demand drives up prices (3); then, as the demand rises, the supply is increased (4). Now the cycle repeats, and generally it eventually reaches equilibrium. Create a spreadsheet model to generate the analysis for these figures where both functions are linear. Incorporate a cobweb graph in your model. Determine conditions that lead to equilibrium being reached (Figure 2.30) or an unstable process being generated (Figure 2.31).

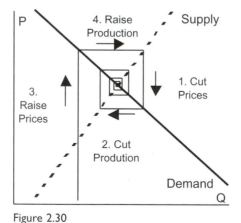

Figure 2.30

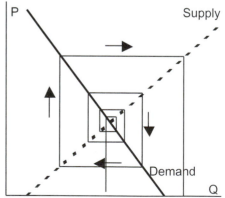

Figure 2.31

Adapted from *The Art of Modeling Dynamic Systems*
by Foster Morrison, John Wiley and Sons, 1991.
Used by permission.

7. Design a similar model to use when the supply and demand functions are not linear, as illustrated in Figure 2.32 with $f(x) = 2/(x + 1)$ and $g(x) = 0.5x^2 + 0.4$.

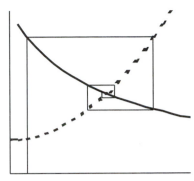

Figure 2.32

Create similar designs using the following pairs of functions:

a. $f(x) = 2e^{-0.5x}$, $g(x) = 0.5x^2$ b. $f(x) = 1 - 0.5\ x$, $g(x) = \ln(1 + x)$

8. A recurrence relation of the form $A(n + 1) = f(A(n))$, $A(0) = c$ is called a *dynamical system*. This topic is discussed throughout the remainder of this chapter, especially in the sections on iterated functions. If $A(n) = c$ for all n, then c is a fixed point of the system. For example, the system $A(n + 1) = -0.5\ A(n) + 3.0$ has 2 as a fixed point. It is easy to evaluate the terms of the recurrence relation on a spreadsheet. For example, using this system and $A(0) = 1$ we obtain the output shown in Figure 2.33. Here the resulting sequence ultimately converges to the fixed point.

n	A(n)
0	1
1	2.5
2	1.75
3	2.125
4	1.9375
5	2.03125
6	1.984375

Figure 2.33

We can also exhibit the behavior of the system by drawing the curve $y = f(x)$ and the line $y = x$ and using them to plot points obtained from the sequence $A(n)$ in a cobweb graph, as shown in Figure 2.34, Figure 2.35, and Figure 2.36. If the process converges to a fixed point, we say that it is an *attracting* fixed point and that the system is stable. If the sequence diverges by continually growing in absolute value, it is called a *repelling* fixed point. If the system satisfies $A(n + m) = A(n)$ for a positive integer m, then the system is called *periodic*. Each of these types can be seen if we create a cobweb design. Create a spreadsheet model to produce cobweb designs, and implement it using the following recurrence relations (a) $A(n + 1) = -0.5A(n) + 3.0$, $A(0) = 0.2$; (b) $A(n + 1) = -A(n) + 5$, $A(0) = 1$; (c) $A(n + 1) = 3A(n) - A^2(n)$, $A(0) = 0.2$ to provide the cobweb designs shown in the figures.

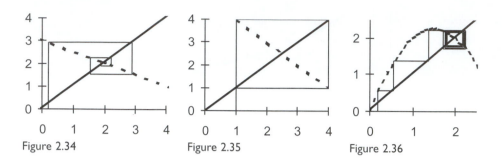

Figure 2.34   Figure 2.35   Figure 2.36

9. Create cobweb designs for the following dynamic systems, examine their natures, and locate fixed points.
   a. $A(n + 1) = -A(n) + 2$, $A(0) = 2$
   b. $A(n + 1) = 1 - 0.5A(n)$, $A(0) = 8$
   c. $A(n + 1) = \cos(A(n))$, $A(0) = 0.4$
   d. $A(n + 1) = 2A(n) - A^2(n) + 1$, $A(0) = 1.5$
   e. $A(n + 1) = \tan^{-1}(A(n))$, $A(0) = 2$
   f. $A(n + 1) = A(n) - 0.3A^2(n) + 1$, $A(0) = 1$

2.3 Population Dynamics

In some of our previous growth models, we have generally treated a population as being homogeneous, with the growth rate as a function of the size of the entire population. Also, birth and death rates have been incorporated into an overall growth rate. However, populations are actually composed of various substrata, with some individuals too young or too old to reproduce and individuals dying at varying rates. This section's model provides a more sophisticated approach to modeling a nation's population that incorporates three states: children, workers, and retired, each with its own death rate, together with a birth component and transition probabilities of moving from one state to another. We can use the model to discover long-term trends and to consider the possible consequences for a nation's retirement plans.

To be able to understand these problems, it is quite useful to understand the principles governing the development of a population. For studying this kind of problem, however, it is not enough to have a model for the development of the total population. We need to know the number of people in different age groups. We will study a very simple model considering only three different age groups. These groups are children (C), working people (W), and retired people (R). We will follow the numbers of people in each group through a longer period, updating the number each year. The changes in population group sizes from one year to the next year are due to the following various reasons:

1. People get older, so some children become workers, and some workers retire.
2. In each age group, people die.
3. New children are born.

We have to find equations that allow us to calculate the numbers in the different age groups for any one year given those numbers for the previous year. To understand the process better we model our problem with the flowchart shown in Figure 2.37.

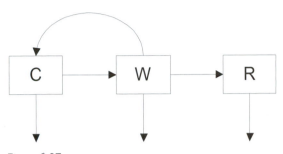

Figure 2.37

In this figure, all horizontal arrows correspond to changes due to people becoming older, vertical arrows correspond to people dying, and the curved arrow corresponds to children being born. Of course the number of children born depends on the current number of adults young enough to have children.

We will assume that for the period we study, the percentage of people from each group making one of the transitions indicated by the arrows is the same for all the years studied. So each arrow is associated with a constant. This constant corresponds to the percentage of group members changing to the next group in 1 year (because of getting older). Demographic tables show that reasonable values for the proportion dying each year are about 0.001 for the children, 0.003 for the working people, and 0.05 for the retired people. These values can be derived from data published by the national statistical offices like the Bureau of Census. These assumptions, of course, simplify the model, but they are good enough for our model.

We have still not defined the age groups that will constitute the children, the workers, and the retired people. We will assume children to be up through age 14, workers up through age 64, and retired from then on. We also have to make reasonable assumptions about the percentages of people passing from one age group to the next one each year. For this problem, a little common sense proves quite helpful.

When the children group comprises all ages from 0 to 14, then there are 15 groups of children with the members of each group having the same year of birth. All the members of the group of the oldest children will become workers the following year, so about 1/15 or 0.066 (or 6.6%) is the proportion of members of the children group changing to the worker group in 1 year (assuming that all these groups are of roughly equal size). A similar argument shows that about 1/50 or 0.02 (or 2%) is the proportion of workers retiring in 1 year. We also need age-specific death rates. These values cannot be found out by reasoning alone. These rates can be found in life expectancy tables, usually published by national statistical agencies. We will use reasonable values that are more or less valid in most Western countries. We add these numbers to our flowchart in Figure 2.38.

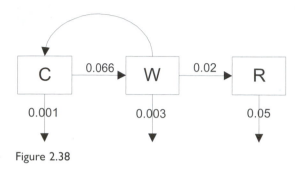

Figure 2.38

Our model is still incomplete, since we have not yet considered the newborn children. Their number should depend upon the number of people in the working age, and again we assume that the number of children born each year is directly proportional to the number of people in the worker group. But how can we find the percentage? Again, common sense is helpful. For simplicity, let us assume that each of the women able to become pregnant will have two children (on the average) while she is from 15 to 64 years of age. So we would have to multiply the number of women with 1/25 (or 0.04) to get the number of births per year. This calculation only takes into account the female members of the worker group. If in addition we assume approximately equal numbers of men and women we have to reduce this factor by 1/2, so our first attempt to get an estimate for this factor is 0.02. Figure 2.39 represents the complete flowchart.

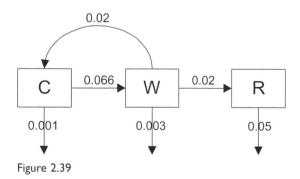

Figure 2.39

We start setting up the spreadsheet table in Figure 2.40.

death rates		
children	workers	retired
0.001	0.003	0.05
transition		
children	workers	
0.066	0.02	
birth rate		
0.02		

Figure 2.40

Now we need starting values for our population. Since we are mainly concerned with the proportions of the different groups in relation to each other, the absolute number is not that important but we have to find numbers representing the proportions in real populations. We will assume that the total population is 1,000,000. If each year the same number of children were born and we could neglect the effect of people dying before reaching the age of 85, we would have a proportion of 15:50:20 between the three age groups. Since these assumptions are not true, we have to try to correct the induced errors.

Each age group becomes smaller from year to year, and death rates are even higher in older age groups than in younger age groups. Therefore, older age groups have smaller numbers of people still alive for each birth year than the younger age groups. A proportion of 15:50:20 would mean percentages of 15/85 or 0.177 (or 17.7%), 50/85 or 0.588 (or 58.8%), and 20/85 or 0.235 (or 23.5%), respectively. Adjusting for these effects of death rates, let us instead assume proportions of 0.19, 0.65, and 0.16. These numbers should not be assumed to be actual data values; they are just "educated guesses" as to what range the actual numbers might lie in.

So we continue to set up our spreadsheet in Figure 2.41.

death rates			
children	workers	retired	
0.001	0.003	0.05	
transition			
children	workers		
0.066	0.02		
birth rate			
0.02			
period	children	workers	retired
1	190000	650000	160000

Figure 2.41

Now we have to calculate the changes for each group. The change in the number of children is given by the following expression:

workers × birth rate – children × transition children – children × death rate children

When we set up the arrow diagram corresponding to this expression it is somewhat complicated, involving six arrows (Figure 2.42).

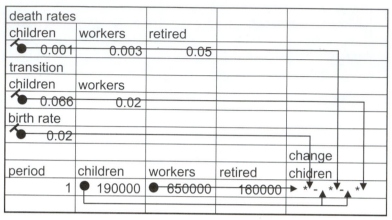

death rates				
children	workers	retired		
0.001	0.003	0.05		
transition				
children	workers			
0.066	0.02			
birth rate				
0.02				
			change	
			chidren	
period	children	workers	retired	
1	190000	650000	160000	*▼ - ▲ *▼ - ▲ *▼

Figure 2.42

Let us note that for all of the periods the current numbers of the population groups will be used, so their references to these cells should be relative (they will always be taken from the same line as the values to be calculated). On the other hand, the transition rates will always be taken from the same cells at the head of our table, so these references should be absolute. The corresponding "arrow tails" in Figure 2.42 already indicate this.

The expression for the change for the workers is

children × trans rate children – workers × trans rate worker
– workers × death rate worker

and is represented by Figure 2.43.

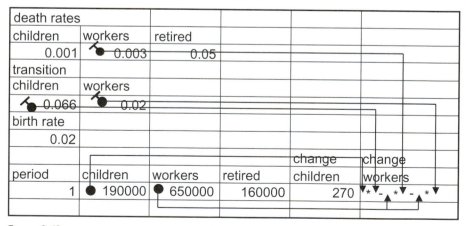

death rates					
children	workers	retired			
0.001	0.003	0.05			
transition					
children	workers				
0.066	0.02				
birth rate					
0.02					
				change	change
				children	workers
period	children	workers	retired		
1	190000	650000	160000	270	*▼ - ▲ *▼ - ▲ *▼

Figure 2.43

The change for the retired can be expressed as

workers × transition rate workers − retired × death rate retired

and is indicated by Figure 2.44.

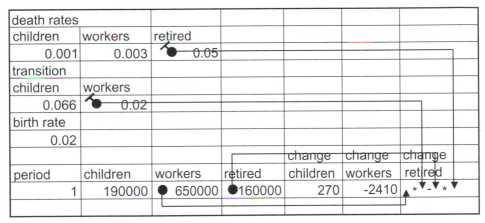

death rates						
children	workers	retired				
0.001	0.003	0.05				
transition						
children	workers					
0.066	0.02					
birth rate						
0.02						
				change	change	change
period	children	workers	retired	children	workers	retired
1	190000	650000	160000	270	-2410	* ▼ - ▼ * ▼

Figure 2.44

Now we want to calculate the population numbers for Period 2. The period number is simply 1 plus the number in the cell above, and the new number of children is the number of children in Period 1 plus the change calculated for Period 1. These two formulas are expressed in Figure 2.45.

death rates						
children	workers	retired				
0.001	0.003	0.05				
transition						
children	workers					
0.066	0.02					
birth rate						
0.02						
				change	change	change
period	children	workers	retired	children	workers	retired
1	190000	650000	160000	270	-2410	5000
+ 1	+					

Figure 2.45

To calculate the new numbers of workers and retired people, we only have to copy the formula for the children to the right, since in arrow notation we have the relations shown in Figure 2.46.

death rates						
children	workers	retired				
0.001	0.003	0.05				
transition						
children	workers					
0.066	0.02					
birth rate						
0.02						
				change	change	change
period	children	workers	retired	children	workers	retired
1	190000	● 650000	● 160000	270	● -2410	● 5000
2	190270	▼ + ◄	▼ + ◄			

Figure 2.46

The formulas for the changes for our three groups in Period 2 can be obtained simply by copying down the formulas directly above them (Figure 2.47). Note that these formulas have been designed to make them "copyable."

death rates						
children	workers	retired				
0.001	0.003	0.05				
transition						
children	workers					
0.066	0.02					
birth rate						
0.02						
				change	change	change
period	children	workers	retired	children	workers	retired
1	190000	650000	160000	270	-2410	5000
2	190270	647590	165000			

Figure 2.47

Now, as in the previous examples, we have to copy down the rest of the formulas. If we want to study the population changes over the next 100 years, then we have to copy these formulas down to the next 100 rows. Because of limited space, the printed table in Figure 2.48 does not extend that far.

death rates					change	change	change
children	workers	retired					
0.001	0.003	0.05					
transition							
children	workers						
0.066	0.02						
birth rate							
0.02							
period	children	workers	retired	children	workers	retired
1	190000	650000	160000	270	-2410	5000
2	190270	647590	165000	203.71	-2336.75	4701.8
3	190473.7	645253.3	169701.8	143.3264	-2269.56	4419.975
4	190617	642983.7	174121.8	88.33236	-2207.9	4153.585

Figure 2.48

To visualize the development of our three population groups, we could simply plot a curve for each of the groups. Having tried this, we know that such a graph does not show the effects very clearly. Therefore, we will use another graph type, the *stacked-line graph*, also called an *area graph*. See Figure 2.49.

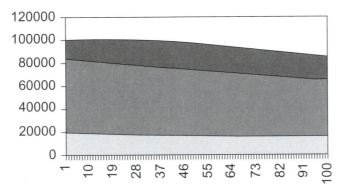

Figure 2.49

The heights of the three different areas at each *x*-value give the number of people in the three population groups (children at the bottom, workers in the middle, retired at the top).

When setting up this graph, we have to be careful about which group to assign to the first data series, which group to the second data series, and so on. Some spreadsheet programs put the area for the first group at the bottom of the graph and some put it at the top, but in most cases this can be adjusted afterward. Before having the final version of the graph, we may have to experiment to find out which data values to assign to which data series.

Since we are studying the implications of financing the retirement plans or pensions, it is very important to know the relative proportions of the different age groups. To do so we introduce a new column, total population (Figure 2.50).

death rates							
children	workers	retired					
0.001	0.003	0.05					
transition							
children	workers						
0.066	0.025						
birth rate							
0.025							
				change	change	change	
period	children	workers	retired	children	workers	retired	total pop
1	190000	650000	160000	270	-2410	5000	
2	190270	647590	165000	203.71	-2336.75	4701.8	
3	190473.7	645253.3	169701.8	143.3264	-2269.56	4419.975	
4	190617	642983.7	174121.8	88.33236	-2207.9	4153.585	

Figure 2.50

This column just adds the values from the three age groups, and we can copy this formula down.

Now we can calculate the percentages for each age group over time by dividing the number in each group by the total population in the respective period. For the children, this is depicted in Figure 2.51 (to get our full table on the page, we have to change the column widths and the number format):

death rates								
children	workers	retired						
0.001	0.003	0.05						
transition								
children	workers							
0.066	0.025							
birth rate								
0.025								
				change	change	change		
period	children	workers	retired	children	workers	retired	total pop	% child
1	190000	650000	160000	270	-2410	5000	1000000	/
2	190270	647590	165000	204	-2337	4702	1002860	
3	190474	645253	169702	143	-2270	4420	1005429	
4	190617	642984	174122	88	-2208	4154	1007723	

Figure 2.51

The formula for the workers is shown in Figure 2.52.

period	children	workers	retired	change children	change workers	change retired	total pop	p child	p work
1	190000	650000	160000	270	-2410	5000	1000000	0.19	
2	190270	647590	165000	204	-2337	4702	1002860		

Figure 2.52

The last two formulas that we created are similar. The second one, however, cannot be obtained by copying the first one since then the reference to the total population number would be moved away from its cell. Making this reference absolute would not help either, because then copying down the formula for later periods would not work. We need a new kind of reference! This reference should stay within the same column all the time, but it should move up and down to another row when the formula containing it is copied to another row. A useful metaphor for this situation is an object sliding along rails; therefore we will use a new kind of arrow (Figure 2.53) to indicate a reference that does not change the *column* it refers to but may change the *row* it refers to. In *Microsoft Excel*, we can create such references by pressing the [F4] key to toggle until the column reference is absolute and the row reference is relative. This is discussed in the Appendix.

period	children	workers	retired	change children	change workers	change retired	total pop	p child	p work
1	190000	650000	160000	270	-2410	5000	1000000	0.19	
2	190270	647590	165000	204	-2337	4702	1002860		

Figure 2.53

To get the percentages for workers and retired people, we only have to copy this formula into the next cell to the right (Figure 2.54).

period	children	workers	retired	change children	change workers	change retired	total pop	p child	p work	p retir
1	190000	650000	160000	270	-2410	5000	1000000	0.19	0.65	
2	190270	647590	165000	204	-2337	4702	1002860			

Figure 2.54

Now we only have to copy down the formulas in the last three columns of this row to get the percentages of the age groups for all the following periods (Figure 2.55).

period	children	workers	retired	change children	change workers	change retired	total pop	p child	p work	p retir
1	190000	650000	160000	270	-2410	5000	1000000	0.19	0.65	0.16
2	190270	647590	165000	204	-2337	4702	1002860	0.1897	0.6457	0.1645
3	190474	645253	169702	143	-2270	4420	1005429	0.1894	0.6418	0.1688
4	190617	642984	174122	88	-2208	4154	1007723	0.1892	0.6381	0.1728

Figure 2.55

Next we will set up an area graph using these percentages instead of the absolute numbers (Figure 2.56).

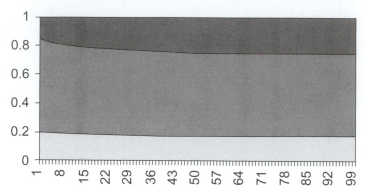

Figure 2.56

This graph shows more clearly the problems underlying the funding of pensions: the percentage of retired people steadily increases. Since the pensions are funded by payments from working people, the change of these percentages is the crucial factor behind financing retirement payments by taxes or social security.

This gives rise to a new number that we might calculate to summarize the problem: divide the number of retired people by the number of workers. This shows "the percentage of one retired person that a working person has to pay for." Alternatively, we could compute the reciprocal value for the number of workers needed to support a retired person.

The new column is easily added: we just need relative references, and we have the formula in Figure 2.57.

period	children	workers	retired	change children	change workers	change retired	total pop	% child	% work	% retir	ret/work
1	190000	650000	160000	270	-2410	5000	1000000	0.19	0.65	0.16	/
2	190270	647590	165000	204	-2337	4702	1002860	0.1897	0.6457	0.1645	
3	190474	645253	169702	143	-2270	4420	1005429	0.1894	0.6418	0.1688	
4	190617	642984	174122	88	-2208	4154	1007723	0.1892	0.6381	0.1728	

Figure 2.57

To get the development of this ratio over time, we simply have to copy the formula down often enough (Figure 2.58).

period	children	workers	retired	change children	change workers	change retired	total pop	% child	% work	% retir	ret/work
1	190000	650000	160000	270	-2410	5000	1000000	0.19	0.65	0.16	0.2462
2	190270	647590	165000	204	-2337	4702	1002860	0.1897	0.6457	0.1645	0.2548
3	190474	645253	169702	143	-2270	4420	1005429	0.1894	0.6418	0.1688	0.263
4	190617	642984	174122	88	-2208	4154	1007723	0.1892	0.6381	0.1728	0.2708

Figure 2.58

Graphing these new data values also shows the problem quite clearly (Figure 2.59).

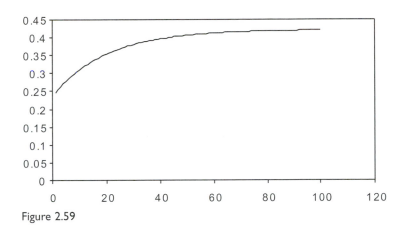

Figure 2.59

Now we can readily experiment with this model. We can study the development of the population with different values for these parameters. In reality, this is not very meaningful because death rates cannot be changed easily and the progress of medicine will have the effect of increasing the percentage of retired people. Nevertheless, we can try to find out how much higher birth rate would be needed to reduce or even compensate for the effect of the percentage of retired people going up. Then politicians could discuss means of increasing the birth rate, since this seems to be the only parameter in our system that can be influenced by political or demographic means.

Another possibility would be to change the starting age of retirement. Then we have to completely readjust our model, because all the rates (transition, birth, and death) would have to be changed (note that we derived their values as a consequence of the number of years a person spends in each of the age groups). So the model can still be used, but all six parameters have to be adjusted according to the new assumptions.

Further Mathematics: Experimenting with Population Dynamics

What is the classical mathematical notation of the model? Let us assume the names in Figure 2.60 for our parameters.

death rates		
children	workers	retired
α	β	γ
transition		
children	workers	
ρ	σ	
birth rate		
μ		

Figure 2.60

Also, let C_n, W_n, and R_n denote the numbers of children, workers, and retired people respectively. Let us additionally denote:

$$\Delta C_n = C_{n+1} - C_n$$
$$\Delta W_n = W_{n+1} - W_n$$
$$\Delta R_n = R_{n+1} - R_n$$

Then we have:

$$\Delta C_n = -\alpha C_n - \rho C_n + \mu W_n$$
$$\Delta W_n = -\beta W_n + \rho C_n - \sigma W_n$$
$$\Delta R_n = -\gamma R_n + \sigma W_n$$

If we are allowed to use linear algebra, and especially matrix notation, we can rewrite the model:

$$\begin{pmatrix} \Delta C_n \\ \Delta W_n \\ \Delta R_n \end{pmatrix} = \begin{pmatrix} -\alpha - \rho & \mu & 0 \\ \rho & -\beta - \sigma & 0 \\ 0 & \sigma & -\gamma \end{pmatrix} \begin{pmatrix} C_n \\ W_n \\ R_n \end{pmatrix}$$

An important question in such a model is whether for given parameters there are stable populations, that is, populations for which the numbers in each group do not change over time. But that means that the changes all must be zero. Therefore the preceding matrix equation must have a solution for

$$\begin{pmatrix} \Delta C_n \\ \Delta W_n \\ \Delta R_n \end{pmatrix} = \begin{pmatrix} -\alpha - \rho & \mu & 0 \\ \rho & -\beta - \sigma & 0 \\ 0 & \sigma & -\gamma \end{pmatrix} \begin{pmatrix} C_n \\ W_n \\ R_n \end{pmatrix} = \begin{pmatrix} 0 \\ 0 \\ 0 \end{pmatrix}.$$

This only can happen (for positive values for the population numbers) when the matrix is singular. So we can try to find sets of values that make the matrix singular.

Using the higher mathematics functions in spreadsheet programs, we even could calculate the determinant and change the values until we get a determinant of 0. This would, however, go far beyond what this book wants to demonstrate and therefore is left to the more advanced reader as an idea for extending this example.

The operator Δ introduced in these equations is called the *difference operator*, and equations using this operator are called *difference equations*.

Construction Summary: Population Dynamics

1. Enter parameter values.
2. Create a counter column.
3. Enter initial population values.
4. Compute the change in children as births less deaths and transitions to workers.
5. Compute the change in workers as influx of children less deaths and transitions to retired.
6. Compute the change in retired as influx from workers less deaths.
7. Compute updated number of children as the previous value plus the change.
8. Copy formulas as indicated.

death rates						
child	worker	retired				
0.001	0.003	0.05				
transition rates						
child	worker		1			
0.066	0.02					
birth rate						
0.020						
				change child	change workers	change retired
period	children	workers	retired			
[2] 1 [3]	190000	650000	160000	[4] 270	[5] -2410	[6] 5000
2 [7]	190270	647590	165000	204	-2337	4702
3	190474	645253	169702	143	-2270	4420
4	190617	642984	174122	88	-2208	4154
5	190705	640776	178275	38	-2151	3902
6	190744	638625	182177	-7	-2099	3664
7	190736	636525	185841	-49	-2051	3438
8	190687	634474	189279	-87	-2008	3226

Exercises

1. Design scroll bars to incorporate into this section's model that allow us to vary the values of parameters effectively.

2. Jar R contains 2 red balls and 1 blue ball. Jar B contains 1 red ball and 3 blue balls. We draw a ball at random from Jar R and replace it. If the ball drawn is red, then we draw again from Jar R; if it is blue, then we draw from Jar B. We continue this process indefinitely (see Sandefur (1993), p. 311). Design a

spreadsheet model to find the probabilities that the nth ball drawn is red or blue. What happens in the long run? In the long run, does it matter from which jar we start? A diagram of the process, created in *Excel*, is shown in Figure 2.61.

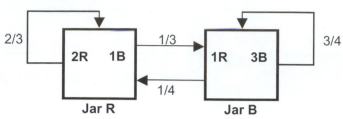

Figure 2.61

3. Create a spreadsheet model for a similar process using three jars (R,B,G) containing various numbers of red, blue, and green balls. Draw succeeding balls from Jar R if a red ball has been drawn, from Jar B if a blue ball has been drawn, and from Jar G if a green ball has been drawn. What happens if each jar contains at least one green ball but Jar G contains only green balls?

4. Exercise 2 and Exercise 3 provide us with examples of Markov chains. Repeat these exercises by employing probability matrices and matrix multiplication. For example, in Exercise 1 use the following matrix:

$$\begin{bmatrix} 2/3 & 1/3 \\ 1/4 & 3/4 \end{bmatrix}$$

5. The Volunteer Car Rental Company rents cars that can be picked up and delivered in either Nashville or Memphis. It currently has 600 cars in Nashville and 200 in Memphis. It is known that 60% of cars rented in Nashville are returned there, with the remaining 40% returned to Memphis. Also, 80% of the cars rented in Memphis are returned there, with 20% returned to Nashville. Design a model to find the distribution of cars at the two sites over a period of many days (see Giordano et al (2003), p. 83.). Does the ultimate distribution depend upon the initial allotments? A diagram of the states, created in *Excel*, is shown in Figure 2.62.

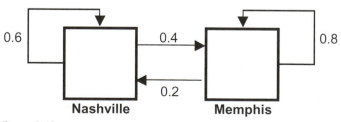

Figure 2.62

6. Between consecutive elections in a certain country, it is known that voters switch parties (Liberal, Conservative, Independent) with probabilities given in Figure 2.63. If in the last election the proportions of voters were Liberal 0.45, Conservative 0.35, and Independent 0.20, determine the proportions in future elections. In the long run, which party will have the most voters? (See Sandefur (1993), p. 313).

		New Party		
		Lib	Con	Ind
Old	Lib	0.67	0.12	0.21
Party	Con	0.17	0.58	0.25
	Ind	0.15	0.10	0.75

Figure 2.63

7. In this exercise, we look at the females of a certain species of animal. Suppose that they live for 3 years and that 50% of the first-year animals die, while 40% of the second-year animals die. During the first year, a female is incapable of reproducing. In the second year, females reproduce an average of 0.7 female each; while in the third year, females produce an average of 3.1 females each. This information is contained in Figure 2.64. Suppose that a certain population starts with 50 females. Create a spreadsheet model to project the number of females in succeeding generations. Use a probability matrix together with matrix multiplication in your design. Design your model so that the variables are parameters, and investigate the values that lead to populations that grow or contract. See Mooney and Swift (1999) for further discussion on similar topics.

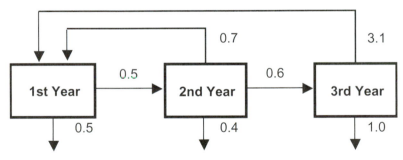

Figure 2.64

8. Two species (Alpha, Beta) of an animal compete for territory. Figure 2.65 gives the probabilities of a given territory changing from one species to another within a 5-year cycle. Suppose that initially the proportions or territories occupied by the indicated species are given by the proportions 0.40, 0.30, 0.25, and 0.05. Create a spreadsheet model using matrix multiplication to project the distribution over succeeding years. Determine the long-range implications. Create an xy-graph of the results. Examples similar to this are found in Chapter 3 of Mooney and Swift (1999).

From\To	Alpha	Beta	Joint	Missing
Alpha	0.68	0.05	0.15	0.12
Beta	0.02	0.8	0.03	0.15
Joint	0.1	0.02	0.82	0.06
Missing	0.06	0.03	0.04	0.87

Figure 2.65

9. Another way to examine the previous models is by computing powers of the transition matrix. Use the compact way of doing this to find powers of the transition matrix of Exercise 8. The first steps of typical output are shown in Figure 2.66, where we have computed successively the square of the transition matrix and the fourth power (the square of the square) of the transition matrix. Observe that the powers eventually reach a steady state.

n = 1			
0.68	0.05	0.15	0.12
0.02	0.80	0.03	0.15
0.10	0.02	0.82	0.06
0.06	0.03	0.04	0.87
n = 2			
0.4856	0.0806	0.2313	0.2025
0.0416	0.6461	0.0576	0.2547
0.1540	0.0392	0.6904	0.1164
0.0976	0.0539	0.0775	0.7710
n = 4			
0.2945	0.1112	0.2923	0.3019
0.0808	0.4368	0.1063	0.3761
0.1941	0.0711	0.5236	0.2113
0.1368	0.0873	0.1389	0.6370

Figure 2.66

2.4 A Predator-Prey Model

In previous models, we have considered the growth of a single population, possibly grouped into subpopulations. Here we create a model for two populations that interact with each other. One population is that of a predator that feeds upon the other population, the prey population. In this model, the sizes of the two populations tend to oscillate. This basic model dates back to the initial work of Lotka and Volterra in the 1920s and 1930s.

The model studies two animal populations occupying the same territory. An example of this could be populations of foxes and rabbits. The two species depend upon each other, but in different ways. The prey population (the rabbits in our example) would grow indefinitely with a constant growth rate, as in the first growth model that we studied, if there were no foxes. The predator population (the foxes) would die out without rabbits, and the rate of change (that is, the percentage of foxes vanishing from one period to the next one) is assumed to be constant. Additionally, the model assumes that the growth rate of the rabbits decreases in proportion to the number of foxes present. A simple explanation for this would be that foxes feed on rabbits, thereby reducing their number. On the other hand, the rate of change in the number of foxes goes up in proportion to the number of rabbits present as we assume that more rabbits cause more foxes to survive since the foxes feed on rabbits.

So in a more formalized way the assumptions are:

$$\frac{\text{change rabbits}}{\text{rabbits}} = \text{initial growth rate rabbits} - \text{prey factor} \times \text{foxes}.$$

The constant prey factor can be interpreted as the percentage of existing rabbits that one fox feeds on per period. This is, of course, a simplification. Over a broader range we cannot assume that given the same number of foxes and double the number of rabbits one fox would feed on twice as many rabbits. However, the model at least qualitatively expresses the fact that more foxes reduce the growth rate of the rabbits and more rabbits increase the growth rate of the foxes.

For the foxes we have the equation

$$\frac{\text{change foxes}}{\text{foxes}} = -\text{initial diminishing rate foxes} + \text{feeding rate} \times \text{rabbits}.$$

The feeding rate expresses the fact that better food conditions—meaning more rabbits to feed on—cause the foxes to have more offspring. The actual change in the number of foxes and rabbits of course is obtained by multiplying both equations with the respective denominator term on the left side.

Given these equations, we can set up the model as a spreadsheet table (Figure 2.67).

growth rate rabbits		0.04		
prey factor		0.0004		
dimin. rate foxes		0.08		
feeding rate		0.0001		
	number		change	
period	rabbits	foxes	rabbits	foxes
1	1000	100		

Figure 2.67

Figure 2.67 shows the initial data. Now let us set up the equations for change. The arrow diagram for the rabbits is shown in Figure 2.68.

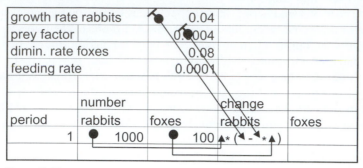

growth rate rabbits		0.04		
prey factor		0.0004		
dimin. rate foxes		0.08		
feeding rate		0.0001		
	number		change	
period	rabbits	foxes	rabbits	foxes
1	1000	100	▲* (◀ - ◀*▲)	

Figure 2.68

Figure 2.69 is the arrow diagram for the change in the number of foxes.

growth rate rabbits		0.04		
prey factor		0.0004		
dimin. rate foxes		0.08		
feeding rate		0.0001		
	number		change	
period	rabbits	foxes	rabbits	foxes
1	1000	100	0	▲* (-▼ +▼*▲)

Figure 2.69

Now we want to generate the numbers for Period 2. We can do this by constructing the formulas shown in the Figure 2.70.

growth rate rabbits		0.04		
prey factor		0.0004		
dimin. rate foxes		0.08		
feeding rate		0.0001		
	number		change	
period	rabbits	foxes	rabbits	foxes
1	1000	100	0	2
+1	+	+		

Figure 2.70

Since the formulas for change refer only to values from the same row, we can simply copy the formulas down (see Figure 2.71).

growth rate rabbits			0.04		
prey factor			0.0004		
dimin. rate foxes			0.08		
feeding rate			0.0001		
	number			change	
period	rabbits	foxes		rabbits	foxes
1	1000	100		0	2
2	1000	102		-0.8	10.1898

Figure 2.71

As in most of our previous examples, the formulas for Period 2 completely describe all of the calculations necessary to get from one period to the next. Thus, we can simply copy this row down as far as we desire. We will need at least 500 periods to study some interesting phenomena. Figure 2.72 shows the initial rows of the copies.

growth rate rabbits		0.04			
prey factor		0.0004			
dimin. rate foxes		0.08			
feeding rate		0.0001			
	number			change	
period	rabbits	foxes		rabbits	foxes
1	1000	100		0	2
2	1000	102		-0.8	2.04
3	999.2	104.04		-1.614707	2.072477
4	997.5853	106.1125		-2.439087	2.096626
5	995.1462	108.2091		-3.267703	2.11166

Figure 2.72

Now let us graph the number of both species against the time axis. For this graph the period values provide the x-axis values (the independent variable), and the numbers of rabbits and foxes are the two y-axis variables (the dependent variables). The graph should look similar to the one in Figure 2.73, in which the upper curve shows the number of rabbits and the lower curve provides the number of foxes.

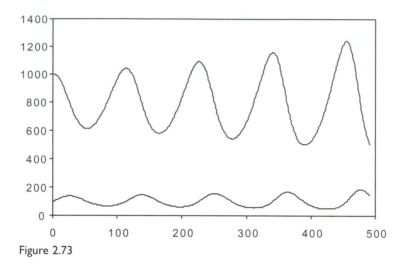

Figure 2.73

We see oscillations with the same period but with a *phase shift*, meaning that the maximal values for rabbits and foxes do not occur at the same time. We also see that the amplitude of these oscillations (the distance between the highest and the lowest point of one "wave") increases over time.

What will happen if somebody wants to change the system behavior? We could assume that in a certain period—let us choose Period 27—a group of hunters decides that there are too many foxes and therefore shoots some of the foxes so that only 67 remain (Figure 2.74).

period	rabbits	foxes	rabbits	foxes
26	796.5656	138.376	-12.22759	-0.047523
27	784.338	67	10.35326	-0.104935
28	794.6913	66.89506	10.52328	-0.035513

Figure 2.74

To study the consequences of this change, all that we need to do is to enter the number 67 into the cell for the number of foxes in Period 27. Looking at the graph again (Figure 2.75), we see some surprising results.

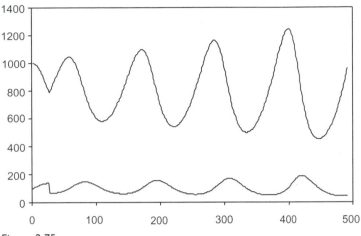

Figure 2.75

The system behavior does not change qualitatively. When we reduce the number of foxes, the number of rabbits goes up immediately; but even with this change, we get essentially the same cycling behavior with the same amplitudes as without this change in the system. So let us put the system back to its original behavior without the external change in the number of foxes in Period 27 (Figure 2.76). Since the contents of the marked cell was changed from the formula to an explicit value, we have only to restore the original formula instead of the value. This can be achieved by just copying the formula from the cell above (Figure 2.77).

period	rabbits	foxes	rabbits	foxes
26	796.5656	138.376	-12.22759	-0.047523
27	784.338	67	10.35326	-0.104935
28	794.6913	66.89506	10.52328	-0.035513

Figure 2.76

period	rabbits	foxes	rabbits	foxes
26	796.5656	138.376	-12.22759	-0.047523
27	784.338	138.3284	-12.02498	-0.21665
28	772.3131	138.1118	-11.7737	-0.382389

Figure 2.77

Since the numbers of foxes and rabbits are highly dependent upon one another in our model, it might be useful to have a graph showing the relation between these two numbers more directly, without an explicit time axis. So for each of the periods we will plot a point with the number of rabbits for the *x*-value and the number of foxes for the *y*-value. And we will connect these points, showing an "orbit" of our system. When we select the values for the axes accordingly we get the graph shown in Figure 2.78.

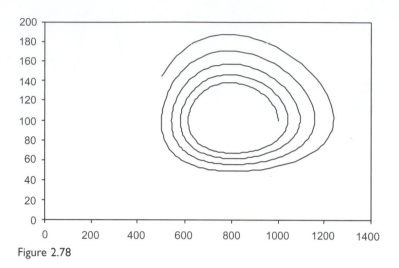

Figure 2.78

This graph makes it easier to understand what happens: the "state" of the system, that is, the number of rabbits and foxes together interpreted as points in the plane, follows a more or less circular orbit. So the numbers of both rabbits and foxes go up and down cyclically but not synchronously. Just following the path of a point along the orbit, we can see how the increase and decrease in both numbers follow a pattern. Now let us make our change to the system again. We manually change the number of foxes in Period 27 to 67. Then the graph of the orbit appears as shown in Figure 2.79.

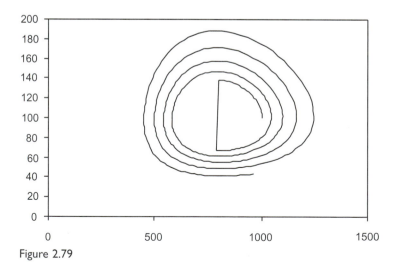

Figure 2.79

The graph shows that essentially we just skip half a circular orbit but stay on the same orbit path. We can experiment with our system by changing the initial conditions. We can, for example, start with 900 rabbits and 100 foxes (but before doing that we should restore the original formula in the "foxes cell" for Period 27 by copying the formula from the cell above).

We get the same kind of circular orbit; but, looking at the minimum and maximum values within the cycles, we see that the amplitude decreases. The center of our circular orbit, however, remains more or less the same. Essentially the same thing happens when we change the starting number of the foxes to 120. Then the amplitude becomes wider again, but we still have the same kind of circular orbits. We get more or less the same effects also when we change the model constants (growth rate, prey rate, and so on) within reasonable bounds. Now it is time for you to do some experiments with different values and watch the results.

Having finished these experiments, let us restore the original values for our four parameters and let us change the initial conditions to 800 rabbits and 100 foxes. The graph is surprising; it consists of just one point.

We have found a stable state of our system. When we start with this configuration, the numbers of foxes and rabbits will remain constant over time. Seeing this we may be tempted to ask the following questions: given our four system parameters, is there always a stable point, and can we find it without experimenting? To reach an answer, we can consider the following facts: A stable point is characterized by the fact that the "system variables" (in our case the number of foxes and rabbits) do not change their values. So the changes (given by our formulas) have to be 0. Now let us look at the equation for the change in the number of rabbits:

$$\frac{\text{change rabbits}}{\text{rabbits}} = \text{initial growth rate rabbits} - \text{prey factor} \times \text{foxes}.$$

The change is 0 when the right-hand side of the equation is 0:

$$\text{initial growth rate rabbits} - \text{prey factor} \times \text{foxes} = 0,$$

which is true when

$$\text{foxes} = \frac{\text{initial growth rate rabbits}}{\text{prey factor}}.$$

A similar equation holds for the change in the number of foxes. This change also has to be 0. Since the change fulfills the following equation:

$$\frac{\text{change foxes}}{\text{foxes}} = -\text{initial diminishing rate foxes} + \text{feeding rate} \times \text{rabbits}.$$

it is 0 when

$$-\text{initial diminishing rate foxes} + \text{feeding rate} \times \text{rabbits} = 0,$$

which is true when

$$\text{rabbits} = \frac{\text{diminishing rate foxes}}{\text{feeding rate}}.$$

Thus, we found a system of two equations for which any set of values for our four system parameters yields a stable state.

Using the numerical values for the parameter values in the original version of our examples, we find the stable point for our system at 100 foxes and 800 rabbits. Looking at some of the cycling graphs again shows that the orbits more or less cycle around the stable point for each configuration. Advanced mathematical theory can show that this is a general principle for this type of equation.

Knowing this, we also see what can happen if we change the system by external influences such as shooting some of the animals. If we want to stabilize the system by bringing it to an orbit close to the stable point, we should try to find the stable point and then try to change the number of one of the animal species just enough to get to such a close orbit. Changing the system too much might result in an orbit with even wider amplitudes than the original one. If we can change only one of the two system variables, timing is very critical. Then we have to find a time when the system variable we cannot change has the value needed for the stable system. Only if we change the other system variable to its stable value at this point in time can we bring the whole system into its stable state.

In our example, this implies that if we are allowed to shoot only foxes and our goal is to stabilize the system, then we have to wait for a time when there are 800 rabbits and more than 100 foxes, and then we have to shoot as many foxes as necessary to have just 100 remaining foxes. The important lesson to be learned from this fact is that if we want to stabilize a system like this one, then we need to have an idea where the stable point might lie. This point can be estimated by looking for the center of the more or less circular orbits. Additionally, timing is critical. When only certain types of changes are possible, we have to take care that we are in a state from which we will manage to reach the stable state under the given constraints.

Assume that our system parameters are given by Figure 2.80.

growth rate rabbits	α
prey factor	β
dimin. rate foxes	γ
feeding rate	δ

Figure 2.80

In classical mathematical notation, we can write our model in the following way:

$$\Delta R_n = R_n \left(\alpha - \beta F_n \right)$$
$$\Delta F_n = F_n \left(-\gamma + \delta R_n \right)$$
$$R_{n+1} = R_n + \Delta R_n$$
$$F_{n+1} = F_n + \Delta F_n$$

This type of equation is called a Volterra-Lotka equation.

Further Mathematics: Difference Equations versus Differential Equations

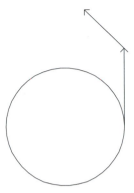

Figure 2.81

Considering the differential equations instead of the difference equations that we have studied in this section, there are some well known results. The solution of the system has closed convex orbits, and there are analytical equations describing these orbits. The solution orbits of our difference equations do not close, because we are approximating convex functions by piecewise linear functions starting in the tangent direction of the convex function. Essentially we are replacing convex curves by the tangents in the initial points. Looking at the graph in Figure 2.81, we immediately see that this results in the approximation curves not being closed.

A curved piece of the orbit is replaced by a straight segment, thereby going outward a little bit. In the next step we again go along a straight line instead of a curve, thereby going more outward and so on. So this example shows that for short-term analysis our model can be quite useful but that qualitatively different behavior can result from replacing the analytic solution of differential equations by the numerical solution of difference equations.

Construction Summary: A Predator-Prey Model

1. Enter parameter values.
2. Create a counter column.
3. Enter the initial numbers of rabbits and foxes.
4. Compute the change in rabbits as
   rabbits(grow$_{rab}$ − factor$_{prey}$ · foxes).
5. Compute the change in foxes as
   foxes(− dim$_{fox}$ + rate$_{feed}$ · rabbits).
6. Compute the updated rabbit population by adding the change to the previous value, copy.
7. Copy as indicated.

growth rate rabbits	0.04		
prey factor	0.0004	1	
dimin. rate foxes	0.08		
feeding rate	0.0001		

		number		change	
period		rabbits	foxes	rabbits	foxes
2 1	3	1000	100	4 0	5 2
2	6	1000.0	102.0	-0.8	2.0
3		999.2	104.0	-1.6	2.1
4		997.6	106.1	-2.4	2.1
5		995.1	108.2	-3.3	2.1
6		991.9	110.3	-4.1	2.1

Exercises

1. Sometimes two species live in the same territory and the presence of each species affects the growth rate of the other. To illustrate this situation, suppose that wolves and coyotes occupy the same territory, a territory that can support 300 wolves and 400 coyotes. The base growth rate for wolves is 0.05 and that for coyotes is 0.07. Without the presence of the other, each would increase from their initial sizes of 10 and 70, respectively, according to our logistic model. However, the growth rate of each is reduced in proportion to the frequency of interaction between the two species, which is proportional to the product of their sizes. Create a model for this situation using $\alpha = 0.00004$ as the constant of proportionality. See page 99 of Meerschaert (1993) for additional discussion of this topic.

2. Design scroll bars for the model of this section to allow us to vary the value of the parameters and initial populations.

3. Design animated graphs to illustrate different aspects of the operation of the predator-prey model. For example, use a scroll bar to cause a trace point to move through the graph as time varies, as in the graph in Figure 2.82, or for the orbit diagram.

4. Use the *Excel* RAND function to generate random points in a square grid in proportion to the number of animals of each type to illustrate the relative denseness of the two populations. One possible output is shown in Figure 2.83, with circles representing foxes and triangles representing rabbits. Vary the time using a scroll bar.

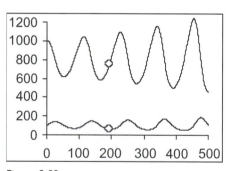

Figure 2.82

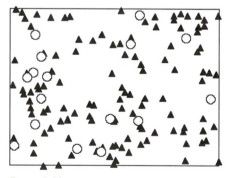

Figure 2.83

2.5 Iterated Maps and Functions

Until now, we have mostly modeled real-world phenomena, or at least modeled systems that try to express some aspects of the physical world mathematically. In this section, we focus on the mathematics of a purely mathematical phenomenon, the iteration of functions.

The main idea of this section is the following: Given a mathematical function $f(x)$, what will happen if we start with a point x, calculate the result of the function, use this value as the new input value again, and so on?

Curves and areas created by similar algorithms have been studied quite extensively during the last few years, and the investigations have led to *chaos theory* and *fractal curves*. We will use spreadsheets to study a number of properties of some of these curves.

Let us start with a simple example: For the function to be applied over and over again, we take the square function, $f(x) = x^2$. We first set up the spreadsheet table shown in Figure 2.84 where we observe the use of the symbol ^ to denote exponentiation within a spreadsheet.

The value generated in the second row is the square of the value in the first row. We want to iterate this process so that the value in Row 3 will be the square of the value in Row 2 and so on. We merely have to copy the formula down into the cells below. We will do at least 200 iterations. Additionally, we want to produce a graph for this sequence of values, so we will need values for the x-axis. To do this, we will just number the values sequentially. The problem is that we want these numbers in the column to the left of our column, but there is no such column. This is no real problem, since all spreadsheet programs allow us to insert a column at any place in the table. For *Excel*, this is discussed in the Appendix.

So we insert a column to the left of our values, making the table shown in Figure 2.85.

Now we add our first column (Figure 2.86).

Figure 2.84

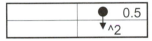

Figure 2.85

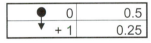

Figure 2.86

The second row contains all of the information for the model. Therefore we copy this row down into at least 200 rows. Figure 2.87 displays only the first few rows of this table. When numbers are very small, they are displayed in scientific notation as in the Figure 2.87.

0	0.5
1	0.25
2	0.0625
3	0.003906
4	1.53E-05
5	2.33E-10

Figure 2.87

We also may write the sequence of numbers we are studying here as

$$x, f(x), f(f(x)), f(f(f(x))), f(f(f(f(x)))), \ldots$$

Now we create a graph with the values in the first column for the x-axis and the values in the second column for the y-axis by selecting the appropriate menu items (Figure 2.88).

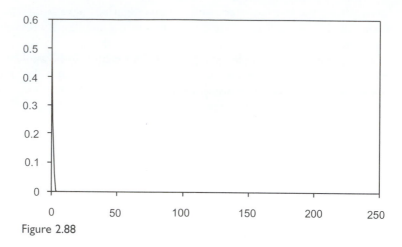

Figure 2.88

Next we will do some experiments with different values for the starting value of our sequence. We can try 0.6, 0.7, and 0.8 instead of 0.5 and look at the graph for each of these starting values. The behavior does not change too much; the graph still goes down to 0 very fast. However, using 1 as a starting value completely changes the graph—now it remains constant at 1. Taking any number larger than 1 makes our sequence explode numerically. Looking at the spreadsheet table, we see that the values become enormously large very quickly; so, in many spreadsheet programs, we do not even get a reasonable picture.

Nevertheless, we see that our system can show radically different behavior that is dependent upon the initial conditions. In some cases the values approach a certain value (0 in our example); in some cases the system has a fixed value and the iteration just stays at this fixed value (1 in our example); and in some cases the system diverges, meaning that the values become arbitrarily large.

The divergent behavior of our system is caused, among other reasons, by the fact that the function $f(x) = x^2$ can become arbitrarily large. As long as we choose starting values between 0 and 1, however, divergence cannot happen. The square of any number between 0 and 1 also lies between 0 and 1, so we do not leave this interval. We already have seen what happens with starting values taken from this interval: If we choose 1 as a starting value, then nothing changes, since the next point also is 1, and so all of the iterated points from this starting point also are 1. For any other starting point from our interval, the iterated points approach 0 very fast. This "uniform" behavior on the interval can be explained very easily. Our function is *monotone*, meaning that if we take starting values x and y with $x < y$, we also have $f(x) < f(y)$. Therefore, if we start, for example, with the value 0.9, we see that the iterated point of 0.9 is 0.81. Additionally we know that each starting value to the left of 0.9 will have its iterated point to the left of 0.81. So we see that the interval [0, 0.9] is compressed into the interval [0, 0.81]. Performing the next iteration step, we see that the iterated point of 0.81 is 0.6561, and again everything between 0 and 0.81 is compressed into the interval [0, 0.6561]. So the fact that our function $f(x) = x^2$ is monotone on the interval [0, 1] makes it clear that we can easily describe the behavior of the iterated points.

With some other functions some more interesting things can happen. For instance, let us study the function $f(x) = 2x(1 - x)$. This function has two properties that are important for our problem: Whenever we give it an input between 0 and 1, the output will also lie between 0 and 1 and it is not monotone on [0, 1].

To study this function, let us create its table of values and its graph with a spreadsheet. Tables of values and graphs of functions are very important tools in mathematics. We will use them quite often in other examples.

First of all we have to decide for what range of x-values we want to get the function values. In our case this is easy; we want to study the function for x-values from 0 to 1. Next, we have to decide how "finely grained" we want the x-axis. Quite often it makes sense to use equally spaced x-values, so we will use a sequence of x-values where each one is 0.025 larger than the previous one. So we will have 41 x-values (yes, it is 41, not 40, since we divide the interval [0, 1] into 40 intervals of length 0.25, and this yields 41 interval endpoints). So we set up our table as shown in Figure 2.89.

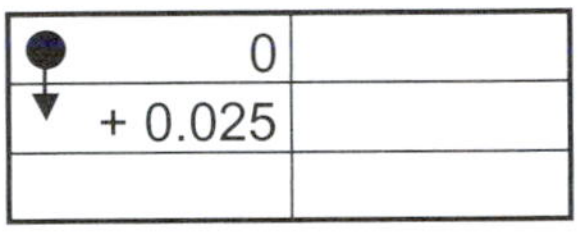

	0	
+ 0.025		

Figure 2.89

The values of our function are generated by the formula in Figure 2.90.

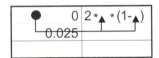

Figure 2.90

We can copy the formula for the function value into the second row since it contains only relative references (Figure 2.91).

	0		0
0.025			

Figure 2.91

The second row is the prototype for all the remaining rows. Therefore we can copy it down as shown in Figure 2.92.

0	0
0.025	0.04875
0.05	0.095
0.075	0.13875
0.1	0.18
0.125	0.21875

Figure 2.92

How many copies of this row do we need? We just noticed that we will need 41 rows in our value table, so we need 39 copies of this row. So when copying we could take care of the number of rows we need by "counting" the copies, that is, do mental bookkeeping about the row numbers. But there is an easier way. We know that the last row of our table has to contain 1 in the first column. So we can create a number of copies slightly higher than what we assume we need, so that the bottom part of our table looks like that shown in Figure 2.93.

0.95	0.095
0.975	0.04875
1	0
1.025	-0.05125
1.05	-0.105
1.075	-0.16125

Figure 2.93

In this table we easily recognize which rows we do not need any more, and with just one "deleting operation" we can get rid of these rows. So all that we need to do is to delete the range shown in Figure 2.94.

0.95	0.095
0.975	0.04875
1	0
1.025	-0.05125
1.05	-0.105
1.075	-0.16125

Figure 2.94

Now we create the graph for our table with 41 rows. This is done by selecting the columns with the numbers for the x-values and for the y-values and using the menu command for creating an xy-graph. Again, for a more detailed account of how to graph numbers in columns, see the Appendix.

Most spreadsheet programs have two types of graphs that seem to serve the same purpose: *line graphs* and *xy-graphs*. For mathematical purposes, the xy-graph almost always is the better choice. The line graph automatically assumes equally spaced x-values, and essentially it treats the x-values only as labels to be used on an equally spaced x-axis. On the other hand, xy-graphs allow one to use x-values with varied spacing since they plot points by their (x,y)-coordinates. They also offer more choices for setting marks for the x-axis. So throughout this book we always will use the xy-graph.

The graph we just produced is shown in Figure 2.95. The actual graph that you get also might have markers on the points representing the xy-value pairs, which is usually not what we desire when displaying a graph of a mathematical function. Spreadsheet programs also offer many possibilities of formatting the graphs, and information about these topics also is contained in the Appendix.

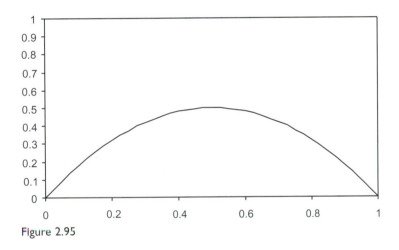

Figure 2.95

The graph you get may not look exactly like the one in Figure 2.95. The y-axis probably stops at the value 0.5. Since the scale can be very important when we are dealing with mathematics, it is quite helpful to be able to control the minimum and maximum values that will be used for the x-axis and the y-axis. You should format your graph in a way that it looks similar to the one in the figure, with both the x-axis and the y-axis ranging from 0 to 1, and with markers set equally spaced with an increment of 0.5. This topic is discussed in the Appendix.

When we were studying the function $f(x) = x^2$, we noticed that it was helpful to know if the function under consideration has the property $f(x) < x$. We also want to know this about our new function, $f(x) = 2x(1 - x)$. One way of doing this is trying to find out graphically by displaying the graphs of two functions simultaneously in one graph. The two functions we want to display are $f(x) = 2x(1 - x)$ and $g(x) = x$.

So we need to create an additional column in our table containing the values for $g(x)$. To make our table more readable we will add column headings. To do this, we insert a new row in the place of our first row (Figure 2.96).

x	f(x)	g(x)
0	0	
0.025	0.04875	
0.05	0.095	
0.075	0.13875	
0.1	0.18	
0.125	0.21875	

Figure 2.96

Since we have $g(x) = x$, the spreadsheet formula for the first term in the column labeled $g(x)$ is straightforward to create (Figure 2.97).

x	f(x)	g(x)
0	0	
0.025	0.04875	
0.05	0.095	
0.075	0.13875	
0.1	0.18	
0.125	0.21875	

Figure 2.97

Additionally, since the input for the last formula only contains one row reference, we copy it down to get the other function values we need (Figure 2.98).

x	f(x)	g(x)
0	0	0
0.025	0.04875	0.025
0.05	0.095	0.05
0.075	0.13875	0.075
0.1	0.18	0.1
0.125	0.21875	0.125

Figure 2.98

In the next step, we create an xy-graph. To create our graph, we select the column labeled x for the x-values and the columns labeled $f(x)$ and $g(x)$ for the y-axis. The process of creating xy-graphs with more than one column of y-values is discussed later in the book as well as in the Appendix. To create our graph, we select all three data columns (Figure 2.99).

x	f(x)	g(x)
0	0	0
0.025	0.04875	0.025
0.05	0.095	0.05
0.075	0.13875	0.075
0.1	0.18	0.1
0.125	0.21875	0.125

Figure 2.99

Of course, this table only shows the topmost part of the table we work with on our computer. The table we need has x-values going up to 1. Selecting the xy-type graph produces a graph similar to the one shown in Figure 2.100.

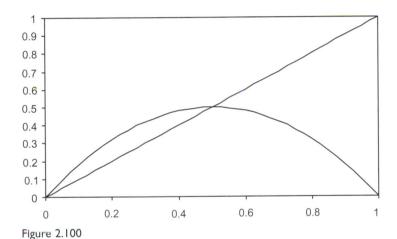

Figure 2.100

The straight line is the graph of $g(x)$ and the parabolic line is the graph of $f(x) = 2x(1 - x)$. This graph shows that for x-values between 0 and 0.5 we have $f(x) > g(x)$ and for x-values between 0.5 and 1 we have $f(x) < g(x)$. So the argument that helped us to understand that all the sequences

$$x, f(x), f(f(x)), f(f(f(x))), f(f(f(f(x)))), \ldots \text{ for } f(x) = x^2$$

and different starting values of x all converge to 0 cannot be applied with our new function $f(x) = 2x(1 - x)$. Using this function we can set up the "iteration sheet" shown in Figure 2.101.

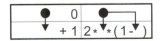

Figure 2.101

Since each input is taken from the previous row, we copy down the second row far enough (say 200 rows) to obtain many iterations. We copy as indicated in Figure 2.102.

0	0.05
1	0.095
2	0.17195
3	0.284766
4	0.407349

Figure 2.102

Using 200 iterated values and creating an xy-graph with the "running numbers" as the x-values and the iterated function values as the y-values, we get the graph shown in Figure 2.103. Notice that instead of plotting lines that connect the (x,y)-coordinates provided by the table of values, we have chosen to plot only markers at each of the points. In *Excel*, we can either select this option at the time that we create an xy-graph or we can right-click on an existing curve in an xy-graph and from the ensuing dialog box choose the options Format Data Series, Patterns and from the options provided choose to display only markers but not lines.

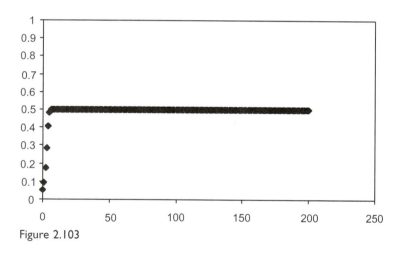

Figure 2.103

Using different starting values for x, we see that the overall behavior of this graph does not change. The graph stabilizes at $y = 0.5$ very quickly for any starting value between 0 and 1. What is so special about the value 0.5? We notice that for our function $f(x) = 2x(1 - x)$ we have $f(0.5) = 0.5$. So, for our function, input 0.5 produces output 0.5. Using the output as input again produces 0.5. If we iterate that process, we always will get the same number that we started with. So if for a certain value of x we have $f(x) = x$, the sequence

$$x, f(x), f(f(x)), f(f(f(x))), \ldots$$

simplifies to

$$x, x, x, x, \ldots$$

So we obtain a constant orbit, meaning that all the iterated values are identical. Points x with the property $f(x) = x$ are called *fixed points* of the function f. Thus, for $f(x) = 2x(1 - x)$ the point $x = 0.5$ is a fixed point. Additionally, wherever we choose the starting point of our sequence of iterated points, the sequence approaches this fixed point.

Using 0 as a starting value, we see that the sequence stays at 0. The value 0 also is a fixed point of our function since $f(0) = 0$. Using any starting point near 0, we see that the sequence of values does not get closer to 0. So with both 0 and 0.5 being fixed points—producing constant iteration sequences or orbits—we see different effects on nearby points. Loosely speaking, we might say that 0.5 seems to attract nearby points whereas 0 seems to push away nearby points.

Further Mathematics: Fixed Points and Calculus

Using some algebra and calculus we can further shed some light onto what is happening: Fixed points are defined by the equation $f(x) = 2x(1 - x) = x$, so all fixed points can be found by solving this equation. The quadratic equation in our example has only two solutions, 0 and 0.5. We have seen these two fixed points "at work." Why is one of them attractive and one of them repulsive? There is one main result dealing with that problem: We have to study the first derivative of the function for the fixed point. If the absolute value of the derivative is less than 1, then the fixed point is attractive. This can be proven using elementary results from calculus. If the absolute value is larger than 1, then the fixed point is repulsive. Nothing can be deduced from the derivative if its value is 1.

Construction Summary: Function Iteration for Square Function

1. Create a counter column.
2. Enter 0.5 as initial value of x.
3. Compute square of the value of the cell above.
4. Copy as indicated.

1		0	2	0.5
		1	3	0.25
		2		0.0625
		3		0.003906
		4		1.53E-05

---

1. Enter 0 as the initial x-value.
2. Increment the previous value of x by 0.025 and copy down.
3. Compute $2x(1 - x)$ and copy down.
4. Reproduce x and copy down.

	x		f(x)		g(x)
1	0	3	0.00000	4	0
2	0.025		0.04875		0.025
	0.050		0.09500		0.050
	0.075		0.13875		0.075

Exercises

1. Imagine the interval from 0 to 1 as a rubber band. Stretching it by a factor of 2 and then folding the right half "on top" of the left half moves point x to point $\min(2x, 2(1 - x)) = 2\min(x, 1 - x)$. Therefore $f_1(x) = 2\min(x, 1 - x)$ gives the position of point x after folding. Distorting the rubber band by moving point α slightly to the right to point β and stretching the interval between 0 and α to the interval between 0 and β and shrinking the interval between α and 1 to the interval between β and 1 is expressed by

$$f_2(x) = \begin{cases} \dfrac{\beta}{\alpha} x, & x \leq \alpha \\ \dfrac{1 - \beta}{1 - \alpha}(x - 1) + 1, & x > \alpha \end{cases}$$

Folding first and then distorting therefore is expressed by $g(x) = f_2(f_1(x))$.

$$g(x) = f_2(f_1(x)) = \begin{cases} \dfrac{\beta}{\alpha} 2\min(x, 1 - x), & \min(x, 1 - x)\alpha \\ \dfrac{1 - \beta}{1 - \alpha}(2\min(x, 1 - x) - 1) + 1, & \min(x, 1 - x) \leq \alpha \end{cases}$$

Investigate the "iteration structure" of this function with $\alpha = \frac{2}{3}$ and $\beta = \frac{3}{4}$.

2.6 More about Iterated Functions

In the previous section, we used the formula $f(x) = 2x(1 - x)$ for iterating the values of a more or less arbitrary starting point. We saw that the shape of the iteration orbit (the sequence of all the points x, $f(x)$, $f(f(x))$, $f(f(f(x)))$, . . .) is quite independent of our choice of the initial x. Now we will investigate what happens when we "fiddle" with the constant 2. We will study the effects of changing this constant (also called a parameter). To do this, we will connect it with a slider. By moving the slider, we will change the value of the constant.

Let us look at our table once again (Figure 2.104).

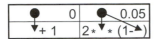

Figure 2.104

The constant 2 is "hardwired" into the formula. What we will study is the behavior of our iterations if we use other numbers instead of 2. With our current table structure, we have to change the formulas in all of the cells that calculate the iterations. This is rather inconvenient, so we will change the formula to make these changes easier. To achieve this, we add two rows at the top of our table and place the label "constant" and the value 2 in the first row (Figure 2.105).

constant		2
	0	0.05
	1	2* ▼ * (1-▲)

Figure 2.105

In the next step, we adjust the formula of the function to reference the constant from the cell in the top row instead of the "hardwired" value of 2 (Figure 2.106).

constant		2
	0	0.05
	1	▼* ▼ * (1-▲)

Figure 2.106

The reference has to be absolute so that we can copy the formula and be able to change the formula for the iterated values by simply changing the value of the constant in the first row. Copying down this expression far enough (again, we should create at least 200 copies of the formulas) according to the table in Figure 2.107 gives us a convenient way of trying different values for the constant in our iteration function.

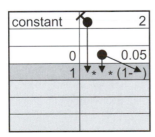

Figure 2.107

Let us recall what we are doing now: We are studying the sequence of values x, $f(x)$, $f(f(x))$, $f(f(f(x)))$, . . . for the function $f(x) = cx(1 - x)$ with the value of $c = 2$. Our table is designed in a way so that we can change the constant in the cell indicated in Figure 2.108.

constant	2
0	0.05
1	0.095
2	0.17195
3	0.284766
4	0.407349

Figure 2.108

We can then easily study the effects of using different values for the constant upon the sequence of the iterated y-values. In the previous section, we saw that for $c = 2$ for any starting point except 0 the sequences approaches 0.5 very quickly. Let us try 2.2 instead of 2.0 as the value of c and look at the resulting xy-graph created from the iterated values as done in the previous section. To make some things more clearly visible in Figure 2.109, we again have slightly changed the format of our xy-graph to show only markers for each point without connecting lines.

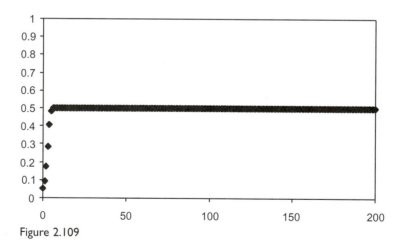

Figure 2.109

In this graph, our iterated sequence again approaches a stable point. Trying different initial values also will have the effect that we already have observed, with the sequence always approaching the same point. Remembering that for $f(x) = cx(1 - x)$

and $c = 2$ the attracting point was the fixed point of $f(x)$, we can now try the same idea for $c = 2.2$. The fixed point(s) of this function can be found by solving the equation $2.2x(1 - x) = x$ or equivalently $-2.2x^2 + 1.2x = 0$ yielding two solutions, $x = 0$ and $x = 12/22 = 0.5454$. We see that again the sequence of iterated points approaches a fixed point of the function.

For what we do in the following steps, it is also important to fix the displayed range on the y-axis. We need for it to run from 0 to 1. Most spreadsheet programs do "autoscaling," that is, select a minimum and a maximum value for the displayed range by using values reasonably close to the minimum and maximum of the y-values in the table. In *Excel*, we do this by right-clicking on the axis and manually setting the maximum and minimum values. Doing this is important for visualizing the effects of changes in the value of c as it ensures that we will always have the same scale on the axis.

When this change is finished the graph appears as shown in Figure 2.110.

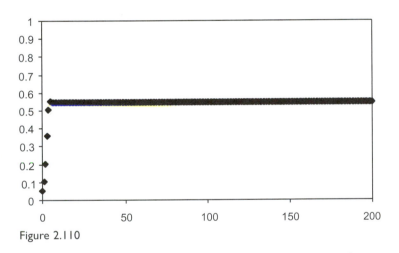

Figure 2.110

Using our value table and this graph, we can examine the effects of changing the value of our constant c still further. Gradually increasing the value of c using 2.3, 2.4, 2.5 . . . at first does not produce anything qualitatively different from what we have seen so far. The graph very soon looks almost like a horizontal line. We also can calculate the values of the fixed points of $f(x) = cx(1 - x)$ for different values of c by solving the equation $cx(1 - x) = x$ or equivalently $cx^2 - (1 + c)x = 0$ giving the solutions $x = 0$ and $x = (c - 1)/c$. We also see that for any starting point x (except 0) the iterated sequence approaches the nonzero fixed point of f. This is, however, not generally true. We only get the typical picture from using values of c below 3. For $c = 3$ (and a corresponding fixed point 0.6667), we get the graph shown in Figure 2.111.

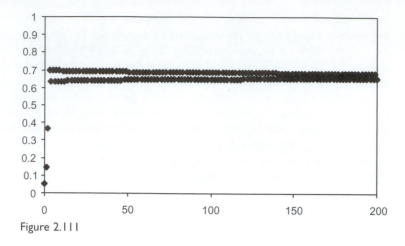

Figure 2.111

It looks as if the "curve" now has two branches. Looking at the value table, we will see that the points "jump up and down," alternating between the branch above 0.6667 and the branch below 0.6667. The graph does not really tell us if eventually we will get a single horizontal line again. Changing the value of c to a value that is only slightly higher than 3, say 3.01, gives a qualitatively different picture (Figure 2.112).

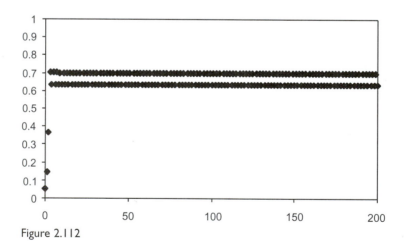

Figure 2.112

Now we have two distinct and separated branches, and the iterated points change from branch to branch with each iteration. What is happening? Changing the constant c even more gives some more interesting graphs. For example, $c = 3.3$ produces the graph shown in Figure 2.113.

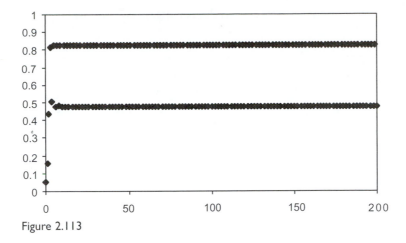

Figure 2.113

Next, with $c = 3.5$ we obtain the graph shown in Figure 2.114.

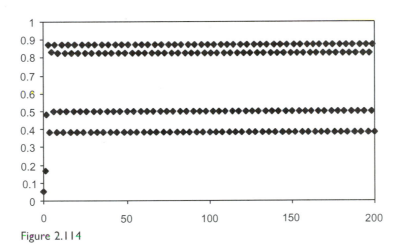

Figure 2.114

Likewise, $c = 3.56$ produces the graph shown in Figure 2.115.

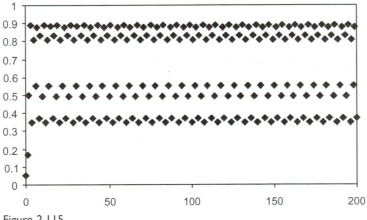

Figure 2.115

Next, the value $c = 3.66$ produces the graph shown in Figure 2.116.

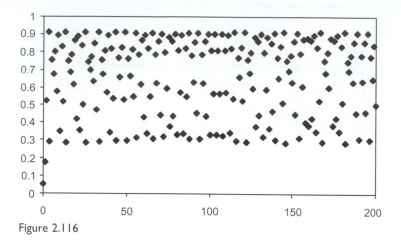

Figure 2.116

We see that for higher values of c the curve acquires more and more branches and finally seems to "sink into chaos." We also notice that by slowly increasing c from time to time we hit values where each one of the branches of our graph splits into two branches, and therefore the number of branches is a power of 2. Once we are in the chaotic region, there seem to be isolated values of c with very strange behavior. For example, with $c = 3.8444$ and a starting value of 0.6, the graph shown in Figure 2.117 is produced.

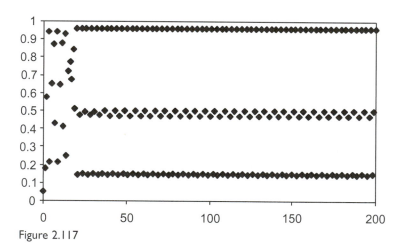

Figure 2.117

This seems as if it is a curve with five branches. Values for c that are slightly below and above 3.8444 produce quite different, chaotic behavior.

We also notice that when using values for c that are higher than 4 the graphs that we obtain behave very irregularly. There is a simple reason for this: With $c > 4$ the function values of $f(x) = cx(1 - x)$ become larger than 1 for values of x between 0 and 1, and therefore we can get divergent behavior.

Our investigations will be enhanced greatly by incorporating a scroll bar into our model to vary the value of the constant c. To create the scroll bar that is illustrated in Figure 2.118, we enter a number into the auxiliary Cell C1 and a formula that divides this value by 100 in Cell B1. We then create a scroll bar that is linked to Cell C1 and set its properties so that it varies between 0 and 400. Then, as we move the slider, we produce a dynamic and animated picture of the effects of what are essentially continuous changes in the parameter c as it varies from 0.00 to 4.00 in increments of size 0.01. Details of the construction of a scroll bar are presented in the Appendix.

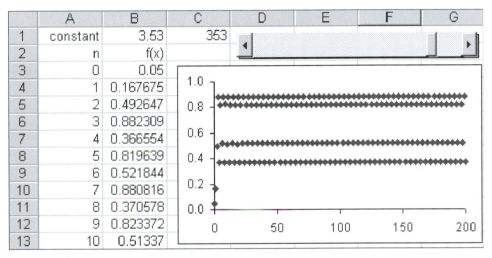

Figure 2.118

Further Mathematics: An Unanswered Question

One of the questions that we have not answered yet is why the behavior of our system changes so dramatically for $c = 3$. Fixed points of f at $(c - 1)/c$ also do exist for values of c between 3 and 4. One way of understanding what is changing at $c = 3$, as in the previous section, uses some algebra and calculus. We noticed that fixed points behave quite differently depending upon the value of the first derivative of the function at the fixed points. Our fixed point 0 (for any value of c larger than 2) has a derivative larger than 1. Using calculus we see that the derivative of $f(x) = cx(1 - x)$ at $x = (c - 1)/c$ has the value $f'((c - 1)/c) = 2 - c$, so for $c > 3$ the absolute value of the derivative at the fixed point is larger than 1. This is the reason that for $c > 3$ the fixed point does not attract the points nearby any more. We can even go further, as $f(x)$ "jumps up and down" between two points for c slightly above 3. This implies that $f(f(x))$ should have a fixed point, since after every second step we get essentially the same value as the start value. We can calculate $f(f(x)) = cx(1-x)(cx^2 - cx + 1)$ and find the fixed points of these functions. This is a fourth degree equation, so solving it is not very easy. It has four solutions: $x_1 = 0$, $x_2 = (c-1)/c$, $x_3 = [\sqrt{c+1}\,(\sqrt{c+1} - \sqrt{c-3})]/2c$, and $x_4 = [\sqrt{c+1}\,(\sqrt{c+1} + \sqrt{c-3})]/2c$. Using the last two fixed points of $f(f(x))$ as input values we see that $f(x_3) = x_4$ and $f(x_4) = x_3$. We can use a spreadsheet to check this with the numeric values. We have seen previously that the behavior of our iterated values changed when the absolute value of the first derivative at the fixed point became larger than 1. This suggests that we try to calculate the first derivative of $f(f(x))$ at our fixed points x_3 and x_4. The first derivative is $(f(f(x)))' = c^2\,(1-2x)(2cx^2 - 2cx + 1)$ and the value of the derivative at both x_3 and x_4 is $2(c+2) - c^2$. So a candidate for the next interesting change of behavior is the value of c where the absolute value of $2(c+2) - c^2$ becomes larger than 1. This happens at $c = \sqrt{6} + 1$, which is approximately 3.45. Changing the value of c in the spreadsheet model and watching the graph shows that this value of c is the value where the graph of the iterated points changes from two branches to four branches. In theory one could try to do the same algebraic calculations for getting still more branches. The problem is that very soon we get equations of rather high degree for which no formula for solving them is available.

Construction Summary: More about Iterated Functions

1. Enter parameter value $c = 2$.
2. Create a counter column.
3. Enter 0.05, the initial value of x.
4. Compute $cx(1-x)$ and copy down.

constant		1	2
2	0	3	0.05
	1	4	0.095
	2		0.17195
	3		0.284766

Exercises

1. Design a scroll bar to incorporate into the model of this section. See the Appendix for details.

2. Investigate the orbits of different points x for the "folding and distorting" function $g(x)$ from the exercise in the last section. Try different values of α and β.

3. [*] Why do the "folding and distorting" function and the quadratic function $f(x) = cx(1-x)$ behave somewhat similarly under iteration?

4. [*] Why does the example stop working for $c > 4$?

2.7 The Feigenbaum Diagram

The Feigenbaum diagram arises in the study of chaotic behavior. We create it by plotting the long-range behavior of a large number of points defined through a particular iterated function of one parameter, c. Initially as c increases all of the points lie on a curve. However, after c reaches a certain value, the sequence splits into points on two (a bifurcation) and then four branches and so on. Eventually the process exhibits chaotic behavior. It is named after its discoverer, physicist Mitchell Feigenbaum, who was born in Philadelphia in 1945.

In this section, we will use the same function as in the previous section, $f(x) = cx(1 - x)$, and study the pattern of the iterated points depending on the value of c. We will use another graphic representation, though. In the previous section, we had a graph showing the sequence of the iterates and for a single value of c. So we could not study the effects of different values of c "side by side." Now we will build a graph giving us an overview of the behavior of the iterates in one single graph. Overlaying sequences of iterates for different values of c does not produce easily "readable" graphs, so we have to find another way of displaying the sequences simultaneously. The basic idea is to use different values of c on the x-axis and display the sequence of iterates "on top" of these c values. In the previous section, we used the number of the iterate for the x-axis; this time we use the value of c for the x-axis.

In the previous section, we noted that the first few iterations might give points that are not very close to the fixed points. Therefore, for a given c we will do 80 iterations but only use Iteration 41 to Iteration 80 for graphing. This will "stabilize" the iteration.

We start with a first column that contains repeated values of the integers from 1 to 40 to help us to decide if we should use the same value of c as in the cell above in the same column or a slightly larger value (Figure 2.119).

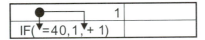

Figure 2.119

In the second column, we start with $c = 2.5$ and enter a formula to reproduce this value of c until the first column is 1. At that time increment the value for c by 0.025. We can interpret the *Excel* IF function used in Figure 2.120 as

IF cell to the left is 1, THEN add 0.025 to cell above, ELSE duplicate the cell above.

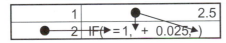

Figure 2.120

We need to copy the formulas in these two columns down until the first column contains 40 and the second column contains 4.0. To do this, we need the formulas in 2,440 rows.

The third column contains the initial value for our iteration in the first row (Figure 2.121). For the remaining rows we need the following formula:

IF count in first column = 1, THEN initial value, ELSE iterate of value above.

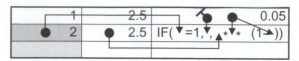

	1	2.5		0.05
	2	2.5	IF(▼=1,▼,,▲*▼* (1➤))	

Figure 2.121

The third column gives us Iterate 1 to Iterate 40. We need one more column for Iterate 41 to Iterate 80. In this column, each value is the iterate of the value above, except in rows where the first column contains 1. In this case, which is the 41st iterate, the value is the iterate of the 40th iterate, to be found in the column to the left and 40 rows lower (Figure 2.122). It is easier to create this formula for the second row and then copy it to the first row later. The new IF formula can be interpreted as

IF count = 1, THEN take the element from column to left, 39 rows farther down,
ELSE iterate of cell above.

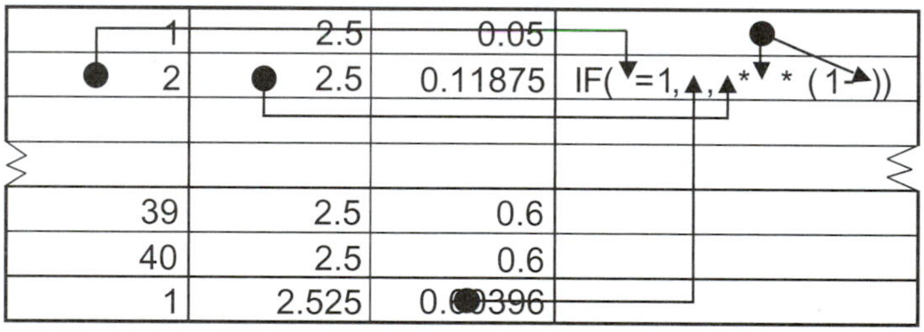

1	2.5	0.05	
2	2.5	0.11875	IF(▼=1,▲,,▲*▼* (1➤))
39	2.5	0.6	
40	2.5	0.6	
1	2.525	0.00396	

Figure 2.122

If we started our worksheet in the very first row, we cannot copy the last formula in the cell above. If we try to do that, then we will get an error, because the first row cannot contain a reference to the row above. The easiest solution to this problem is to insert an additional empty first row and then copy the formula we just created into the cell above (Figure 2.123).

1	2.5	0.05	
2	2.5	0.11875	IF(▼=1,▲,▲*▼*(1▲))
39	2.5	0.6	
40	2.5	0.6	
1	2.525	0.60896	

Figure 2.123

In copying the formulas in Column 3 and Column 4 down far enough, we will already have all of the numbers that we need to create the Feigenbaum diagram. The only problem is that we want to graph the nonadjacent Column 2 and Column 4 in an xy-chart. It is possible to do this, but an easier solution is to move Column 3 and Column 4 one column to the right, then duplicate the values from the new Column 5 in Column 3 with a formula and copying down the formula (Figure 2.124).

1	2.5		0.05	0.6
2	2.5		0.11875	0.6
3	2.5		0.261621	0.6

Figure 2.124

Selecting Column 2 and Column 3 and creating an xy-chart (or scatter plot) gives the graph shown in Figure 2.125.

This is the well-known *Feigenbaum diagram* illustrating complex behavior arising from a relatively simple equation.

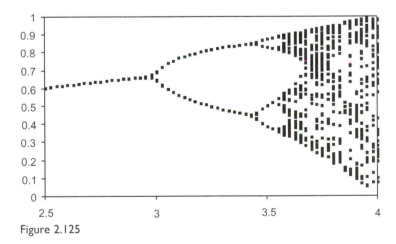

Figure 2.125

1. Enter 1 to start count.
2. Generate 1 if cell above is 40, else increment count by 1.
3. Enter initial values for x, y.
4. Add 0.025 to cell above if count is 1, else duplicate cell above.
5. Duplicate initial value if count is 1, else compute next iterate.
6. Copy Column 4 of 39 rows below if count is 1, else next iterate.
7. Duplicate value in right column.
8. Copy as indicated.

[1] 1	[3] 2.5	[7] 0.6	[3] 0.05		0.6
[2] 2	[4] 2.5	0.6	[5] 0.1188	[6]	0.6
3	2.5	0.6	0.2616		0.6
4	2.5	0.6	0.4829		0.6
5	2.5	0.6	0.6243		0.6
6	2.5	0.6	0.5864		0.6
40	2.5		0.6		0.6
1	2.525	0.60396	0.05		0.60396

Exercises

1. Modify the spreadsheet for the Feigenbaum diagram to use the "folding and distorting" function $g(x)$ from the exercise in the last section. Use a fixed value for α and combine it with different values of β (so that β plays the role of the constant c in the current section).

2.8 Sierpinski's Triangle

The Sierpinski triangle is a fractal that was created in 1916 by the Polish mathematician Waclaw Sierpinski (1882–1969). In its original form, a process without any element of randomness creates it. We use a related but different construction method involving a random process.

In the previous two sections, we investigated iterative applications of a function of one variable and found that some surprising patterns can emerge when we study the sequence of points x, $f(x)$, $f(f(x))$, $f(f(f(x)))$, ... Now we will study a similar phenomenon in the plane.

We start with the vertices of an equilateral triangle: $(0,0)$, $(1,0)$, and $(1/2, \sqrt{3}/2) = (0.5, 0.866)$. Then we choose an arbitrary starting point within the triangle, for example, $(0.3, 0.2)$. The spreadsheet arrangement in Figure 2.126 will be helpful.

			0	0	1	0	0.5	SQRT(3)/2
0.3	0.2							

Figure 2.126

Now we imagine standing at (0.3, 0.2) and walking half the distance to Point (1, 0). Then our destination is

$$(0.3, 0.2) + 0.5 \cdot ((1,0) - (0.3, 0.2)) = 0.5 \cdot (0.3, 0.2) + 0.5 \cdot (1,0).$$

For any two points A and B, the point halfway from A to B is the midpoint of the line segment joining the two points. Therefore its coordinates can be calculated as the arithmetic mean of the coordinates of Point A and Point B.

The formula used for the x-coordinate of the midpoint between the points $(0.3, 0.2)$ and $(0,0)$ is illustrated in Figure 2.127.

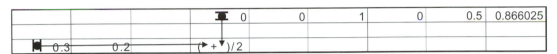

Figure 2.127

The formula for the y-coordinate of the midpoint is illustrated in Figure 2.128.

			0	0	1	0	0.5	0.866025
0.3	0.2			(▶ + ▼)/2				

Figure 2.128

We are using mixed references here to be able to reuse the formulas for the destination points $(1,0)$ and $(\sqrt{3}/2)$ later by just copying. The formulas look very similar, but the second one cannot be created by copying the first one. The reference to the first column is "column absolute" and therefore will not change into a reference to the second column, which is what we need. Therefore, the second formula has to be created separately.

Copying our two midpoint formulas as shown in Figure 2.129 calculates the "halfway points" from our given point $(0.3, 0.2)$ to the destinations $(1, 0)$ and $(\sqrt{3}/2)$.

			0	0	1	0	0.5	0.866025
0.3	0.2		0.15	0.1	0.65	0.1	0.4	0.533013

Figure 2.129

Now we will choose one of the three possible destinations randomly. To do that, we calculate an integer random number in the range from 1 to 3. Spreadsheet programs usually have a function that returns a real (noninteger) random number in the range 0 to 1. In *Excel*, this function is called RAND(). When using the random function, one has to type the empty parentheses at the end, since this is *Excel*'s way of understanding that this is not a named range but a function. This function never will return 0 or 1. Instead, the result always will be between these limits. By multiplying the result by 3, we will get a number between 0 and 3; and by taking its integer part, we will get integer values of 0, 1, or 2. By adding 1, we will get integer values 1, 2, or 3. INT() is the *Excel* function that returns the integer part of a number, so INT(3*RAND())+1 is an expression that returns an integer random number in the range from 1 to 3. Entering this formula into our sheet as shown in Figure 2.130 allows us to choose the next point at random from the three candidates. We will use a new spreadsheet function, OFFSET, to select our destination. The function OFFSET takes three arguments—a cell, a vertical offset, and a horizontal offset. If both offsets are 0, then OFFSET returns the value from the given cell. If the vertical offset is 1 and the horizontal offset is 0, then OFFSET returns the value from the cell below the indicated cell. If the vertical offset is 0 and the horizontal offset is 3, OFFSET returns the value from the cell three cells to the right of the indicated cell and so on. Therefore, entering the formula shown in Figure 2.131 into our sheet selects the *x*-coordinate of the next point at random from the *x*-coordinates of the three candidate points. The last argument to OFFSET yields 0 if the random number is 1, 2 if the random number is 2, and 4 if the random number is 3; so it gives just the offset that we need. Using a column-absolute reference allows us to copy the formula to the right, giving us the *y*-coordinate of the same candidate we just used for the new *x*-coordinate (Figure 2.132).

			0	0	1	0	0.5	0.866025
0.3	0.2	INT(3*RAND())+1	0.1	0.65	0.1	0.4	0.533013	

Figure 2.130

			0	0	1	0	0.5	0.866025
0.3	0.2	3	0.15	0.1	0.65	0.1	0.4	0.533013
OFFSET(,0,2*(-1))								

Figure 2.131

			0	0	1	0	0.5	0.866025
0.3	0.2	3	0.15	0.1	0.65	0.1	0.4	0.533013
0.4	0.533013							

Figure 2.132

Copying the remaining cells from the starting point to the next row (see Figure 2.133) creates the setting that allows us to apply the same mechanism again. To finish this project, we copy the last line down 2,000 rows and then select the first two columns and turn them into an xy-graph (only points, no connecting lines). This gives a picture similar to the one shown in Figure 2.134.

			0	0	1	0	0.5	0.866025
0.3	0.2	3	0.15	0.1	0.65	0.1	0.4	0.533013
0.4	0.533013	2	0.2	0.266506	0.7	0.266506	0.45	0.699519

Figure 2.133

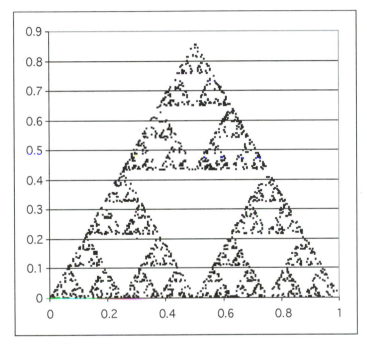

Figure 2.134

This graph is quite interesting. It looks very regular despite the fact that the selection of points is random. It is an example of a self-similar shape. The small triangle (half the size of the original one) in the left lower corner of the graph looks like a scaled down version of the whole triangle, and so does the right lower triangle and the upper triangle.

Recalculating the spreadsheet (in *Excel*, this can be accomplished by pressing function key F9) creates new random numbers. Trying this, we see that the graph does not change noticeably. Changing the starting point of our sequence does not change the graph either.

Rewriting the algorithm creating this graph can help us understand the shape a little better. We have three transformations in the plane:

$$t_1(x, y) = \left(\frac{x}{2}, \frac{y}{2} \right)$$

$$t_2(x, y) = \left(\frac{x + 1}{2}, \frac{y}{2} \right)$$

$$t_3(x, y) = \left(\frac{x}{2} + \frac{1}{4}, \frac{y + \sqrt{3}/2}{2} \right)$$

These transformations can be described as shrinking with different centers. Thus, t_1 shrinks toward $(0, 0)$ by a factor of 0.5; t_2 shrinks toward $(1, 0)$ by a factor of 0.5; and t_3 shrinks toward $(\frac{1}{2}, \sqrt{3}/2) = (0.5, 0.866)$, also by a factor of 0.5. Starting with an arbitrary point (x_0, y_0) we define $(x_{n+1}, y_{n+1}) = t_j (x_n, y_n)$ where j is chosen at random from the numbers 1, 2 or 3.

Plotting the sequence (x_n, y_n) produces a graph that can be described in the following way: Start with the equilateral triangle with vertices $(0, 0)$, $(1, 0)$, and $(\frac{1}{2}, \sqrt{3}/2)$ $= (0.5, 0.866)$. Color the triangle black (Figure 2.135).

Figure 2.135

Now apply t_1, t_2, and t_3 to the triangle. This produces the smaller triangles, shrunk toward the vertices of the original triangle. Replace the original triangle by the three smaller triangles (Figure 2.136).

Figure 2.136

Now apply t_1, t_2, and t_3 to the new shape (creating three smaller copies of the shape) and replace the shape by the three shrunk shapes (Figure 2.137).

Figure 2.137

We repeat this procedure once again, yielding the shape in Figure 2.138.

Figure 2.138

This shape is very similar to the dot pattern we produced with the spreadsheet. The main difference is that the spreadsheet method used random numbers, whereas the second method is completely deterministic (that is, nonrandom). Mathematical theory created by Michael Barnsley (see Barnsley (1988)) and others studies the connections between these two methods (both involving a set of shrinking transformations, in our case t_1, t_2, and t_3) and explains why these two methods produce very similar results. Comparing these graphs with the ones in the exercises about Pascal's triangle and the ones about cellular automata shows that there is a striking similarity.

Construction Summary: Sierpinski's Triangle

	x	y	k	x₁	y₁	x₂	y₂	x₃	y₃
				[1] 0	0	1	0	0.5	0.86603
[2]	0.300	[2] 0.200	[3] 3	[4] 0.150	[5] 0.100	0.650	0.100	0.400	0.533
[6]	0.400	0.533	2	0.200	0.267	0.700	0.267	0.450	0.700
	0.700	0.267	2	0.350	0.133	0.850	0.133	0.600	0.566
	0.850	0.133	1	0.425	0.067	0.925	0.067	0.675	0.500
	0.425	0.067	1	0.213	0.033	0.713	0.033	0.463	0.466

1. Enter the (x,y)-coordinates of the three vertices of a triangle.
2. Enter the (x,y)-coordinates of any interior point of the triangle.
3. Generate a random integer, k, in the range 1, 2, 3 via =INT(3*RAND()+1).
4. Compute the x-coordinate of the midpoint of the new point and the x-coordinate of the first point in the top row.
5. Compute the y-coordinate of the midpoint of the new point and the y-coordinate of the first point in the top row. Then copy the block of (x,y)-coordinates to the right.
6. Use the =OFFSET function to produce the x-coordinate of the kth point of the row above.
7. Copy the formula for x into y and then copy that block down. Copy the formulas in the other columns down.

Excercises

1. Modify the spreadsheet for the Sierpinski triangle in the following way:

   Instead of the three vertices of a triangle use the four vertices of a square. Try to guess what the graph will look like, and be ready for some surprises.

2. What do you get when instead of the three corners of a triangle you use only the two endpoints of a line segment?

3. What do you get when instead of walking half the distance to the selected corner you walk two-thirds of the distance, or one-third of the distance? Modify the spreadsheet in a way you can arbitrarily change this "shrinkage factor."

3

Physics Applications

CHAPTER OUTLINE

3.1   Simple Projectile Motion

A projectile is fired (or a stone is thrown) with a given initial velocity and a given initial angle to the surface, which we assume is horizontal. We want to calculate the flight path, or trajectory, of the projectile. This path can be described by giving the horizontal distance of the projectile from the starting point together with its height above the surface at any given time after firing. In this model, we neglect effects due to air resistance.

We can visualize the initial situation using a velocity vector (Figure 3.1). The initial velocity is in the direction of the given angle with the ground, and the magnitude of the velocity, given in meters per second, corresponds to the length of the vector.

Let us start our example using an initial angle of 40 degrees and an initial velocity of 30 meters per second. In our spreadsheet, we calculate the position of the projectile for the times 0 second, 1 second, 2 seconds, . . . after firing. Instead of talking about distance and height, we talk about the *x*-coordinate and the *y*-coordinate of the projectile. Consequently, we can label our graph differently (see Figure 3.2).

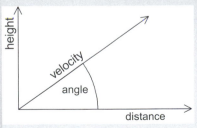

Figure 3.1

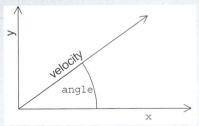

Figure 3.2

We will have our projectile start at Position (0,0). Therefore, we set up the first lines of the spreadsheet as shown in Figure 3.3.

time	x	y
0	0	0

Figure 3.3

For the next line in this sheet, we want to calculate the position of the projectile at time 1 second. Simplifying the physics, we will assume that the projectile travels along a straight line for the whole first second of its flight. We know that the length of this path is 30 meters since the initial velocity is 30 meters per second. We also know that the path has an angle of 40 degrees. For our spreadsheet, we need to know the change in the x-coordinate and the change in the y-coordinate. Using basic trigonometry, we can summarize the relations between the x-change and y-change and the velocity and angle as shown in Figure 3.4.

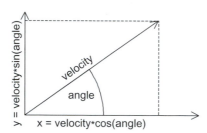

Figure 3.4

We now modify our spreadsheet to calculate the x-change and the y-change for the initial velocity. To do this, we add a few lines at the beginning of our sheet (Figure 3.5).

velocity	angle	
30	40	
x-change	y-change	
time	x	y
0	0	0

Figure 3.5

As a next step, we calculate the initial x-change and y-change in meters. The trigonometric functions SIN and COS are built-in functions in *Microsoft Excel* and most other spreadsheet programs. However, these functions require their arguments to be in radians, not in degrees. Therefore, we have to create the two formulas shown in Figure 3.6 and Figure 3.7, respectively.

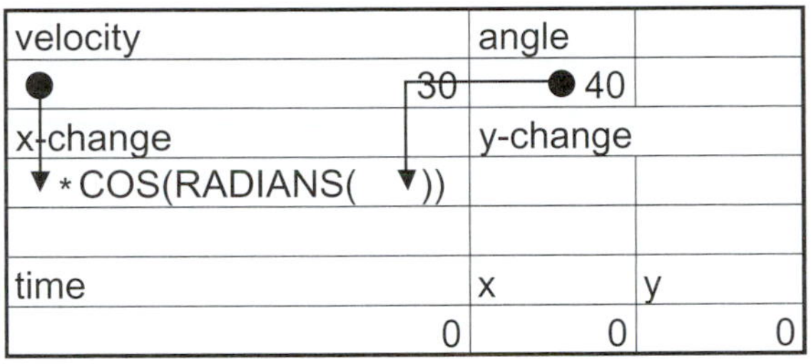

velocity		angle	
●	30	● 40	
x-change		y-change	
▼ * COS(RADIANS(▼))			
time		x	y
	0	0	0

Figure 3.6

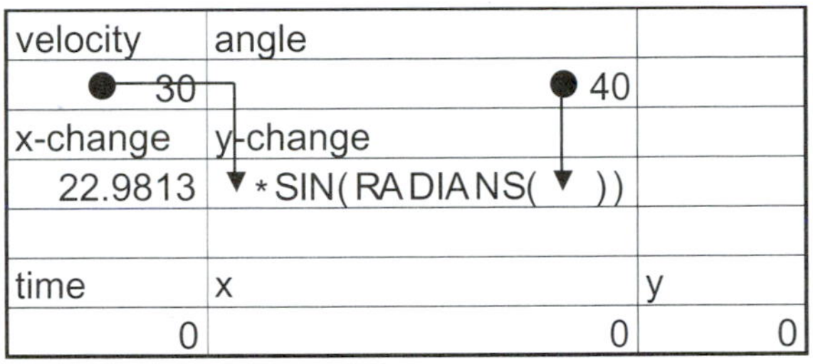

velocity	angle		
● 30		● 40	
x-change	y-change		
22.9813	▼ * SIN(RADIANS(▼))		
time	x		y
0		0	0

Figure 3.7

Now we add two columns for the horizontal and vertical components of the velocity of the projectile (Figure 3.8).

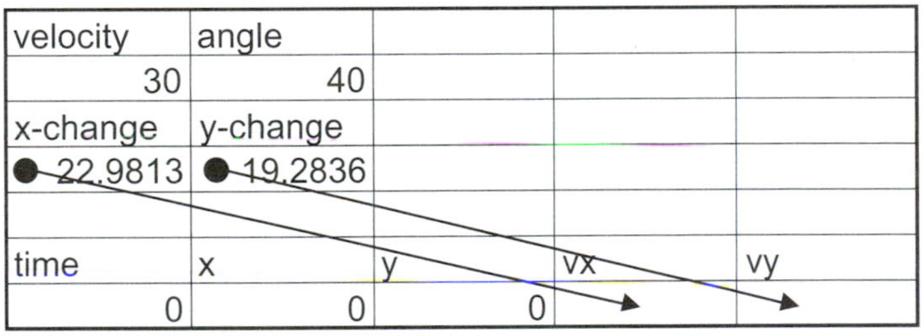

velocity	angle			
30	40			
x-change	y-change			
● 22.9813	● 19.2836			
time	x	y	vx	vy
0	0	0		

Figure 3.8

Since the column labeled vx gives the change of the x-coordinate in 1 second (that is how velocity is defined), we can calculate the x-coordinate of our projectile at Time 1 second by adding the x-change to the previous value of x (see Figure 3.9).

velocity	angle			
30	40			
x-change	y-change			
22.9813	19.2836			
time	x	y	vx	vy
● 0 ●	┌ 0	0 ●	22.9813	19.2836
1 +▼	▼ +▼			

Figure 3.9

Calculating the new y-coordinate is similar to calculating the new x-coordinate; everything is just one column to the right. Therefore, we only need to copy the formula for the x-coordinate to the right (see Figure 3.10).

velocity	angle			
30	40			
x-change	y-change			
22.9813	19.2836			
time	x	y	vx	vy
0	0	0	22.9813	19.2836
1	22.9813	19.2836		

Figure 3.10

To compute the position at time 2, we need the velocity for the time between 1 second and 2 seconds. Because of the effect of gravity, the velocity vector for this time will have a flatter angle, so that our simplified orbit will look somewhat like that shown in Figure 3.11.

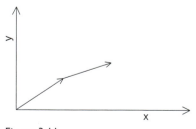

Figure 3.11

To study the change in the velocity vector in detail, we first insert a shifted dashed copy of the initial velocity vector with the same starting point as the second velocity vector (Figure 3.12).

The dashed vertical arrow in this graph gives the change in the y-component of the velocity between the first point and the second point. Newton's theory of gravity, the basis of our calculations, tells us that this change is constant in time given by the gravitational constant 9.81 meters/second/second. Therefore, the velocity vector for the third second can be drawn by applying exactly the same vertical change to the second velocity vector, and so on (see Figure 3.13).

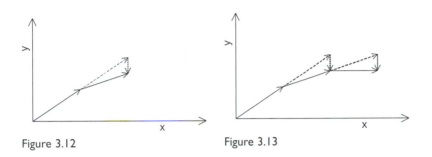

Figure 3.12   Figure 3.13

This change in velocity is called *acceleration*. As a consequence of Newton's theory, we can always calculate the velocity for the next second by adding a constant change to the current velocity. This change is strictly vertical, so the x-component of the velocity does not change. The vertical component of change is 9.81, the acceleration for free fall when we measure distance in meters and time in seconds. Since acceleration points downward in our example, it is negative and we subtract 9.81 from the previous vertical component of velocity.

Now we can approximate the velocity at Time 1 second. The x-component does not change, while the y-component is obtained by subtracting 9.81 from the cell above (see Figure 3.14).

velocity	angle			
30	40			
x-change	y-change			
22.9813	19.2836			
time	x	y	vx	vy
0	0	0	● 22.9813	● 19.2836
1	22.9813	19.2836	▼	▼ -9.81

Figure 3.14

The row for Time 1 second now contains formulas that compute values for time, location, and velocity from the previous row. Therefore we copy this row down, and we see that after 5 seconds the projectile would be lower than its starting point. Assuming that the motion is taking place above a horizontal surface, this means that the projectile will hit the surface at some time between 4 seconds and 5 seconds (see Figure 3.15).

velocity	angle			
30	40			
x-change	y-change			
22.9813	19.2836			
time	x	y	vx	vy
0	0	0	22.9813	19.2836
1	22.9813	19.2836	22.9813	9.4736
2	45.9627	28.7573	22.9813	-0.3364
3	68.9440	28.4209	22.9813	-10.1464
4	91.9253	18.2745	22.9813	-19.9564
5	114.9067	-1.6819	22.9813	-29.7664
6	137.8880	-31.4482	22.9813	-39.5764

Figure 3.15

The graph of the trajectory, created as an *xy*-chart, is shown in Figure 3.16.

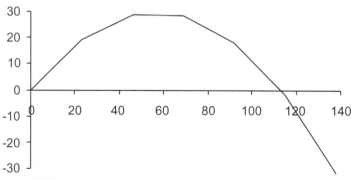

Figure 3.16

This shows that computing the location of the projectile using 1-second intervals gives us a very coarse approximation. So to see if we can get a smoother graph, we will calculate the trajectory with shorter periods between the computed locations. What do we need to change in our spreadsheet model to do this?

We can observe that different values for the time interval between successive points will influence the resulting curve. Therefore, we put this value in a cell so that we can change it easily. We start with a modified version of the sheet (Figure 3.17).

velocity	angle			
30	40			
x-change	y-change			
22.9813	19.2836			
delta t	0.1			
time	x	y	vx	vy
0	0	0	22.9813	19.2836

Figure 3.17

The new entry, *delta t*, is the time interval between successive points. We will choose a value of 0.1 second for this interval. In mathematics it is quite common to use the expression *delta* (the name of the Greek letter Δ) for differences. Therefore, the time value for the next row is computed by adding *delta t* to the time value from the previous row. Since we will copy this formula down later, the reference to the time difference has to be absolute (Figure 3.18).

velocity	angle			
30	40			
x-change	y-change			
22.9813	19.2836			
delta t	0.1			
time	x	y	vx	vy
0	0	0	22.9813	19.2836
+				

Figure 3.18

Since only one-tenth of a second passes between two successive points, and since velocity gives the amount of change of location in a full second, we compute the new location by adding to the old location the velocity multiplied by the time difference (Figure 3.19). Again, the reference to the time interval has to be absolute for later copying.

velocity	angle			
30	40			
x-change	y-change			
22.9813	19.2836			
delta t	0.1			
time	x	y	vx	vy
0	●	0	0 ● 22.9813	19.2836
0.1	▼+ ▼ * ▼			

Figure 3.19

Since the columns for the *y*-location and the *y*-velocity are one column to the right of these for the *x*-location and the *x*-velocity, we can copy this formula to the right to get the computation for the next *y*-location (Figure 3.20).

velocity	angle			
30	40			
x-change	y-change			
22.9813	19.2836			
delta t	0.1			
time	x	y	vx	vy
0	0	0	22.9813	19.2836
0.1	2.2981	1.9284		

Figure 3.20

Next, the *x*-velocity is constant while the *y*-velocity changes by 0.1 of the change for a full second. Therefore we insert the two formulas shown in Figure 3.21.

velocity	angle			
30	40			
x-change	y-change			
22.9813	19.2836			
delta t	0.1			
time	x	y	vx	vy
0	0	0	● 22.9813	● 19.2836
0.1	2.2981	1.9284	▼	- ▼ * 9.81+ ▼

Figure 3.21

We now have completed the computation of a row from the previous row, so we can copy everything down to compute the trajectory with our smaller time step (Figure 3.22).

velocity	angle			
30	40			
x-change	y-change			
22.9813	19.2836			
delta t	0.1			
time	x	y	vx	vy
0	0	0	22.9813	19.2836
1	2.2981	1.9284	22.9813	18.3026
2	4.5963	3.7586	22.9813	17.3216
3	6.8944	5.4908	22.9813	16.3406
4	9.1925	7.1249	22.9813	15.3596
5	11.4907	8.6608	22.9813	14.3786
6	13.7888	10.0987	22.9813	13.3976

Figure 3.22

We need to extend this table for many more rows, because in our earlier calculations we observed that the projectile would have flown for nearly 5 seconds. The trajectory, again created as an *xy*-chart from the location columns, is shown in Figure 3.23.

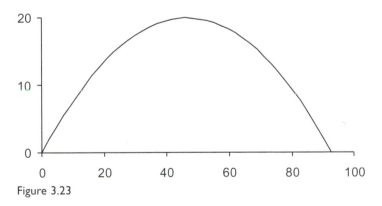

Figure 3.23

The trajectory now looks much smoother. We can change the time step from 0.1 second to 0.01 second to see if it changes the trajectory once again. Doing this and extending the range of data to be graphed show that the graph changes slightly but not nearly as much as it did when we went from a time step of 1 second to a time step of 0.1 second. Therefore, using one-tenth of a second in this example may seem to provide a reasonable degree of accuracy. However, this conclusion should be reexamined in light of the exercises at the end of this section or when compared with the analytical solution. In particular, with a time step of 0.1 second, the distance from the starting point to the point at which the projectile lands is overestimated. Other difficulties will arise as well. On the other hand, real-world projectiles do not behave exactly according to the analytical model either, so both our spreadsheet model and the analytical solution of the problem are approximations of the real trajectory.

Further Mathematics: A Calculus-Based Solution

The development of our model also provides the framework for a calculus-based solution of the problem. If we denote by Δy and Δv the changes in height and velocity, respectively, over the time unit Δt that we use in our model for calculating the y-component of the path, then we have used the relations $\Delta y = v\Delta t$ and $\Delta v = -g\Delta t$ to calculate the changes in y and v. If we divide each of these by Δt, we obtain $\Delta y/\Delta t = v$ and $\Delta v/\Delta t = -g$. If we then let the size of Δt decrease toward 0 and take the corresponding limits, then we obtain

$$\frac{dy}{dt} = \lim_{\Delta t \to 0} \frac{\Delta y}{\Delta t} = v \text{ and } \frac{dv}{dt} = \lim_{\Delta t \to 0} \frac{\Delta v}{\Delta t} = -g$$

Integrating the latter equation and using the initial velocity as v_0, we obtain

$$v = -gt + v_0.$$

Substituting this into the first equation, we obtain

$$dy/dt = v = -gt + v_0.$$

By integrating this and using the initial height as y_0, we obtain

$$y = -0.5\, gt^2 + v_0 t + y_0.$$

Construction Summary: Simple Projectile Motion

1. Enter parameter values.
2. Compute initial velocity components.
3. Create time counter column.
4. Enter initial x,y values.
5. Reproduce the initial velocity components, copy right.
6. Compute new x-position, copy right and down.
7. Reproduce the x-component of velocity, copy down.
8. Compute new y-component of velocity, copy down.

velocity		angle			
1	30	1	40		
x-change		y-change			
2	22.981	2	19.284		
delta t		1	0.1		
time		x	y	vx	vy
3	0.0	4	0	4	0	5	22.981	19.284
	0.1	6	2.298	1.928	7	22.981	8	18.303
	0.2		4.596	3.759		22.981	17.322
	0.3		6.894	5.491		22.981	16.341

Exercises

1. In an elementary physics or calculus course, the y-component of the flight of a projectile is given by

$$y = -0.5\, gt^2 + v_0 t + y_0,$$

   where

   g is the gravitational constant 9.81 meters/second/second,

   v_0 is the initial velocity in the vertical direction, and

   y_0 is the initial height (which is 0 in our model).

   Include a column in the model of this section to compute the height of the projectile using this equation. Then drag the column into the graph. Compare both the numerical and graphical aspects of this result with our basic model using various time step sizes.

2. Modify our model to locate the time and (x,y) coordinates of the point of maximum elevation, and the time and the distance traveled before the projectile returns to the surface. Create a data table to generate this information for various angles. Determine the angle that enables the projectile to travel the maximum distance. Also find the resulting distance. Observe how the time step size influences the results.

3. Modify the model for a projectile that is fired from a given height above the surface.

4. A large convention hall has a known length and ceiling height. What is the minimal velocity needed to throw a ball the length of the hall without hitting the ceiling or the floor? Take into account the height above the surface from which the ball is thrown. Obtain the measurement of a hall with which you are familiar and determine if it is physically possible for someone to throw a ball the length of the hall. Note that the fastest baseball pitchers can throw a ball approximately 100 miles per hour, or about 44.7 meters per second.

5. How fast must a baseball pitcher throw a pitch so that it will cross home plate in the strike zone? What will be the amount of change in the elevation of the path? Use reasonable estimates for the height of the pitcher and batter. In a standard baseball field, the pitcher's rubber is 60.5 feet from the plate and sits on a mound that is 10 inches high. Also see the book *The Physics of Baseball* by Robert Adair (2002) for various other throwing and hitting situations to analyze through our model.

3.2 Projectile Motion with Air Resistance

In the previous section, we investigated the flight path of a projectile by taking only gravity into account. We did not consider the effects of air resistance. The effect of gravity can be described in the following way: Adding a constant vector "bends" the velocity vector downward. Air resistance has an additional effect: adding a constant vector pointing exactly opposite to the velocity vector, thereby shortening (or reducing the magnitude of) the resulting velocity vector. The length of this opposing vector is directly connected with the length of the velocity vector. The longer the velocity vector, the longer the opposing vector. This expresses the fact that air resistance increases with speed. A graph displaying the vectors is shown in Figure 3.24.

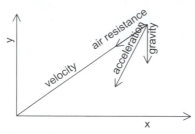

Figure 3.24

The formula for the length of the opposing vector is not as well established as the formula for gravity. In our model, we will assume that this length is proportional to the square of the velocity. Therefore, doubling velocity increases the air resistance by a factor of 4. There is an additional complication: Air resistance also depends on the size, shape, and surface of the moving projectile.

Let us write $\vec{v} = (v_x, v_y)$ for the velocity vector and $\vec{a} = (a_x, a_y)$ for the acceleration vector (the change in velocity per second). Since in our example velocity is changed by gravity and by air resistance, acceleration can be calculated by adding the two vectors expressing the effects of gravity and air resistance. The gravity vector, which we used in the previous section, is constant, $\vec{g} = (0, -g) = (0, -9.81)$. We are using the standard gravitational constant $g = 9.81 \ m/s^2$ for gravity near the surface of the earth. The length, or magnitude, of the velocity vector is $|\vec{v}| = \sqrt{v_x^2 + v_y^2}$ (see Figure 3.25).

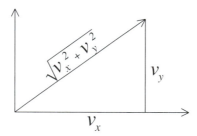

Figure 3.25

Now let $c > 0$ be an arbitrary constant. Then the vector

$$-c\,|\vec{v}|\,\vec{v} = -c\vec{v}\sqrt{v_x^2 + v_y^2} = -c\sqrt{v_x^2 + v_y^2}(v_x, v_y) = \left(-cv_x\sqrt{v_x^2 + v_y^2}, -cv_y\sqrt{v_x^2 + v_y^2}\right)$$

points in the opposite direction of $\vec{v}$ and its length is proportional to the square of the length of $\vec{v}$. Therefore, it can be used to express the effect of air resistance. Since acceleration, the change in velocity, is the sum of the gravity and air resistance acceleration vectors, we have

$$(a_x, a_y) = (0, -g) + \left(-cv_x\sqrt{v_x^2 + v_y^2}, -cv_y\sqrt{v_x^2 + v_y^2}\right)$$

$$= \left(-cv_x\sqrt{v_x^2 + v_y^2}, -g - cv_y\sqrt{v_x^2 + v_y^2}\right).$$

When we compute the flight path, $\vec{v}$ is known, but we need a value for the constant c. Since this constant depends on the moving projectile, we need a practical way to compute c under realistic conditions. To find a reasonable value, we use the following argument: If there were no air resistance, a falling object's velocity would increase indefinitely until it impacted the earth, since gravity is constant and therefore always causes an increase in velocity. Air resistance, however, always points in the reverse direction of the flight direction. For a free-falling object, the air resistance vector therefore points upward and increases in magnitude with increasing velocity. Therefore, at a certain velocity, it will have the same length as the gravity vector, which is pointing down. At this velocity, the net change of velocity is 0, and therefore the free-falling object has reached its terminal speed. The terminal speed for falling objects can be measured; and when we know the terminal speed, we can calculate the constant c for the object under consideration. Let $\vec{V} = (0, -V)$ be the terminal velocity. We use $-V$ as the y-component of the velocity vector because when written this way the terminal speed can be expressed as a positive number, measured in meter per second. Since at the terminal velocity the velocity does not change any more, acceleration at this velocity is 0. Since at the terminal velocity in free fall we have $v_x = 0$ and $v_y = -V$, then

$$(0, 0) = \left(-cv_x\sqrt{v_x^2 + v_y^2}, -g - cv_y\sqrt{v_x^2 + v_y^2}\right) = (0, -g + cV^2).$$

Therefore we have $c = \dfrac{g}{V^2}$.

Now we can calculate the air resistance constant when we know the terminal velocity for the falling object. For a parachutist in free fall, this terminal velocity has been measured as approximately 300 kilometers per hour. Since we will perform all of our computations in meters and seconds, for a parachutist we can assume a terminal velocity of 85 meters per second (1 meter per second corresponds to 3.6 kilometers per hour). A golf ball can reach a similar velocity in free fall. Generally speaking, the terminal velocity depends on the size, shape, and smoothness of the surface of the object.

To calculate the path of a projectile with air resistance now, we do not have to start from scratch. Instead, we can modify the spreadsheet that we set up for projectile movement without air resistance, as shown in Figure 3.26.

Since this spreadsheet assumes constant acceleration, the formulas for calculating velocity are no longer valid for our current model. Therefore, we clear all the cells for velocity except the first ones, which calculate the initial velocity for the given angle (see Figure 3.27).

velocity	angle			
30	40			
x-change	y-change			
22.98133	19.28363			
delta t	0.1			
time	x	y	vx	vy
0	0	0	22.98133	19.28363
0.1	2.298133	1.928363	22.98133	18.30263
0.2	4.596267	3.758626	22.98133	17.32163
0.3	6.8944	5.490788	22.98133	16.34063
0.4	9.192533	7.124851	22.98133	15.35963
0.5	11.49067	8.660814	22.98133	14.37863
0.6	13.7888	10.09868	22.98133	13.39763

Figure 3.26

velocity	angle			
30	40			
x-change	y-change			
22.98133	19.28363			
delta t	0.1			
time	x	y	vx	vy
0	0	0	22.9813	19.2836
0.1	2.298133	1.928636		
0.2	2.298133	1.928636		
0.3	2.298133	1.928636		
0.4	2.298133	1.928636		
0.5	2.298133	1.928636		
0.6	2.298133	1.928636		

Figure 3.27

The columns for x and y at the moment do not describe a projectile orbit since the computation depends on values in the velocity columns. But later, when we have constructed the formulas for velocity, the formulas for x and y will use the new values for velocity and the values for x and y will be computed correctly again. Therefore, we do not delete the formulas but keep in mind that for some time the displayed values should be ignored.

In this spreadsheet, acceleration (the change of velocity) was constant and, therefore, we needed no columns to display the change in acceleration over time. In our current spreadsheet, acceleration does change over time, so we need additional columns (see Figure 3.28). We will compute the acceleration caused by gravity and the acceleration caused by air resistance separately. Therefore, we add the following columns: *ax* and *ay* for total acceleration, *gx* and *gy* for the acceleration by gravity, and *rx* and *ry* for the acceleration caused by air resistance.

velocity	angle										
30	40										
x-chg	y-chg										
22.981	19.284										
delta t	0.1										
time	x	y	vx	vy	ax	ay	gx	gy	rx	ry
0		0	0	22.9813	19.284					
0.1	2.2981	1.9284								
0.2	2.2981	1.9284								
0.3	2.2981	1.9284								
0.4	2.2981	1.9284								
0.5	2.2981	1.9284								
0.6	2.2981	1.9284								

Figure 3.28

Since the acceleration by gravity is constant, we can immediately enter the corresponding values and formulas, as shown in Figure 3.29.

velocity	angle										
30	40										
x-chg	y-chg										
22.981	19.284										
delta t	0.1										
time	x	y	vx	vy	ax	ay	gx	gy	rx	ry
0	0	0	22.9813	19.284			0	-9.81		
0.1	2.2981	1.9284								
0.2	2.2981	1.9284								
0.3	2.2981	1.9284								
0.4	2.2981	1.9284								
0.5	2.2981	1.9284								
0.6	2.2981	1.9284								

Figure 3.29

To compute the acceleration caused by air resistance, we first need to compute the constant c, mentioned in the discussion at the beginning of this section, from the terminal velocity. Therefore, we insert the values and formulas shown in Figure 3.30.

Using this constant, we can begin to compute the effect of air resistance. Since air resistance depends upon the absolute value of the current velocity (or the length of the velocity vector), we add an additional column, v, for computing this value (Figure 3.31).

The newly created formula belongs to the last column, the column labeled v, not the one before it. Using the constant and the value in the v-column, we can calculate rx and ry (see Figure 3.32).

velocity	angle		term v	const								
30	40		● 85	9.81/▲^2								
x-chg	y-chg											
22.981	19.284											

Figure 3.30

velocity	angle		term v	const								
30	40		85	0.0014								
x-chg	y-chg											
22.981	19.284											
delta t	0.1											
time	x	y	vx	vy	ax	ay	gx	gy	rx	ry	v	
0	0	0	29.981	19.284				0	-9.81		sqrt(▲^2 + ▲^2)	
0.1	2.2981	1.9284						0	-9.81			
0.2	2.2981	1.9284						0	-9.81			
0.3	2.2981	1.9284						0	-9.81			
0.4	2.2981	1.9284						0	-9.81			
0.5	2.2981	1.9284						0	-9.81			
0.6	2.2981	1.9284						0	-9.81			

Figure 3.31

velocity	angle		term v	const								
30	40		85	0.0014								
x-chg	y-chg											
22.981	19.284											
delta t	0.1											
time	x	y	vx	vy	ax	ay	gx	gy	rx	ry	v	
0	0	0	29.981	19.284				0	-9.81	-▼* ▲*	←	▶◀ 30
0.1	2.2981	1.9284						0	-9.81			
0.2	2.2981	1.9284						0	-9.81			

Figure 3.32

This formula contains an absolute reference to the constant and a mixed reference to the length of the velocity vector. Setting it up this way allows us to copy the formula to the right as shown in Figure 3.33, computing the value for *ry*.

The total acceleration is the sum of the acceleration from gravity and the acceleration from air resistance. Therefore, we add the formula for *ax* shown in Figure 3.34.

Copying this formula to the right (see Figure 3.35) computes *ay*.

velocity	angle		term v	const								
30	40		85	0.0014								
x-chg	y-chg											
22.981	19.284											
delta t	0.1											
time	x	y	vx	vy	ax	ay	gx	gy	rx	ry	v	
0	0	0	22.981	19.284			0	-9.81	-0.936	-0.785	30	
0.1	2.2981	1.9284					0	-9.81				
0.2	2.2981	1.9284					0	-9.81				

Figure 3.33

velocity	angle		term v	const								
30	40		85	0.0014								
x-chg	y-chg											
22.981	19.284											
delta t	0.1											
time	x	y	vx	vy	ax	ay	gx	gy	rx	ry	v	
0	0	0	22.981	19.284	+		0	-9.81	-0.936	-0.785	30	
0.1	2.2981	1.9284					0	-9.81				
0.2	2.2981	1.9284					0	-9.81				

Figure 3.34

velocity	angle		term v	const								
30	40		85	0.0014								
x-chg	y-chg											
22.981	19.284											
delta t	0.1											
time	x	y	vx	vy	ax	ay	gx	gy	rx	ry	v	
0	0	0	22.981	19.284	-0.936	-9.025	0	-9.81	-0.936	-0.785	30	
0.1	2.2981	1.9284					0	-9.81				
0.2	2.2981	1.9284					0	-9.81				

Figure 3.35

Now we need to compute *vx* and *vy* for Time 0.1. These values are computed by adding *ax* and *ay* multiplied by the time step to the previous values. The setup of our spreadsheet allows us to copy the formulas for *x* to the columns for *vx* and *vy* (see Figure 3.36), because the previous value and the increment have the same relative position.

velocity	angle		term v	const							
30	40		85	0.0014							
x-chg	y-chg										
22.981	19.284										
delta t	0.1										
time	x	y	vx	vy	ax	ay	gx	gy	rx	ry	v
0	0	0	22.981	19.284	-0.936	-10.6	0	-9.81	-0.936	-0.785	30
0.1	2.2981	1.9284	22.888	18.224				0	-9.81		
0.2	2.2981	1.9284						0	-9.81		

Figure 3.36

All of the other values for Time 0.1 can be computed by copying the formulas from the row for Time 0 (Figure 3.37).

velocity	angle		term v	const							
30	40		85	0.0014							
x-chg	y-chg										
22.981	19.284										
delta t	0.1										
time	x	y	vx	vy	ax	ay	gx	gy	rx	ry	v
0	0	0	22.981	19.284	-0.936	-10.6	0	-9.81	-0.936	-0.785	30
0.1	2.2981	1.9284	22.888	18.224	-0.909	-10.53	0	-9.81	-0.909	-0.724	29.257
0.2	2.2981	1.9284						0	-9.81		

Figure 3.37

Copying the row for Time 0.1 down as far as needed finishes our spreadsheet (Figure 3.38).

Graphing the columns with *x* and *y* produces a graph of the projectile's flight path. Using different values for the terminal velocity allows us to study the effects of air resistance on flight paths. This graph does not show the distance traveled versus time; it shows the shape of the trajectory.

velocity	angle		term v	const							
30	40		85	0.0014							
x-chg	y-chg										
22.981	19.284										
delta t	0.1										
time	x	y	vx	vy	ax	ay	gx	gy	rx	ry	v
0	0	0	22.981	19.284	-0.936	-10.6	0	-9.81	-0.936	-0.785	30
0.1	2.2981	1.9284	22.888	18.224	-0.909	-10.53	0	-9.81	-0.909	-0.724	29.257
0.2	4.5869	3.7508	22.797	17.171	-0.883	-10.48	0	-9.81	-0.883	-0.665	28.54
0.3	6.8666	5.4678	22.708	16.123	-0.859	-10.42	0	-9.81	-0.859	-0.61	27.85
0.4	9.1374	7.0802	22.623	15.081	-0.835	-10.37	0	-9.81	-0.835	-0.557	27.189
0.5	11.4	8.5883	22.539	14.045	-0.813	-10.32	0	-9.81	-0.813	-0.506	26.557
0.6	13.654	9.9927	22.458	13.013	-0.791	-10.27	0	-9.81	-0.791	-0.459	25.955

Figure 3.38

Construction Summary: Projectile Motion with Air Resistance

velocity	angle		term		const						
[1] 30	[1] 40		[1] 85		0.0014						
x-chg	y-chg			[2]							
[3] 23.0	[3] 19.3										
delta t [1]	0.1										
time	x	y	vx	vy	ax	ay	gx	gy	rx	ry	v
[4] 0.0	[5] 0.0	[5] 0.0	[6] 23.0	19.3	[10] -0.9	-10.6	[7] 0	[7] -9.81	[9] -0.9	-0.8	[8] 30.0
0.1	[11] 2.3	1.9	22.9	18.2	-0.9	-10.5	0	-9.81	-0.9	-0.7	29.3
0.2	4.6	3.8	22.8	17.2	-0.9	-10.5	0	-9.81	-0.9	-0.7	28.5
0.3	6.9	5.5	22.7	16.1	-0.9	-10.4	0	-9.81	-0.9	-0.6	27.9

1. Enter parameter values.
2. Compute constant as g/v^2.
3. Compute x- and y-components of initial velocity vector.
4. Use column to generate time in steps of *delta t*.
5. Enter initial coordinates of projectile.
6. Enter a formula to reproduce the initial x-component of velocity, copy to right.
7. Enter 0 and −9.81 as constant x- and y-components of gravity acceleration.
8. Compute length of the velocity vector.
9. Enter formula for acceleration effect of air resistance, copy to right.
10. Enter formula to sum the x-components of acceleration, copy to right.
11. Compute next value of x, copy right through three columns for y and velocity.
12. Copy down as indicated.

Exercises

1. Add the graph for the flight path to our model.

2. Modify your model to compute the flight path of the projectile using a smaller value of *delta t*.

3. Create a graph to plot the (x,y)-coordinates of the path of the projectile, creating dots at regular time intervals, as in the graph in Figure 3.39.

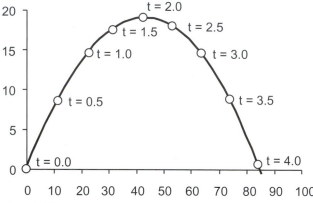

Figure 3.39

4. Build a spreadsheet model to determine both the flight path computed without air resistance and the one with air resistance in one spreadsheet. Graph both flight paths in one graph and compare the flight paths. What do you observe about the shape of the resulting curve?

5. Using the ideas and techniques from Exercise 3 of the previous section, compare the distance traveled before the projectile returns to the surface in both models.

6. Modify the model for a projectile that is fired from a given height above the surface.

7. Design a spreadsheet model to observe the motion of a free-falling body taking into account air resistance and using various values of *c*. This model will illustrate steps in the development of the concept of a limiting velocity. Create separate graphs of distance and velocity as functions of time.

3.3 Planetary Orbits: Half-Step Method

We will now set up a spreadsheet model to compute a planetary orbit, or the orbit of a satellite. To do so, we will use a simplified version of Newton's theory of gravity. This simplified version covers the case with one very heavy object (the sun in the case of a planetary orbit, and the earth in the case of a satellite orbit) and one much lighter object (the planet or the satellite in our case).

The sun is placed at the origin of a coordinate system, and the planet moves around the sun. At any time t, the planet's coordinate vector $\vec{x}(t) = (x(t), y(t))$ gives its position and its velocity vector $\vec{v}(t) = (v_x(t), v_y(t))$ gives its velocity. The fundamental principle of Newton's theory can be expressed in the following way: a vector pointing from the planet to the sun gives the change in velocity, the acceleration. The effect of this change is that the velocity vector is constantly bent toward the sun. The chart in Figure 3.40 (which is not true to scale) shows the change of velocity caused by this principle.

Figure 3.40

The length of this vector is inversely proportional to the square of the distance from the sun. Therefore, the length (but not the direction) of this vector is constant along circles around the sun; and if we compare two positions where the second position has twice the distance from the sun of the first position, the length of the acceleration vector at the second position is only one-fourth of the length at the second position.

To produce a formula from this description, we note that for $r = \sqrt{x^2 + y^2}$ the vector

$-\dfrac{1}{r}(x, y)$ is a vector of length 1 pointing in the opposite direction of the vector (x, y).

For an arbitrary constant c, multiplying this vector with $\dfrac{c}{r^2}$ yields a vector of length $\dfrac{c}{r^2}$.

Therefore, $-\dfrac{c}{r^3}(x, y)$ is the expression that yields the acceleration for our model.

In this spreadsheet model, we will not use real dimensions. Instead, we will use a coordinate system for which the numbers for distances and time lie in the range between 0 and 10^4. We start by setting up the spreadsheet with values for the initial position and the initial velocity, including a value known to work as the gravity constant c (Figure 3.41). These values imply that initially the planet is at the distance 100 from the sun and that it is moving perpendicular to the radius from the sun to the planet.

grav		100						
t	x		y	vx	vy	ax	ay	r
0	100		0	0	1			

Figure 3.41

In the next step, we compute the distance between the sun and the planet. Although this is straightforward to do in this special case, we will create a general formula that works in any possible position. The newly created formula is entered in the last column (Figure 3.42).

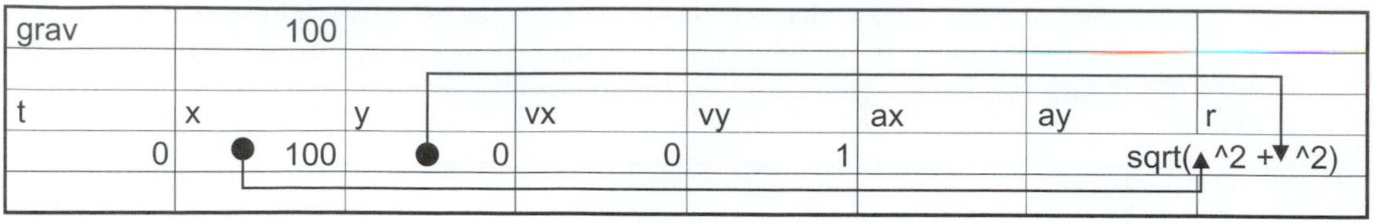

grav		100							
t	x	y	vx	vy	ax	ay	r
0	100	0	0	1			sqrt(▲^2 +▼^2)

Figure 3.42

Using this value that gives the distance of the planet from the sun together with the gravity constant, we can compute the acceleration. We first determine its *x*-component (Figure 3.43).

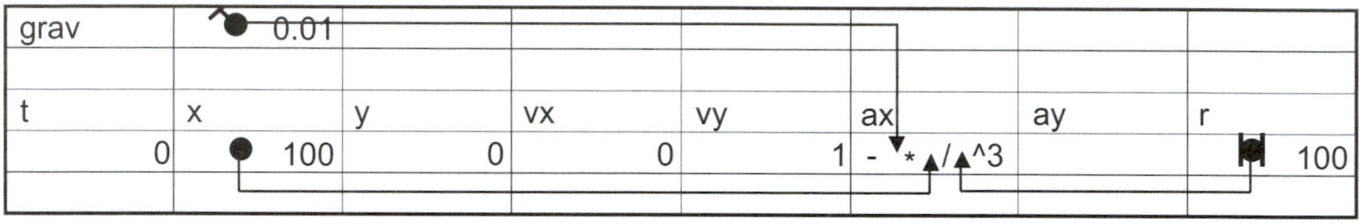

grav		0.01						
t	x	y	vx	vy	ax	ay	r
0	100	0	0	1	- ▼*▲/▲^3		100

Figure 3.43

Since we set up the formula to reference the column containing the distance from the sun with an absolute column reference, we can copy the formula to the right to compute the *y*-component of the acceleration vector (Figure 3.44).

grav		100						
t	x	y	vx	vy	ax	ay	r
0	100	0	0	1	-0.01	0	100

Figure 3.44

We will compute the orbit with a time step of 1. Therefore, we can calculate time and *x* in the next row as shown in Figure 3.45.

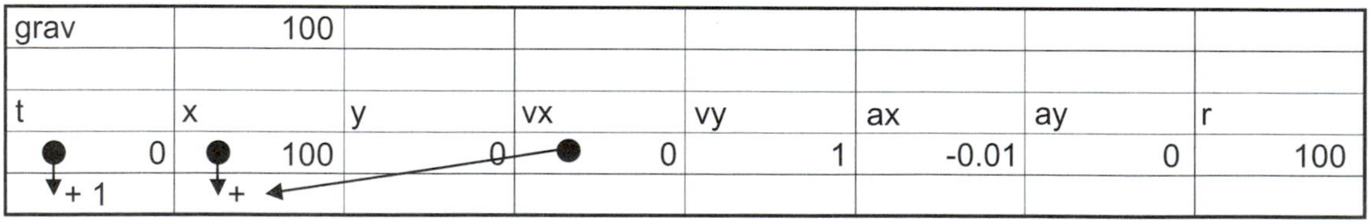

grav		100						
t	x	y	vx	vy	ax	ay	r
0	100	0	0	1	-0.01	0	100
▼+ 1	▼+						

Figure 3.45

Using the layout of the table we can copy the formula for *x* to the right (Figure 3.46).

grav		100								
t	x		y		vx		vy		ax		ay		r
0	100		0		0		1		-0.01		0		100
1	100		1		-0.01		1						

Figure 3.46

Since the formulas in the last three columns only use values from the same row and an absolute reference, they can be copied down (Figure 3.47).

grav		100							
t	x		y		vx		vy	ax	ay	r
0	100		0		0	1	-0.01	0	100
1	100		1		-0.01	1	-0.009999	-1E-04	100.005

Figure 3.47

The row for Time 1 contains the complete set of formulas to compute the present state of the planetary system from the previous state. Therefore, we can copy it down to see a full orbit (Figure 3.48).

grav		100							
t	x		y		vx		vy	ax	ay	r
0	100		0		0	1	-0.01	0	100
1	100		1		-0.01	1	-0.009999	-1E-04	100.005
2	99.99		2		-0.019999	0.9999	-0.009996	-0.0002	100.01
3	99.97		2.9999		-0.029995	0.9997	-0.009993	-0.0003	100.015
4	99.94001		3.9996		-0.039987	0.9994	-0.009988	-0.0004	100.02
5	99.90002		4.999		-0.049975	0.999	-0.009983	-0.0005	100.025
6	99.85004		5.998001		-0.059958	0.998501	-0.009976	-0.000599	100.03
7	99.79009		6.996502		-0.069934	0.997902	-0.009969	-0.000699	100.0351

Figure 3.48

To get a few revolutions of the planet around the sun, we need to copy the formulas down until at least to Time 2,000. When the formulas are copied far enough, we create an *xy*-chart from the *x*- and the *y*-columns (Figure 3.49).

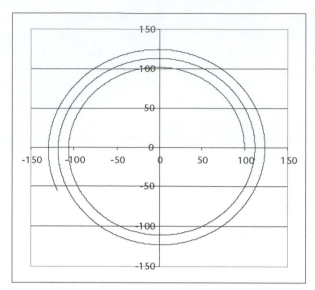

Figure 3.49

For our example, it is important that the scales for the x-axis and the y-axis are equal. Otherwise, we will not be able to see when the orbit is circular and when it is elliptical. Circular orbits are important because the velocity vector has a constant length, and therefore one can compute the time for a full orbit from the radius of the orbit.

We notice that we do not get closed orbits. We could try to improve the accuracy of the computation by changing the time step like we did in the projectile project. For the current project, however, there is a better way.

Looking at the equations of our model, we see that $\vec{x}(1) = \vec{x}(0) + \vec{v}(0)$ and $\vec{v}(1) = \vec{v}(0) + \vec{a}(0)$. These equations can be rewritten as $\vec{v}(0) = \vec{x}(1) - \vec{x}(0)$ and $\vec{a}(0) = \vec{v}(1) - \vec{v}(0)$. So we have considered the distance traveled between Time 0 and Time 1 as the velocity at Time 0. If we were to measure distances and compute velocities from these distances, it would make sense to consider the distance traveled between Time 0 and Time 1 as the velocity at Time 0.5. So, while we originally start with the layout shown in Figure 3.50, we will now change the way that we calculate location, velocity, and acceleration. Let us assume that we know or compute location and acceleration at Times 0, 1, 2, 3. . . and velocity at Times 0.5, 1.5, 2.5, 3.5 . . . .

$\vec{x}(0)$	$\vec{v}(0)$	$\vec{a}(0)$
$\vec{x}(1)$	$\vec{v}(1)$	$\vec{a}(1)$
$\vec{x}(2)$	$\vec{v}(2)$	$\vec{a}(2)$

Figure 3.50

Arranging these in a table ordered by time values produces the pattern in Figure 3.51. Using this table, we can compute $\vec{x}(1) = \vec{x}(0) + \vec{v}(0.5)$ and $\vec{v}(1.5) = \vec{v}(0.5) + \vec{a}(0)$, or, more generally, $\vec{x}(t + 1) = \vec{x}(t) + \vec{v}(t + 0.5)$ and $\vec{v}(t + 0.5) = \vec{v}(t - 0.5) + \vec{a}(t)$.

To save space in the table, we will rearrange the cells as shown in Figure 3.52.

$\vec{x}(0)$		$\vec{a}(0)$
	$\vec{v}(0.5)$	
$\vec{x}(1)$		$\vec{a}(1)$
	$\vec{v}(1.5)$	
$\vec{x}(2)$		$\vec{a}(2)$

Figure 3.51

$\vec{x}(0)$	$\vec{v}(0.5)$	$\vec{a}(0)$
$\vec{x}(1)$	$\vec{v}(1.5)$	$\vec{a}(1)$
$\vec{x}(2)$	$\vec{v}(2.5)$	$\vec{a}(2)$

Figure 3.52

Using the calculation method that we just described, a new *x*-value is computed by adding the element above and the value from the two cells above and to the right. A new *v*-value is computed by adding the values from the two cells above and above and to the right. To implement this in our spreadsheet, we have to change the formula for velocity from that shown in Figure 3.53 to that shown in Figure 3.54.

grav		100							
t		x	y	vx	vy		ax	ay	r
	0	100	0	●	0	1	● -0.01	0	100
	0	100	1	▼+	◄		-0.009999	-1E-04	100.005

Figure 3.53

grav		100							
t		x	y	vx	vy		ax	ay	r
	0	100	0	●	0	1	-0.01	0	100
	0	100	1	▼+	◄		●.009999	-1E-04	100.005

Figure 3.54

Now our labeling is somewhat imprecise. The columns for vx and vy do not contain the values for the time given in the column for t. Instead, they contain the values for $t + 0.5$. Therefore, we relabel the columns to make things clear (Figure 3.55).

grav	100						
t	x(t)	y(t)	vx(t + 0.5)	vy(t + 0.5)	ax(t)	ay(t)	r(t)
0	100	0	●	0	1	-0.01	0	100
1	100	1	▼+	←		●.009999	-1E-04	100.005

Figure 3.55

Now we can copy the new formula for vx (which also works for vy) to the right and down (Figure 3.56).

grav	100						
t	x(t)	y(t)	vx(t + 0.5)	vy(t + 0.5)	ax(t)	ay(t)	r(t)
0	100	0	0	1	-0.01	0	100
1	100	1	-0.009999	0.9999	-0.009999	-1E-04	100.005
2	99.99	1.9999	-0.019995	0.9997	-0.009996	-0.0002	100.01
3	99.97001	2.9996	-0.029987	0.9994	-0.009993	-0.0003	100.015
4	99.94002	3.999	-0.039975	0.999001	-0.009988	-0.0004	100.02
5	99.90004	4.998001	-0.049958	0.998501	-0.009983	-0.000499	100.025
6	99.85009	5.996502	-0.059934	0.997902	-0.009976	-0.000599	100.03
7	99.79015	6.994404	-0.069902	0.997203	-0.009969	-0.000699	100.035

Figure 3.56

After that change in our workbook, or spreadsheet model, the chart also changes (Figure 3.57).

We get a closed orbit. This modified method for computing the orbit is called the *half-step method*. Now we can change the initial position and the initial velocity of the planet and study the effects on the shape of the orbit.

A final remark: The differential equation underlying this system is known to be very sensitive to changes in parameters and to changes in the time step chosen for the numerical iteration. Therefore, changing the gravitational constant or choosing an initial radius very different from the values in our project may produce some strange results that are definitely not elliptical orbits.

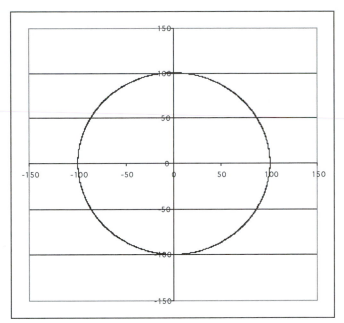

Figure 3.57

Construction Summary: Planetary Orbits (Half-Step Method)

grav	1	100									
	t		x	y		vx	vy		ax	ay	r
2	0	3	100	3	0	3	0	3	1	5	-0.010	0.000	4	100.000
	1	6	100.00	1.000	7	-0.010	1.000		-0.010	0.000	100.005
	2		99.99	2.000		-0.020	1.000		-0.010	0.000	100.010
	3		99.97	3.000		-0.030	0.999		-0.010	0.000	100.015
	4		99.94	3.999		-0.040	0.999		-0.010	0.000	100.020

1. Enter parameter value for the gravitational constant, c.
2. Generate a counter column.
3. Enter the initial position and velocity vectors.
4. Find r, the distance from the sun, as the square root of $x^2 + y^2$, copy down.
5. Compute the x-component of the acceleration vector as $-cx/r^3$, copy right and down.
6. Update x as the previous x-value plus previous velocity v_x. Copy right and down.
7. Update v_x as the previous v_x plus current acceleration a_x. Copy right and down.

Exercises

1. Change the initial position (keep y fixed at 0 and change only the x). For any given initial x, try to find an initial velocity for which the orbit is circular. Try to find a formula for v_y that always produces a circular orbit. In our example, the velocity vector for the initial point $(x, 0)$ is perpendicular to the location vector. Under these circumstances, a circular orbit can be identified in the following way: Find the minimum value and the maximum value of the x-coordinates. If the minimum is the negative of the maximum, the orbit is circular.

2. Using the velocity found in Exercise 1, try to find a formula for the time for one full circular orbit of the planet expressed as a function of the distance from the sun. Using the velocity (which is constant for a circular orbit) and the circumference of the orbit (which can be computed by the formula for a circle), you can compute the orbital period. You might say that now you have rediscovered Kepler's third law.

3. [*] Try to find an initial velocity vector large enough that it does not produce a closed orbit. In that case, the "planet" is not a planet but leaves the solar system. Some space probes had orbits like that. Does this velocity depend on the distance from the sun? Make the initial velocity point directly away from the sun and study the same problem.

3.4 The Shape of the Moon's Orbit

Let us study the following problem: We have a planet revolving around a sun and a moon revolving around the planet. Suppose that the orbit of the planet around the sun is a circle and that the orbit of the moon around the planet also is a circle. Also assume that both orbits lie in the same plane. We want to graph the orbit of the moon with respect to the sun.

We choose our time scale such that a full orbit of the planet takes one time unit, and we assume the radius of the planet orbit to be 1. To characterize the system, we need to know r, the radius of the moon's orbit around the planet, and m, the number of full orbits of the moon in one full planet orbit, which in analogy to the earth-moon system we will call the number of months per planet year.

Our situation is illustrated in Figure 3.58.

The vector from the sun to the planet, $\vec{x}_p$, moves along a circle with Radius 1 and completes one full circle in the time interval from 0 to 1. Therefore, $\vec{x}_p(t) = (\cos(2\pi t), \sin(2\pi t))$ gives the position of the planet at Time t. The vector, $\vec{x}_{mrel}$, from the planet to the moon, moves along a circle with Radius r, and it completes m full circles in the time interval from 0 to 1. Therefore, $\vec{x}_{mrel}(t) = (r\cos(2\pi mt), r\sin(2\pi mt))$ gives the position of the moon relative to the planet at Time t. As a consequence, the orbit of the moon relative to the sun is

$$\vec{x}(t) = \vec{x}_p(t) + \vec{x}_{mrel}(t) = (\cos(2\pi t) + r\cos(2\pi mt), \sin(2\pi t) + r\sin(2\pi mt)).$$

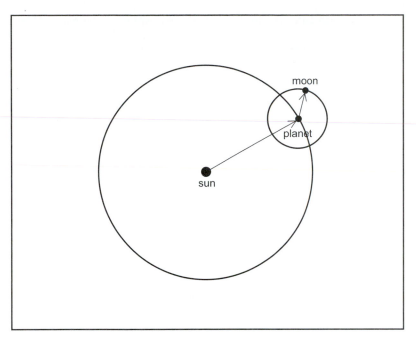

Figure 3.58

Let us set up a spreadsheet using the parameter values $r = 0.3$ and $m = 12$ (Figure 3.59).

r	0.3					
months	12					
t	x planet	y planet	xr moon	yr moon	x moon	y moon
0						

Figure 3.59

Now we enter the formula for the *x*-position of the planet (Figure 3.60).

r	0.3					
months	12					
t	x planet	y planet	xr moon	yr moon	x moon	y moon
0	COS(2*PI()* ▼)					

Figure 3.60

We also enter the formula for the *y*-position of the planet (Figure 3.61).

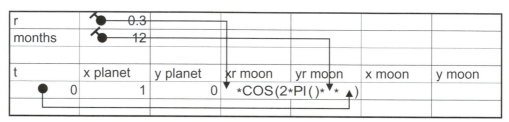

r	0.3					
months	12					
t		x planet	y planet	xr moon	yr moon	x moon	y moon
	0		SIN(2*PI()*)				

Figure 3.61

In the next step, we enter the formula for the relative *x*-position of the moon relative to the planet (Figure 3.62).

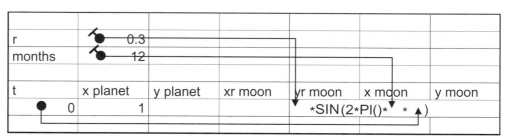

r		0.3					
months		12					
t		x planet	y planet	xr moon	yr moon	x moon	y moon
	0	1		0	*COS(2*PI()* *)		

Figure 3.62

We also enter the formula for the relative *y*-position of the moon relative to the planet (Figure 3.63).

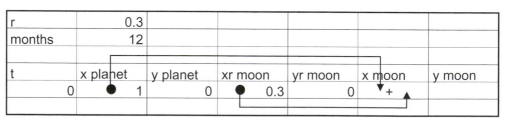

r		0.3					
months		12					
t		x planet	y planet	xr moon	yr moon	x moon	y moon
	0	1			*SIN(2*PI()* *)		

Figure 3.63

The *x*-position of the moon relative to the sun is computed by adding the respective *x*-positions (Figure 3.64).

r	0.3					
months	12					
t		x planet	y planet	xr moon	yr moon	x moon	y moon
	0	1	0	0.3	0	+	

Figure 3.64

We can copy this formula for the *y*-position (Figure 3.65).

r	0.3					
months	12					
t	x earth	y earth	xr moon	yr moon	x moon	y moon
0	1	0	0.3	0	1.3	0

Figure 3.65

In the ensuing step we create the formulas for the next position. To get enough points for a smooth curve, we will create 1,000 points. Therefore, the time step is 0.001. For the next row, we only need to create the formula for the time column and copy down the formulas for the positions of the planet and the moon (Figure 3.66).

r	0.3					
months	12					
t	x earth	y earth	xr moon	yr moon	x moon	y moon
0	1	0	0.3	0	1.3	0
+ 0.001						

Figure 3.66

As the last step in creating the worksheet, we only have to copy the complete row down far enough, until the time 1.000 (Figure 3.67).

r	0.3					
months	12					
t	x earth	y earth	xr moon	yr moon	x moon	y moon
0	1	0	0.3	0	1.3	0
0.001	0.99998	0.006283	0.299148	0.022598	1.299128	0.028881
0.002	0.999921	0.012566	0.296596	0.045068	1.296517	0.057634
0.003	0.999822	0.018848	0.292358	0.067281	1.29218	0.08613
0.004	0.999684	0.02513	0.286459	0.089112	1.286144	0.114243
0.005	0.999507	0.031411	0.278933	0.110437	1.27844	0.141848

Figure 3.67

Creating an *xy*-chart from the column *x* moon and the column *y* moon produces the graph shown in Figure 3.68.

We see that for our current parameter values, with a moon orbit of Radius 0.3 and a moon orbital period that is 1/12 of the planet orbital period, the orbit has loops. Changing these values produces different graphs. For example, $r = 0.2$ and $m = 4$ produce a

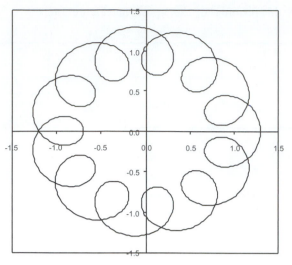

Figure 3.68

loop-free orbit, while $r = 0.01$ and $m = 9$ even produce a convex orbit that looks somewhat like a regular polygon with smooth edges.

Now we have to find values that describe the system sun-earth-moon. The distance from the sun to the earth varies between 147,500,000 kilometers and 152,500,000 kilometers, while the distance from the earth to the moon varies between 364,000 kilometers and 403,000 kilometers. The moon orbits the earth once in 27.3 days, while the earth orbits the sun in 1 year, or 365.25 days. Therefore, the moon orbits around the earth about 13 times in 1 year.

Simplifying the actual situation somewhat, we can say that $r = 0.0025$ and $m = 13$. This produces a loop-free convex orbit that is almost indistinguishable from a circle. Therefore, we see that the moon's orbit around the sun does not contain any loops and is even convex.

In fact, the orbits of planets and moons are generally ellipses of different eccentricities, e, where $0 < e < 1$, rather than circles, which have an eccentricity of $e = 0$. As an exercise we are challenged to create a model in which orbits are ellipses with foci at the sun and the planet.

Construction Summary: Shape of the Moon's Orbit

1. Enter parameter values.
2. Generate a column for Point t.
3. Compute the (x,y)-components of the planet.
4. Compute the (x,y)-components of the moon relative to the planet.
5. Compute the moon's x-component.
6. Copy as indicated.

r	1	0.3				
month	1	12				
	t	x earth	y earth	xr moon	yr moon	x moon	y moon
2	0 3	1 3	0 4	0.3 4	0 5	1.3	0
	0.001	1.000	0.006	0.299	0.023	1.299	0.029
	0.002	1.000	0.013	0.297	0.045	1.297	0.058
	0.003	1.000	0.019	0.292	0.067	1.292	0.086

Exercises

1. The rotations for the planet and the moon in our model are in the same direction (counterclockwise in our graph). Modify the worksheet so that the planet orbits counterclockwise around the sun and the moon orbits clockwise around the planet.

2. Try to find a relation between m and r that ensures that that the orbit does not contain loops.

3. Try to find a relation between m and r that ensures that the orbit is convex.

4. Investigate the nature of the curves produced for various values of r and m. Discover how our model can be used to produce a variety of classical curves, such as cardioids. What happens if m is not an integer? How do we need to modify our model to provide full illustrations of the resulting curves?

5. An *epicycloid* is a curve traveled by a point on a smaller disk gliding around a larger disk (spirographs produce epicycloids). How can our planetary orbit model produce epicycloids?

6. Design a spreadsheet model for planetary motion in which orbits are ellipses, with the sun and the planet as foci. Hint: The parametric equation of an ellipse of eccentricity e and center at the origin can be given by

$$x = a\cos(2\pi mt), \text{ and } y = b\sin(2\pi mt),$$

where

$$c = ae,$$

and

$$b^2 = a^2 - c^2 \text{ (Figure 3.69)}.$$

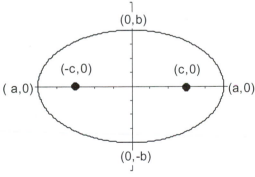

Figure 3.69

If the ellipse is translated so that the foci are at $(-2c,0)$ and $(0,0)$, then

$$x = a\cos(2\pi mt) - c \text{ (Figure 3.70)}.$$

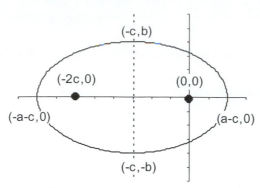

Figure 3.70

7. Consult references containing data on the planets, moons, and principal comets of our solar system; incorporate them into an orbit model. You may wish to include several orbits within the same graph.

8. Enhance your orbit models to incorporate a scroll bar to animate the motion of the planet and moon around the sun. Possible output from our primary model is provided in Figure 3.71.

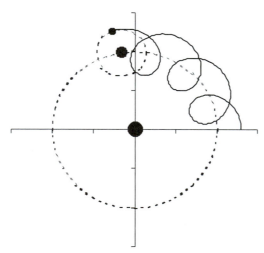

Figure 3.71

3.5   Heat Propagation

A spreadsheet can be an excellent tool for visualizing and analyzing mathematical examples from physics and the other sciences, as well as for implementing numerical algorithms in these fields. Here we provide two examples from physics involving heat flow. The latter example is also applicable to the study of other topics, such as groundwater flow in geology.

1. One-Dimensional Model

Suppose that we have a thin, straight metal rod whose temperature is initially uniformly at the room temperature of 10° C in a cool laboratory. Next we provide a constant heat source of 60° C at one end of the rod and another constant heat source of 100° C at the other end. What will happen to the temperature of the rod after a period of time? Clearly, it will heat up, but will all of the points on it eventually have the same temperature, or is there another way that we can describe the long-term state of the rod's temperatures?

We begin building our model by considering the rod to be subdivided into a number of segments. In order to present the concepts of the model in a condensed manner, we will divide the rod into seven sections in addition to the two ends for a total of nine segments. However, we could easily increase the number of segments for a more precise analysis. We will study the changes in temperature over a series of periods of time.

We enter the initial temperatures of the segments of the rod, together with the temperatures of the heat sources, into the cells of a row, as shown in Figure 3.72. In addition, we will use a column to the left to count time periods. We enter 0 into the top row of that column and insert our standard counting formula into the cell below it, that is, "add 1 to the cell above." Also, into the end cells of the second row, we enter simple formulas to reproduce the temperatures at the two heat sources.

Period									
0	60	10	10	10	10	10	10	10	100
1									

Figure 3.72

Next, we need to decide how to obtain an approximation of the temperature in the next period of the first interior segment on the rod. One way to think about the process is that in a given period each segment will exchange with its two neighboring segments a fixed percentage, α, of the difference in their temperatures.

The idea can be seen in Figure 3.73, in which we assume that in every period each cell shares 40 percent of the difference in the temperature with its two neighboring cells, with the heat flowing from a warmer cell to a cooler one. The temperature of each cell is provided inside the box.

Thus, since the first cell is 30 degrees warmer than the cell to its right, in the next period it gives up 12 degrees (0.4 × 30) to the second one. Because the second cell is 20 degrees warmer than the cell to its right, it gives up 8 degrees to that cell. Thus, the second cell has a net increase of 4 degrees (12 − 8). Finally, because the third cell is 40 degrees cooler than the cell on its right, it receives 16 degrees from that one, for a net increase of 24 degrees (8 + 16). Without further information, we cannot adjust the temperatures of the other two cells. The new temperatures appear in the lower part of the figure.

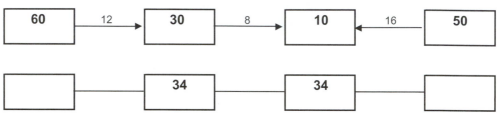

Figure 3.73

We can actually show the heat exchange with all of the arrows going in one direction (see Figure 3.74) if we interpret negative difference as meaning flow in the opposite direction.

Figure 3.74

Thus, we can display the general case using the state diagram in Figure 3.75, together with either the upper or the lower set of arrows, which are equivalent. This will help us in our designing of the spreadsheet equations. It should be mentioned that in keeping the temperatures of the endpoints of the rod constant we are employing an outside heat source to supply the heat required to adjust for the heat exchanged with its neighbor.

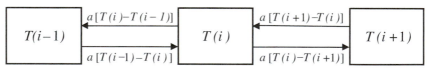

Figure 3.75

Thus, if we denote the temperature of Point i in Period n by $T_n(i)$, then its temperature in period $n + 1$ will be

$$T_{n+1}(i) = T_n(i) + \alpha[T_n(i-1) - T_n(i)] - \alpha[T_n(i) - T_n(i+1)]$$

or

$$T_{n+1}(i) = T_n(i) + \alpha[T_n(i-1) - T_n(i)] + \alpha[T_n(i+1) - T_n(i)].$$

Thus, to the previous temperature of Point i we add a fixed percentage, α, of the difference between its temperature and the temperature of the segment to its left, and the same percentage of the difference in its temperature and the temperature of the segment to its right.

Since the resulting formula is rather lengthy to enter into the spreadsheet, we can use algebra to simplify it somewhat to obtain

$$T_{n+1}(i) = (1 - 2\alpha)T_n(i) + \alpha[T_n(i-1) + T_n(i+1)].$$

We now enter this formula into the cell adjacent to the left endpoint, being sure that the reference to α is absolute. Here we use the value $\alpha = 0.4$. We then copy this expression across the row to complete the temperature distribution for the period $n = 1$ (Figure 3.76).

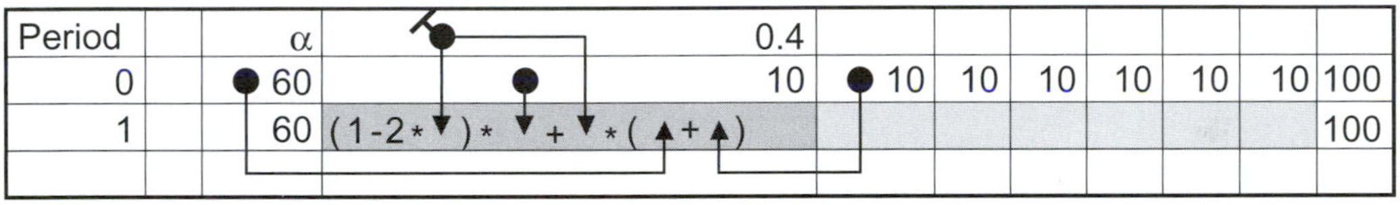

Figure 3.76

We can now copy the formulas in that row downward to generate the temperature distribution for any number of additional periods (Figure 3.77).

Period		α	0.4							
0		60	10	10	10	10	10	10	10	100
1		60	30	10	10	10	10	10	46	100
2		60	34	18	10	10	10	24.4	53.2	100
3		60	38	21.2	13.2	10	15.76	30.16	60.4	100

Figure 3.77

In Figure 3.78, we indicate the output from some of the first 157 periods of the process with $\alpha = 0.4$. If we examine the entire output in more detail, we will see that initially there is a sizable amount of change in some of the cell temperatures as the heat flows from the relatively hot endpoints to the initially cooler interior points. Eventually the temperatures effectively cease to change, and we say that the heat distribution has reached steady state. From the output shown in the figure, we can see that we have essentially reached this point by the 156th period. Here we might notice that the final distribution increases linearly from the temperature at the left endpoint to the temperature at the right endpoint.

If we experiment with various values for the exchange rate, α, we will see that when the algorithm converges, it always converges to the same steady-state distribution. This can be verified mathematically as well. However, you can also find that for some values of α the process diverges.

Period	α	0.4							
0	60	10	10	10	10	10	10	10	100
1	60	30	10	10	10	10	10	46	100
2	60	34	18	10	10	10	24.4	53.2	100
3	60	38	21.2	13.2	10	15.76	30.16	60.4	100
10	60	47.6	37.63	32.32	33.1	41.05	55.96	76.42	100
11	60	48.57	39.46	34.74	35.93	43.83	58.18	77.67	100
30	60	59.89	60.56	62.66	66.64	72.65	80.55	89.88	100
156	60	65	70	75	80	85	90	95	100
157	60	65	70	75	80	85	90	95	100

Figure 3.78

Next, we will present one way to construct a graph of the changes in the temperature distribution of the system. To do this, we insert a few empty rows above the current top of our spreadsheet model and enter the number, n, that represents a selected time period (here, Period $n = 10$ has been selected). We want our model to reproduce the temperatures for this period into this row. We will describe one method for doing this.

To help us identify the rod's segments in the graph, we also generate a counter for the segments ranging from 0 for the left endpoint up to 8 for the right endpoint (Figure 3.79).

		0	1	2	3	4	5	6	7	100
10	60	47.6	37.63	32.23	33.1	41.05	55.96	76.42		100
Period	α	0.4								
0	60	10	10	10	10	10	10	10		100
1	60	30	10	10	10	10	10	46		100

Figure 3.79

It is easy to generate the desired row. In the first column (see Figure 3.80), we enter the formula OFFSET (*start,row,col*). This function returns the value of the cell that is offset by *row* rows and *col* columns from a starting location, *start*. Here we start from the first segment in the initial row and go *n* rows and 0 columns to find the temperature for that segment. We can copy this formula across the row to generate a copy of the selected row. However, we must ensure that the reference to the row number is absolute.

At this point, we can also create a scroll bar in our model and connect it to the cell that determines the period. This adds an animated feature to our graph, allowing us to show a more continual change in the temperature distribution from one time period to the next.

			0	1	2	3	4	5	6	7	8
10		OFFSET(▲ , ▼,0)									
Period			α	0.4							
0			60	10	10	10	10	10	10	10	100
1			60	30	10	10	10	10	10	46	100
2			60	34	18	10	10	10	24.4	53.2	100

Figure 3.80

We now select the two new rows and use them to generate a column graph in which the heights of the columns represent the temperatures of the various segments of the rod during a given time period. The initial temperature distribution is shown in Figure 3.81. We use $\alpha = 0.4$ in this figure and the next two figures.

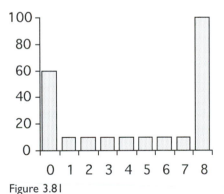

Figure 3.81

The temperatures for the 10th period and the 200th period are shown in Figure 3.82 and Figure 3.83, respectively. Notice that Figure 3.83, which shows the steady-state temperature distribution, indicates a linear relationship. The linear nature at steady state can also be seen by using a line graph.

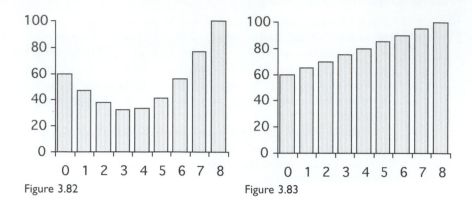

Figure 3.82   Figure 3.83

Finally, there is another approach that we can use to create an extremely compact version of our model and, in the process, lay the groundwork for the ideas of the two-dimensional heat model that follows. This time we create only the first two rows of our model. Then, instead of extending the model down for further rows, we record a *macro* (see the Appendix for a discussion of this topic) in which we select and copy the entries of the second row and then paste their values (only the values, not the formulas) into the top row. This moves the second row's temperatures into the top row and, in turn, generates the temperature distribution of the next period in the second row.

After we turn off the macro recorder, we can design a button to execute the macro. We can then repeatedly generate the temperature distributions for successive periods by clicking on the button. This model may be used to generate the graph constructed previously. We see the model's initial screen display in Figure 3.84. This time we use $\alpha = 0.3$.

Period		α	0.3							
0		60	10	10	10	10	10	10	10	100
1		60	25	10	10	10	10	10	37	100

Figure 3.84

After invoking the macro 10 times, we now see the distribution of temperatures for the 10th period (Figure 3.85).

Period		α	0.3							
10		60	44.9	32.59	25.4	25.13	32.93	49.09	72.43	100
11		60	45.74	34.13	27.48	27.55	35.44	51.25	73.7	100

Figure 3.85

With this value of α, it takes a little longer to reach steady state, as seen in Figure 3.86. As a challenge, try to determine for which value of the parameter α the method converges to steady state the fastest, and for which values it diverges.

Period	α	0.3								
209	60	65	70	75	79.99	85	90	95	100
210	60	65	70	75	80	85	90	95	100

Figure 3.86

The case $\alpha = 0.5$ is often used in calculations in science. When we simplify our recurrence relation for this case, our recurrence relation becomes

$$T_{n+1}(i) = \frac{T_n(i-1) + T_n(i+1)}{2}$$

so that the next temperature of a segment of the rod is the average of the current temperatures of its two neighboring points. The more general expression involving the parameter α that we have used gives a weighted average.

Construction Summary: Heat Flow in a Rod

1. Create a counter column.
2. Enter the boundary temperatures and copy down.
3. Enter the initial temperatures of the interior cells in the top row.
4. Compute the next temperature of the leftmost interior cell as the average temperature of the two adjoining cells above.
5. Copy to the right and down as indicated.

Period	α	0.5						
1	0	2	40	3	20.0	20.0	20.0	20.0	2	100
	1		40	4	30.0	20.0	20.0	60.0		100
	2		40		30.0	25.0	40.0	60.0		100
	3		40		32.5	25.0	40.0	70.0		100
	4		40		32.5	27.5	50.0	70.0		100
	5		40		33.8	27.5	50.0	75.0		100
	6		40		33.8	30.3	56.6	75.0		100
	7		40		35.2	30.3	56.6	78.3		100

Three columns are hidden to condense the display.

2. Two-Dimensional Model

Next we expand our study of heat flow by considering a two-dimensional model. Suppose that the temperature throughout a uniform thin plate is initially a constant 0° C. In addition, suppose that the top and the bottom boundaries of the plate are insulated so that temperatures at those points will always remain at 0° C. Next, heat sources are applied to both of the left and right boundaries of the plate so that the temperature on those edges increases linearly from 0° C at the top and bottom of the plate until they reach 50° C at the center of each vertical edge. The heat sources will remain constant at all times. Our challenge is to determine what happens to the temperature distribution of the plate over a range of time.

To analyze the situation, we create a two-stage model similar to the last one produced for our one-dimensional model in the preceding section. Again, we will approximate changes in the temperature distribution on the plate over successive time periods. This time we subdivide the plate into small rectangular segments, represented by a rectangular block of cells in the spreadsheet (see Figure 3.87).

First, we enter the initial temperature distribution into the cells of the top half of the model. The boundary temperatures will remain constant throughout. The bottom half of the model then computes an approximation for the temperatures in the next time period. Thus, in the figure, the distribution of the initial, or the 0th, period's temperature distribution is shown in the upper block of cells, with the next, or 1st, period's distribution computed in the lower block. We will need to enter only one for-

										α	0.15
0	0	0	0	0	0	0	0	0	0	n	0
10	0	0	0	0	0	0	0	0	10		
20	0	0	0	0	0	0	0	0	20		
30	0	0	0	0	0	0	0	0	30		
40	0	0	0	0	0	0	0	0	40		
50	0	0	0	0	0	0	0	0	50		
40	0	0	0	0	0	0	0	0	40		
30	0	0	0	0	0	0	0	0	30		
20	0	0	0	0	0	0	0	0	20		
10	0	0	0	0	0	0	0	0	10		
0	0	0	0	0	0	0	0	0	0		
0	0	0	0	0	0	0	0	0	0	n	+1
10									10		
20									20		
30									30		
40									40		
50									50		
40									40		
30									30		
20									20		
10									10		
0	0	0	0	0	0	0	0	0	0		

Figure 3.87

mula in the upper-left interior cell in the bottom half and then copy it throughout the rest of the interior cells. The formula for the first cell references the same location and the four neighboring cells in the top half.

The construction of the formulas for the next period in the bottom block mimics the method that we used in the one-dimensional case. As before, we assume that in a given period each segment will exchange with its four neighboring segments (left, right, above, and below) a fixed percentage, α, of the difference in their temperatures. Thus, if we denote the temperature of point (i,j) in Period n by $T_n(i, j)$, then its temperature in period $n + 1$ will be

$$T_{n+1}(i, j) = T_n(i,j) + \alpha[T_n(i - 1, j) - T_n(i, j)] + \alpha[T_n(i + 1,j) - T_n(i, j)]$$
$$+ \alpha[T_n(i, j + 1) - T_n(i, j)] + \alpha[T_n(i, j - 1) - T_n(i, j)].$$

Since the resulting formula is rather lengthy to enter into the spreadsheet, we can use algebra to simplify it somewhat to obtain

$$T_{n+1}(i) = (1 - 4\alpha)T_n(i,j) + \alpha[T_n(i - 1, j) + T_n(i + 1, j) + T_n(i, j + 1) + T_n(i, j - 1)].$$

Again, we can view the new temperature as a weighted average of the previous temperatures of the point and its four neighbors. We simply enter this formula in the upper-left interior cell and then copy it throughout the remainder of the interior cells. The basic step in our model is illustrated in Figure 3.88 where we have used $\alpha = 0.15$.

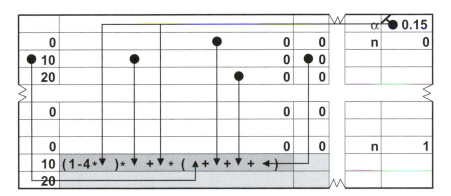

Figure 3.88

We next follow the same approach that we used in the last version of our one-dimensional model. We turn on the macro recorder, select and copy the entire bottom distribution (including the value of the time period, n), and paste the values only in place of the previous contents of the block of cells in the top part.

If we then attach this macro to a button that will serve as a trigger for activating it, then each time that we click on the button we will generate the new temperature distribution for the next time period. The initial display is shown in Figure 3.89, and the result of 200 updates of our model is shown in Figure 3.90. Long before this time, we will see that the temperature distribution essentially has reached steady state.

It is also convenient to design an initialization macro and button that will place a 0 in each of the interior cells of the upper block of cells and set the initial counter to 0.

Figure 3.89 — first block ($\alpha = 0.15$, $n = 0$):

										α	0.15
0	0	0	0	0	0	0	0	0	0	n	0
10	0	0	0	0	0	0	0	0	10		
20	0	0	0	0	0	0	0	0	20		
30	0	0	0	0	0	0	0	0	30		
40	0	0	0	0	0	0	0	0	40		
50	0	0	0	0	0	0	0	0	50		
40	0	0	0	0	0	0	0	0	40		
30	0	0	0	0	0	0	0	0	30		
20	0	0	0	0	0	0	0	0	20		
10	0	0	0	0	0	0	0	0	10		
0	0	0	0	0	0	0	0	0	0		

Figure 3.89 — second block ($n = 1$):

0	0	0	0	0	0	0	0	0	0	n	1
10	1.5	0	0	0	0	0	0	1.5	10		
20	3	0	0	0	0	0	0	3	20		
30	4.5	0	0	0	0	0	0	4.5	30		
40	6	0	0	0	0	0	0	6	40		
50	7.5	0	0	0	0	0	0	7.5	50		
40	6	0	0	0	0	0	0	6	40		
30	4.5	0	0	0	0	0	0	4.5	30		
20	3	0	0	0	0	0	0	3	20		
10	1.5	0	0	0	0	0	0	1.5	10		
0	0	0	0	0	0	0	0	0	0		

Figure 3.89

Figure 3.90 — first block ($\alpha = 0.15$, $n = 250$):

										α	0.15
0	0	0	0	0	0	0	0	0	0	n	250
10	8.46	7.16	6.24	5.78	5.78	6.24	7.16	8.46	10		
20	16.7	13.9	12	11.1	11.1	12	13.9	16.7	20		
30	24.3	19.8	16.9	15.4	15.4	16.9	19.8	24.3	30		
40	30.7	24.2	20.2	18.4	18.4	20.2	24.2	30.7	40		
50	34.4	26.1	21.5	19.4	19.4	21.5	26.1	34.4	50		
40	30.7	24.2	20.2	18.4	18.4	20.2	24.2	30.7	40		
30	24.3	19.8	16.9	15.4	15.4	16.9	19.8	24.3	30		
20	16.7	13.9	12	11.1	11.1	12	13.9	16.7	20		
10	8.46	7.16	6.24	5.78	5.78	6.24	7.16	8.46	10		
0	0	0	0	0	0	0	0	0	0		

Figure 3.90 — second block ($n = 251$):

0	0	0	0	0	0	0	0	0	0	n	251
10	8.46	7.16	6.24	5.78	5.78	6.24	7.16	8.46	10		
20	16.7	13.9	12	11.1	11.1	12	13.9	16.7	20		
30	24.3	19.8	16.9	15.4	15.4	16.9	19.8	24.3	30		
40	30.7	24.2	20.2	18.4	18.4	20.2	24.2	30.7	40		
50	34.4	26.1	21.5	19.4	19.4	21.5	26.1	34.4	50		
40	30.7	24.2	20.2	18.4	18.4	20.2	24.2	30.7	40		
30	24.3	19.8	16.9	15.4	15.4	16.9	19.8	24.3	30		
20	16.7	13.9	12	11.1	11.1	12	13.9	16.7	20		
10	8.46	7.16	6.24	5.78	5.78	6.24	7.16	8.46	10		
0	0	0	0	0	0	0	0	0	0		

Figure 3.90

From the numerical display shown in the two preceding figures, it may be hard to visualize what the final distribution looks like. However, if we create a three-dimensional surface graph, with the heights representing the temperatures, we will see that a saddle surface has been generated (Figure 3.91).

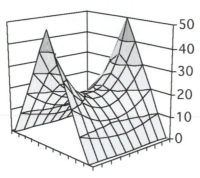

Figure 3.91

A view of an implementation of the model with a three-dimensional graph using buttons to initialize or update the process is shown in Figure 3.92. The construction of these graphs is discussed in the Appendix.

A	B	C	D	E	F	G	H	I	J	K	L	M
1		INITIAL			UPDATE							
2											α	0.15
3	0	0	0	0	0	0	0	0	0	0	n	25
4	10	7.11	4.67	2.95	2.07	2.07	2.95	4.67	7.11	10		
5	20	14.1	9.19	5.76	4.03	4.03	5.76	9.19	14.1	20		
6	30	20.8	13.3	8.24	5.72	5.72	8.24	13.3	20.8	30		
7	40	26.6	16.5	10.1	6.92	6.92	10.1	16.5	26.6	40		
8	50	30	18	10.8	7.36	7.36	10.8	18	30	50		
9	40	26.6	16.5							40		
10	30	20.8	13.3	8						30		
11	20	14.1	9.19	5						20		
12	10	7.11	4.67	2						10		
13	0	0	0							0		
14												
15	0	0	0							0	n	26
16	10	7.2	4.8							10		
17	20	14.2	9.4							20		
18	30	20.9	13.5							30		
19	40	26.7	16.8							40		
20	50	30.2	18.3	11.1	7.7	7.7	11.1	18.3	30.2	50		
21	40	26.7	16.8	10.4	7.3	7.3	10.4	16.8	26.7	40		
22	30	20.9	13.5	8.5	6.0	6.0	8.5	13.5	20.9	30		
23	20	14.2	9.4	6.0	4.3	4.3	6.0	9.4	14.2	20		
24	10	7.2	4.8	3.1	2.2	2.2	3.1	4.8	7.2	10		
25	0	0	0	0	0	0	0	0	0	0		

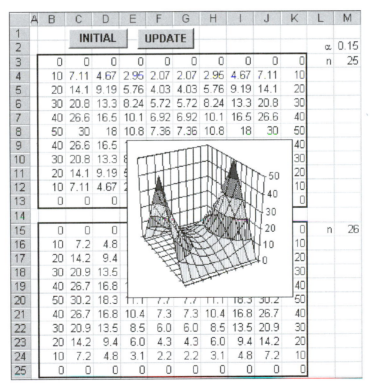

Figure 3.92

On your spreadsheet, it may also be possible to employ conditional formatting features to use the appearance of the screen display to show the contours of the temperature regions, as shown in Figure 3.93. Here, different shading is used to indicate different temperature ranges. This topic is discussed in the Appendix.

										α	0.15
0	0	0	0	0	0	0	0	0	0	n	250
10	8.46	7.16	6.24	5.78	5.78	6.24	7.16	8.46	10		
20	16.7	13.9	12	11.1	11.1	12	13.9	16.7	20		
30	24.3	19.8	16.9	15.4	15.4	16.9	19.8	24.3	30		
40	30.7	24.2	20.2	18.4	18.4	20.2	24.2	30.7	40		
50	34.7	26.1	21.5	19.4	19.4	21.5	26.1	34.7	50		
40	30.7	24.2	20.2	18.4	18.4	20.2	24.2	30.7	40		
30	24.3	19.8	16.9	15.4	15.4	16.9	19.8	24.3	30		
20	16.7	13.9	12	11.1	11.1	12	13.9	16.7	20		
10	8.46	7.16	6.24	5.78	5.78	6.24	7.16	8.46	10		
0	0	0	0	0	0	0	0	0	0		

Figure 3.93

In the special case of $\alpha = 0.25$, our formula can be simplified to the following one:

$$T_{n+1}(i, j) = \frac{T_n(i, j - 1) + T_n(i + 1, j) + T_n(i, j + 1) + T_n(i - 1, j)}{4}$$

where $T_n(i, j)$ refers to the temperature in the nth iteration at Point (i, j). Thus, in this case, the temperature of a point in the next iteration is the average of the current temperatures of its four neighbors.

As in the one-dimensional case, it is a good challenge to find an optimal value for the parameter α and to find out for which values it leads to divergence.

1. Enter the boundary temperatures in the boundary cells (bold) of the top block.
2. Enter 0 as the initial temperature in the upper-left interior cell and copy throughout the interior.
3. Enter 0 as the initial iteration counter.
4. Create the same boundary temperatures in the boundary cells (bold) of the bottom block.
5. Compute the next temperature at the upper-left interior cell as the average of the four surrounding cells from above and copy throughout the interior.
6. Compute the next iteration counter as 1 plus the previous counter.
7. Create a macro to copy the values of the lower display into the upper display.

Top block:

1	0	0	0	0	0	0	n	3	0
10	2 0.0	0.0	0.0	0.0	10				
20	0.0	0.0	0.0	0.0	20				
30	0.0	0.0	0.0	0.0	30				
40	0.0	0.0	0.0	0.0	40				
50	0.0	0.0	0.0	0.0	50				
40	0.0	0.0	0.0	0.0	40				
30	0.0	0.0	0.0	0.0	30				
20	0.0	0.0	0.0	0.0	20				
10	0.0	0.0	0.0	0.0	10				
0	0	0	0	0	0				

Bottom block:

4	0	0	0	0	0	0	n	6	1
10	5 2.5	0.0	0.0	2.5	10				
20	5.0	0.0	0.0	5.0	20				
30	7.5	0.0	0.0	7.5	30				
40	10.0	0.0	0.0	10.0	40				
50	12.5	0.0	0.0	12.5	50				
40	10.0	0.0	0.0	10.0	40				
30	7.5	0.0	0.0	7.5	30				
20	5.0	0.0	0.0	5.0	20				
10	2.5	0.0	0.0	2.5	10				
0	0	0	0	0	0				

Exercises

1. Design two buttons for each model in this section. One button is used to initialize the temperatures of the interior cells. The second one copies the values of the bottom block of cells and the counter into the upper block in order to carry out one iteration of the process.

2. Use the two-dimensional model of this section to obtain the steady-state condition that results from other boundary conditions. Display the output using both three-dimensional graphs and conditional formatting (see Appendix) of the numerical output.

   a. Start at 100° C in the upper-left corner, vary the temperature distribution linearly from 100° C down to 0° C along the left and the top sides, and maintain the temperature at a constant 0° C along the right and the bottom sides.

b. Start with 50° C in the middle of all four sides and decrease the heat distribution linearly until it is 0° C in each of the four corners.

3. Create condensed spreadsheet models for some mathematical algorithms that compute the first iteration of an algorithm. Then record a macro to copy the values of the bottom row(s) of the display back into upper row(s) to generate further values of the algorithm. Link the macros to user-designed buttons to improve the user interface of the model. Use this approach to compute Fibonacci numbers and factorials (both shown in Figure 3.94), as well as to implement two additional algorithms of your choice. Possible examples include the computation of x^n via the recurrence relation $x \cdot x^{n-1}$, the annual growth of a savings account at a given interest rate, the unit fraction expansion of a fraction (see Exercise 3 of Decimal Expansions in Chapter 4), or a cumulative sum such as $1 + x + x^2 + \ldots$

n	Fib(n)
1	1
2	1
3	2

n	Fib(n)
2	1
3	2
4	3

n	n!
1	1
2	2

n	n!
2	2
3	6

Figure 3.94

4. Suppose that the heat distribution in a long, straight rod is initially 0 throughout. If the two ends of the rod are held at 0°C and a heat source concentrated at the center of the rod is held at 50°C, what happens to the heat distribution in the rod after a long period of time? How does the diffusion coefficient affect the time required to reach equilibrium? Create a spreadsheet model to investigate these questions. The initial distribution is shown in Figure 3.95.

0	0	0	0	0	0	0	0	0	0	50	0	0	0	0	0	0	0	0	0	0

Figure 3.95

4

Number Theory

CHAPTER OUTLINE

4.1   Numbers in Different Bases

In this section, we examine different ways of representing integers. In everyday work, we customarily write numbers in the decimal, or base 10, notation using the 10 digits 0, 1, 2, . . . , 9. However, digital computers store numbers in the binary, or base 2, notation because the only digits that are needed, 0 and 1, can be represented naturally in an electronic circuit as "on" and "off." In addition, computer scientists sometimes employ the hexadecimal, or base 16, notation. We will create models that allow us to change from base 10 to another base and then back again.

> When we write a number, for example, 3,127, this is shorthand for
>
> $$3{,}127 = 3 \cdot 1{,}000 + 1 \cdot 100 + 2 \cdot 10 + 7 = 3 \cdot 10^3 + 1 \cdot 10^2 + 2 \cdot 10^1 + 7 \cdot 10^0.$$
>
> This representation is called the *decimal expansion*, or the expansion in base 10, of the number.

It is possible to use other bases instead of 10. Computers internally work with base 2 numbers. In base 2, $1101_2 = 1 \cdot 2^3 + 1 \cdot 2^2 + 0 \cdot 2^1 + 1 \cdot 2^0 = 8 + 4 + 1 = 13_{10}$. When two different bases are used within one equation, the base is usually indicated as an index, or subscript, (and the base itself is written in base 10). When the index is omitted, the base in most cases is understood to be 10.

Suppose that b is a positive integer greater than 1. Then, in writing numbers in base b, the digits that we need are those for the numbers 0,1, . . . , $b - 1$. For bases less than 10, we use the usual digits (for example, for base 8 we use 0,1, . . . , 7). For bases larger than 10, we need digits for the numbers 11, 12, and so on. Usually, the letters a–z or A–Z are used, with the interpretation a = A = 10, b = B = 11, ... This convention limits us to bases up to 35, but it is very uncommon to work with larger bases.

We will now set up a spreadsheet that converts a number given in base 10 to another base $b \leq 35$. The central idea in our spreadsheet model is that the last digit of the number n in base b is the remainder that is produced when dividing n by b.

As our first illustration, we will convert 2105_{10} to base 8. Before we do that, however, we set up a worksheet that calculates the digits of a number in base 10. Then we will simply change the base, and the number representation will change accordingly.

The last digit of the base 10 representation of 2105_{10} is the remainder that is produced in dividing 2105 by 10, which is 5. Spreadsheet programs have a built-in function for computing remainders, and in *Microsoft Excel* this function is called MOD.

To compute the next digit (going from the last digit to the next to the last digit), we subtract the last digit from our number and divide the resulting difference by 10 (our base). The last digit of this new number is the next to the last digit of the original number. Therefore, we can transform our algorithm for expanding a number in base 10 to an algorithm for calculating the last digit of a sequence of numbers derived from the original number.

So let us set up a worksheet for our problem. First, we calculate the last digit. Since we will reuse the formula that calculates the remainder, we create the reference to the base as an absolute reference (Figure 4.1).

In the next step, we compute the next number in our sequence by subtracting the last digit from the original number and dividing the resulting difference by the base (Figure 4.2).

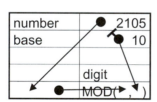

Figure 4.1

Figure 4.2

Again we make the reference to the base an absolute reference. The next digit that we need to calculate is the remainder of the "new" number when it is divided by the base. Since we have already created this formula in calculating the last digit of 2105, we need only to copy it down (Figure 4.3).

number	2105
base	10
	digit
2105	5
210	

Figure 4.3

Copying the last two lines of our worksheet down farther gives all the digits for our given number. To be able to compute the digits for larger numbers also, we copy the formulas down farther than needed for our "current" number (Figure 4.4). Proceeding in reverse order from the end, we can read off the digits manually: 2105.

number	2105
base	10
	digit
2105	5
210	0
21	1
2	2
0	0
0	0
0	0

Figure 4.4

Now the last column contains the digits, but in reverse order. Changing the base immediately produces the digits for another base (8 in our case). The notation for this is $2105_{10} = 4071_8$. We can check this by observing that

$$4071_8 = 4 \cdot 8^3 + 0 \cdot 8^2 + 7 \cdot 8^1 + 1 \cdot 8^0 = 2048 + 0 + 56 + 1 = 2105.$$

number	2105
base	8
	digit
2105	1
263	7
32	0
4	4
0	0
0	0
0	0

Figure 4.5

Now we want to "assemble" the digits to produce the standard representation of the number in base 8, which is 4071. *Excel* has a string concatenation operator &. A formula referencing two cells with the symbol & between the two references will just put the text content of the two cells together. The rightmost bottom cell of our worksheet contains the first digit of the new representation. To create a "nice" representation, however, we want to suppress leading 0s. Therefore, we enter the formula shown in Figure 4.6.

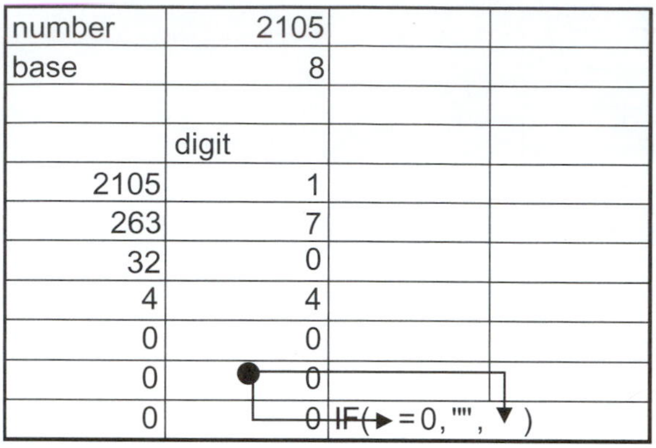

number	2105	
base	8	
	digit	
2105	1	
263	7	
32	0	
4	4	
0	0	
0	0	
0	0	IF(▸ = 0, "", ▾)

Figure 4.6

This will produce an empty string if the first digit is 0 and the correct digit if it is nonzero.

To continue, we now need to "paste" the second digit after the first digit, but again under some circumstances the second digit might be a leading 0 and therefore should be suppressed. It is possible for us to formulate the condition as follows: If all the digits to the left of the current one are leading zeros and the current digit is zero, then it also is a leading digit and therefore should be suppressed. This is computed by the formula shown in figure 4.7.

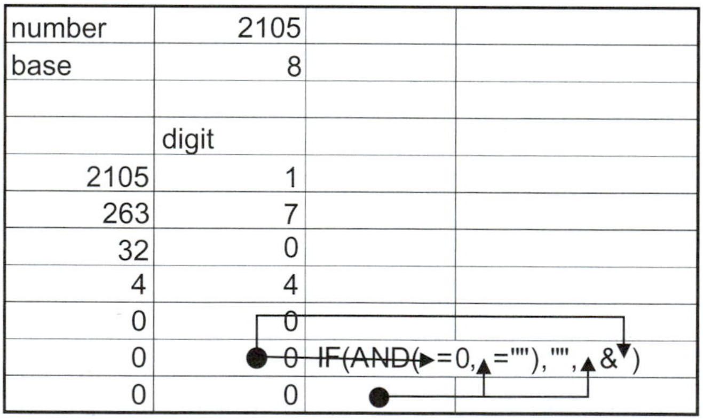

number	2105	
base	8	
	digit	
2105	1	
263	7	
32	0	
4	4	
0	0	
0	0	IF(AND(▸ = 0, ▴ = ""), "", ▴ & ▾)
0	0	

Figure 4.7

The last formula works not only for the second digit but for all further digits. Therefore we can copy it up (Figure 4.8).

number		2105	
base		8	
	digit		
2105		1	4071
263		7	407
32		0	40
4		4	4
0		0	
0		0	
0		0	

Figure 4.8

Now we have the representation of 2105 in base 8. Changing the base or the number in the model will convert the number to a given base. Well, not always.

We still cannot produce a nice representation when the base is larger than 10, because our base 10 number system does not possess single-digit symbols for digits larger than 10. To be able to do that, we will use the LOOKUP function. To begin, we change the number to 2,975 and the number base to 16. Then, we set up a table for translating decimal numbers into digits, including letters for digits with values larger than 9. To enter 0 as a text and not as a number into a cell, one has to press the apostrophe key before pressing the 0 key. So one enters '0, and the cell displays 0 (see Figure 4.9).

number		2975				0	0
base		16				1	1
						2	2
	digit					3	3
2975		15	11915			4	4
185		9	119			5	5
11		11	11			6	6
0		0				7	7
0		0				8	8
0		0				9	9
0		0				10	A
						11	B
					33	X
					34	Y
					35	Z

Figure 4.9

We see that the first digit after the leading zeros has a value of 11 and therefore should be written as B. Therefore, we delete the formulas in the column currently producing an incorrect number representation (the column with 11915 as its first non-empty cell), and we enter the formula shown in Figure 4.10.

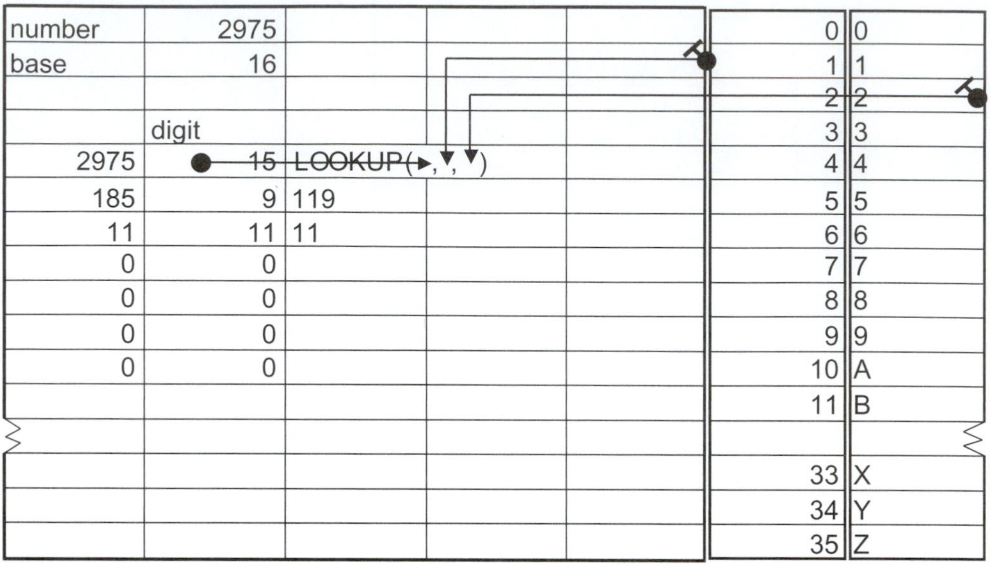

number	2975					0	0
base	16					1	1
						2	2
	digit					3	3
2975	●	15	LOOKUP(→,↓,↓)			4	4
185		9	119			5	5
11		11	11			6	6
0		0				7	7
0		0				8	8
0		0				9	9
0		0				10	A
						11	B
						33	X
						34	Y
						35	Z

Figure 4.10

The second and the third references in the newly created formula are *range* references. They refer to the whole range containing the numbers 0 to 35 and the symbols 0 to Z, respectively. Copying down this formula (Figure 4.11) produces the sequence of digits for the representation of 2975 in base 16, also called hexadecimal representation.

number	2975				0	0
base	16				1	1
					2	2
	digit				3	3
2975	15	F			4	4
185	9	9			5	5
11	11	B			6	6
0	0	0			7	7
0	0	0			8	8
0	0	0			9	9
0	0	0			10	A
					11	B
					33	X
					34	Y
					35	Z

Figure 4.11

Using the same technique as before, we can suppress a leading zero if necessary (Figure 4.12).

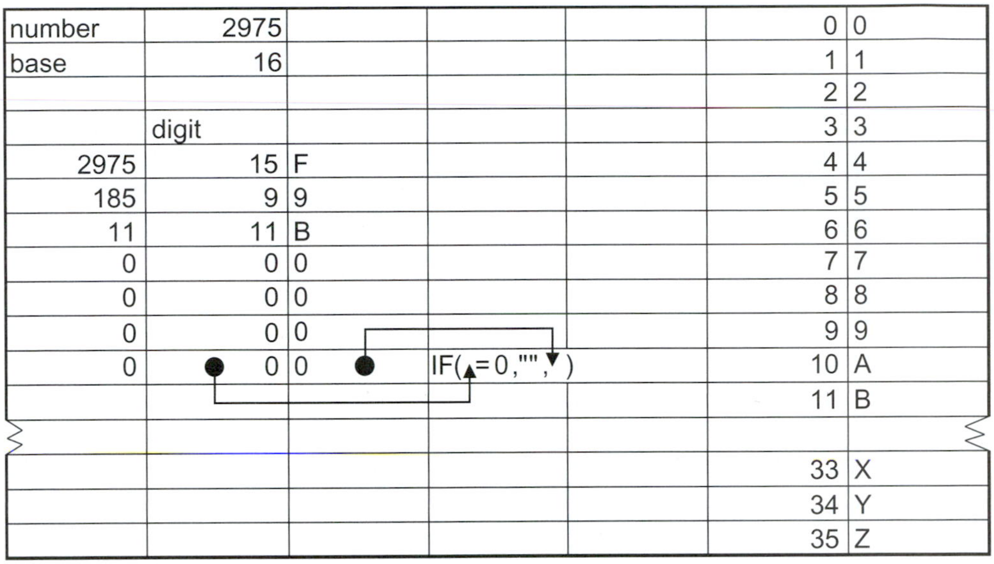

number	2975				0	0
base	16				1	1
					2	2
	digit				3	3
2975	15	F			4	4
185	9	9			5	5
11	11	B			6	6
0	0	0			7	7
0	0	0			8	8
0	0	0			9	9
0	● 0	0	●	IF(▲=0,"",▼)	10	A
					11	B
					33	X
					34	Y
					35	Z

Figure 4.12

Using the same technique as before, we can combine the digits to the number representation (Figure 4.13).

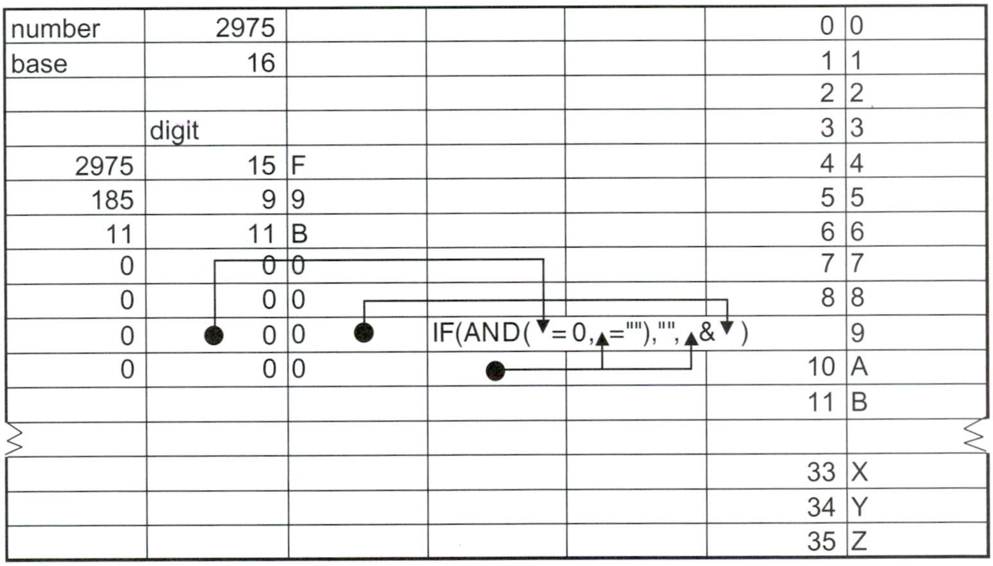

number	2975				0	0
base	16				1	1
					2	2
	digit				3	3
2975	15	F			4	4
185	9	9			5	5
11	11	B			6	6
0	0	0			7	7
0	0	0			8	8
0	● 0	0	●	IF(AND(▼=0,▲=""),"",▲&▼)		9
0	0	0			10	A
					11	B
					33	X
					34	Y
					35	Z

Figure 4.13

Copying this formula up produces the hexadecimal representation of 2975, B9F (see Figure 4.14). Thus,

$$2975_{10} = B9F_{16}$$

number	2975				0	0
base	16				1	1
					2	2
	digit				3	3
2975	15	F	B9F		4	4
185	9	9	B9		5	5
11	11	B	B		6	6
0	0	0			7	7
0	0	0			8	8
0	0	0			9	9
0	0	0			10	A
					11	B
					33	X
					34	Y
					35	Z

Figure 4.14

We can now enter any base up to 35 and our worksheet will convert decimal numbers into the given base. If we want to be able to compute the representation for very large numbers, we cannot simply copy down our formulas because the formula in the last row is different from the formulas above it. It should not, however, be very difficult to modify the worksheet accordingly.

So far, we started with a number given in base 10 and converted it to another base, b. Now we will perform the opposite conversion, converting a number from another base, b, to base 10.

In our model, we will convert $A73B_{12}$ to base 10. As in the worksheet for converting from base 10 to another base, we will need a table for converting "large digits" to numerical values. In the last project, we converted decimal values to "string digits." Now we convert "string digits" to decimal values. Therefore we start with the table shown in Figure 4.15.

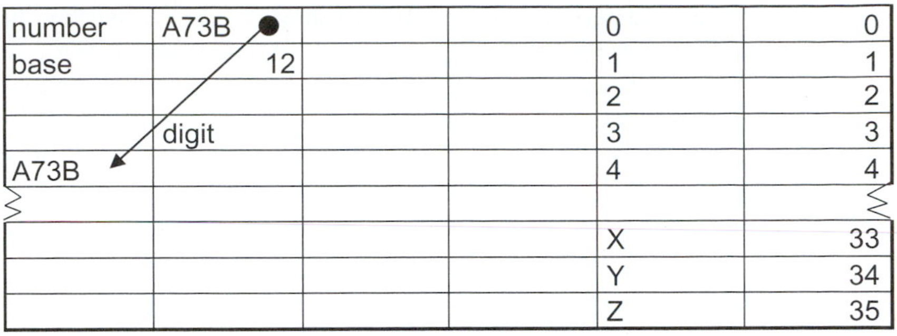

number	A73B			0	0
base	12			1	1
				2	2
	digit			3	3
A73B				4	4
				X	33
				Y	34
				Z	35

Figure 4.15

As a first step, we take the last digit from the number. It is the rightmost character of the string representing the number; therefore, the *Excel* function RIGHT gives us this number (Figure 4.16).

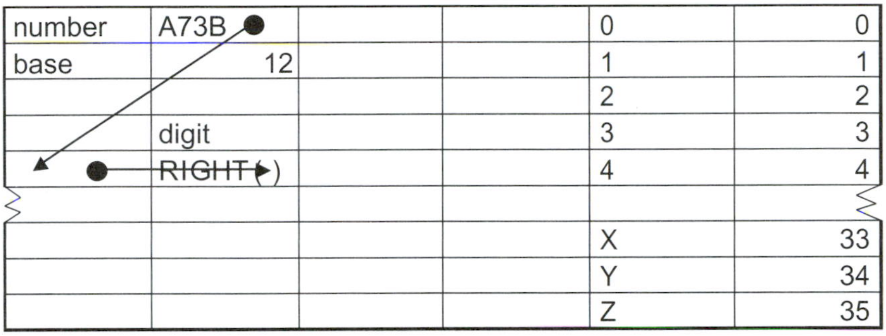

number	A73B			0	0
base	12			1	1
				2	2
	digit			3	3
	RIGHT( )			4	4
				X	33
				Y	34
				Z	35

Figure 4.16

Now we have to calculate the decimal value of this digit, and this is done using the LOOKUP function (see Figure 4.17).

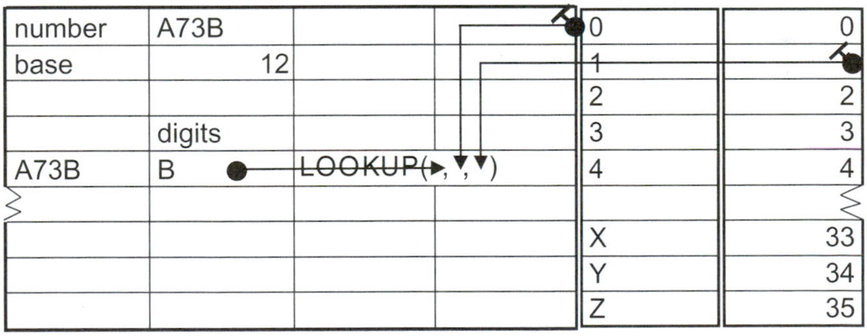

number	A73B			0	0
base	12			1	1
				2	2
	digits			3	3
A73B	B	LOOKUP(, ,)		4	4
				X	33
				Y	34
				Z	35

Figure 4.17

As a next step, we "cut off the last digit" of the current number we are converting. To accomplish that, we use the LEFT function, which gives the leftmost character(s) of a character string, and the LEN function, which computes the length of a string. Starting from our original number string, we compute the string consisting of its leftmost characters, but with length one shorter than the original string (Figure 4.18).

number	A73B			0	0
base	12			1	1
				2	2
	digits			3	3
A73B ●—— B		11		4	4
LEFT(▾,LEN(▾)-1)					
				X	33
				Y	34
				Z	35

Figure 4.18

The next to the last digit of the original number is the last digit of the new number. Therefore, we can just copy down the formulas for the last digit and its value (Figure 4.19).

number	A73B			0	0
base	12			1	1
				2	2
	digits			3	3
A73B	B		11	4	4
A73	3		3	5	5
				6	6
				X	33
				Y	34
				Z	35

Figure 4.19

Copying down the three formulas should compute all digits. If we do that, however, we will get errors. The reason for this is that we cannot take the leftmost digits from a string with no digits. So we have to modify our formula for the value of a digit (Figure 4.20).

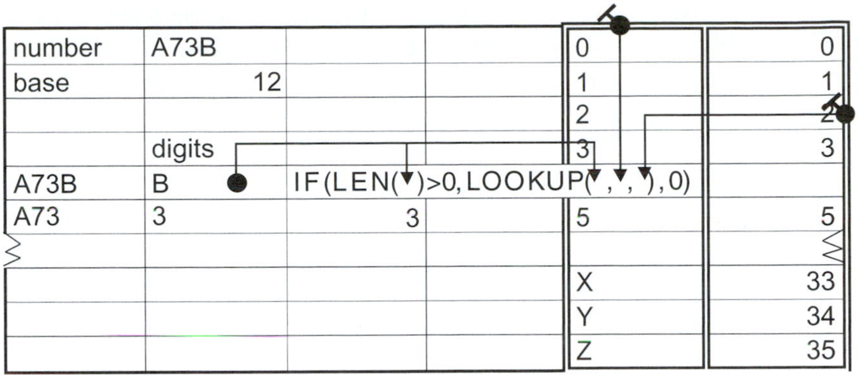

number	A73B			0	0
base	12			1	1
				2	2
	digits			3	3
A73B	B	●	IF(LEN(▼)>0,LOOKUP(▼,▼,▼),0)		
A73	3		3	5	5
				X	33
				Y	34
				Z	35

Figure 4.20

We also have to modify the formula for "cutting off" the last digit (Figure 4.21). Now we can copy the adjusted formulas down their columns (Figure 4.22).

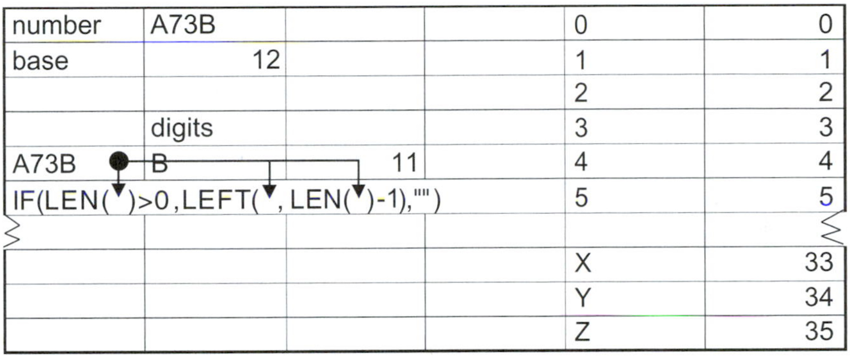

number	A73B			0	0
base	12			1	1
				2	2
	digits			3	3
A73B ●	B		11	4	4
IF(LEN(▼)>0,LEFT(▼,LEN(▼)-1),"")				5	5
				X	33
				Y	34
				Z	35

Figure 4.21

number	A73B			0	0
base	12			1	1
				2	2
	digits			3	3
A73B	B		11	4	4
A73	3		3	5	5
A7	7		7	6	6
A	A		10	7	7
			0	8	8
			0	9	9
			0	A	10
				X	33
				Y	34
				Z	35

Figure 4.22

We have computed the decimal values for all the digits, but we still have to multiply them by the appropriate powers of 12. There is a straightforward way of doing this. If we take a decimal number and append one digit, then the value of the new number is the value of the old number multiplied by 12 plus the value of the new digit. So we start computing the value for the last number in our sequence created by cutting off digits from the right one by one. Using this idea, we start with the formulas in Figure 4.23.

number	A73B			0	0
base	12			1	1
				2	2
	digits			3	3
A73B	B	11		4	4
A73	3	3		5	5
A7	7	7		6	6
A	A	10		7	7
		0		8	8
		0 + *		9	9
		0 → 0 0	A	10
				X	33
				Y	34
				Z	35

Figure 4.23

Finally, coping these formulas up gives a calculate worksheet that the decimal values we were looking for (Figure 4.24).

number	A73B			0	0
base	12			1	1
				2	2
	digits			3	3
A73B	B	11	18335	4	4
A73	3	3	1527	5	5
A7	7	7	127	6	6
A	A	10	10	7	7
		0	0	8	8
		0	0	9	9
		0	0	A	10
				X	33
				Y	34
				Z	35

Figure 4.24

Construction Summary: Simple Expansion

1. Enter parameter values.
2. Reproduce the given number at the top of the left column.
3. Compute the first digit in the new base using =MOD, and copy down.
4. Find the next number: Subtract the digit from previous and divide by the base. Copy down.
5. Starting below the last nonzero digit, create the empty string "".
6. Use =IF to generate the empty string until the first nonzero digit. Thereafter, concatenate the digit onto the right of the string below. Copy upward.

number	**1**	2105	
base		8	
		digit	
2 2105	**3**	1	4071
4 263		7	407
32		0	40
4		4	4
0		0	
0		0	**6**
0		0	**5**

Construction Summary: Integer Expansion with Target Base > 10

1. Enter parameter values.
2. Create a conversion table.
3. Reproduce the given number at the top of the left column.
4. Compute the first digit using the =MOD function, and copy down.
5. Find the next number: Subtract the digit, divide by the base, and copy down.
6. Convert digit to other base, copy.
7. Starting below the last nonzero digit, create the empty string "".
8. Use =IF to generate the empty string until the first nonzero digit. Thereafter, concatenate the digit onto the right of the string below. Copy upward.

number	2975	**1**			**2**	0	0	**2**
base	16					1	1	
						2	2	
	digit					3	3	
3 2975	**4** 15	F	**6**	B9F		4	4	
5 185	9	9		B9		5	5	
11	11	B		B		6	6	
0	0	0				7	7	
0	0	0				8	8	
0	0	0			**8**	9	9	
0	0	0			**7**	10	A	

Construction Summary: Integer Expansion with Source Base Different from 10

1. Enter parameter values.
2. Create a conversion table.
3. Reproduce the given number at the top of the left column.
4. Generate the rightmost digit.
5. Convert the digit to base 10.
6. Use =LEFT and =LEN to generate the string of the remaining left digits. Copy down in first three columns.
7. Generate 0 as below the last digit.
8. Multiply the cell below by base, add the value of new digit, and copy upward.

number	A73B						0	[2]	0
base	12	[1]					1		1
							2		2
	digits						3		3
A73B	3	B	4	5	11	18335	4		4
A73	6	3			3	1527	5		5
A7		7			7	127	6		6
A		A			10	10	7		7
					0	0	8		8
					0	0	9		9
					0	0	A		10
					0	[8] 0	B		11
					0	[7] 0	C		12

Exercises

1. With some experimentation, try to find a "rule" how one can convert numbers between base 2 and base 4, and between base 2 and base 16.

2. Generalize the "rule" from the previous exercise to cover conversion between a base b and base b^2 and, more generally, between base b and base b^k.

3. [*] Why does the problem get harder when the number base is larger than 36?

4. [*] Can you give an easy rule for divisibility by 2 when the number is written in base 2? What about a rule for divisibility rule by 4 in base 2?

4.2 Decimal Expansions

In this section, we continue the work of the previous section and determine the digits that appear in the representations of fractions in the decimal and other bases. This will enable us to generate and investigate a wide range of patterns in the resulting sequences of digits.

In this example, we will see how determining the decimal representation for a fraction can lead to some interesting mathematical investigations. We recall a standard process from arithmetic by finding the decimal expansion of 3/7. Since both the traditional notation and the techniques used are culturally dependent, we will outline the U.S. and European versions in Figure 4.25 and Figure 4.26, respectively. Note that, in these two systems, the roles of the symbols , and . are reversed.

```
   0.428              7
7)3.000          3,000 : 7 = 0,428
  2 8                2 8
   20                 20
   14                 14
   60                 60
   56                 56
    4                  4
```

Figure 4.25   Figure 4.26

Actually, we can present essentially the same steps of the algorithm in a slightly different fashion as a series of one-step divisions as shown in Figure 4.27 or as shown in Figure 4.28.

```
   4      2      8        7           7           7
7)30   7)20   7)60    30 : 7 = 4   20 : 7 = 2   60 : 7 = 8
  28     14     56       28           14           56
   2      6      4        2            6            4
```

Figure 4.27   Figure 4.28

We will implement the equivalent steps for this process in a spreadsheet.

In essence, in the first stage of the algorithm, we first multiply the numerator (3) by 10 and divide the result ($10 \times 3 = 30$) by the denominator (7). The integer part of the quotient is the first digit (4) in the decimal expansion. Then we simply repeat the process using the current remainder, $30 - (4 \times 7) = 2$, as the new numerator.

In implementing this process in a spreadsheet, it is advantageous to generalize our work by using a separate cell to contain the value 10 (the base of the decimal system) as a parameter of the model. We do this so that later we can do the same operation with different bases. In this example, we will only work with fractions in which the numerator and denominator are positive and in which the numerator is less than the denominator.

We begin by entering the labels and parameters that are shown in Figure 4.29. Then in the first column we use our standard procedure to generate values of a counter for the digits. The second column generates the successive digits in the decimal expansion, with the corresponding remainders computed in the third column. To begin, we reproduce the numerator as the initial remainder at the top of the third column.

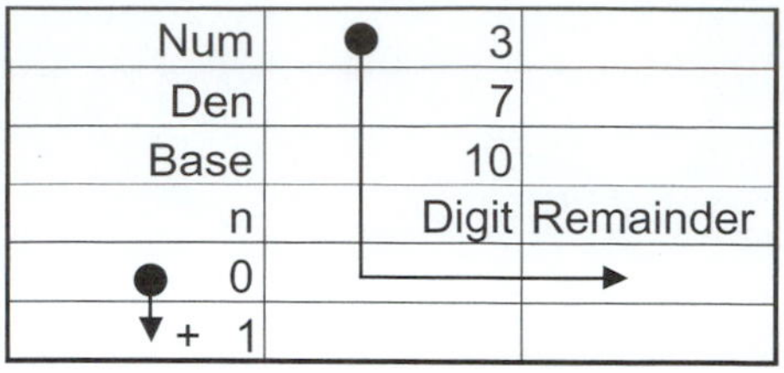

Figure 4.29

Next, in the second column, we compute the first digit of the decimal expansion as the integer quotient produced when we multiply the previous remainder by the base 10 and then divide by the denominator, that is, $\text{INT}(10 \times 3/7) = \text{INT}(30/7) = 4$. In designing the formula, we observe that both the base (10 in this example) and the denominator (7 in this example) must be absolute locations (Figure 4.30).

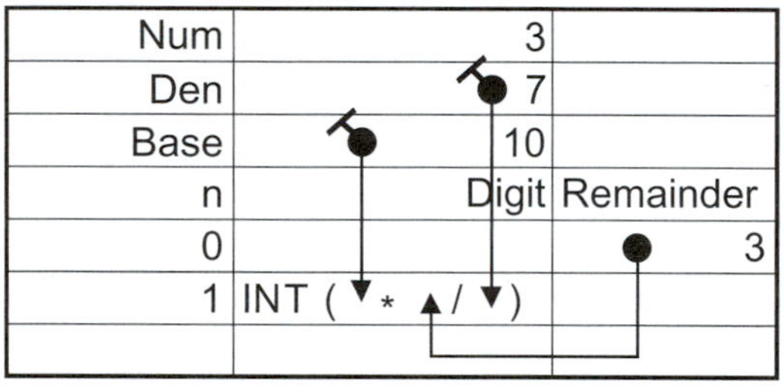

Figure 4.30

Now, we find the next remainder by subtracting the product of the current quotient (that is, the current digit, 4) and the denominator (7) from the product of the previous remainder (3) and the base 10, or $(3 \times 10) - (4 \times 7) = 30 - 28 = 2$.

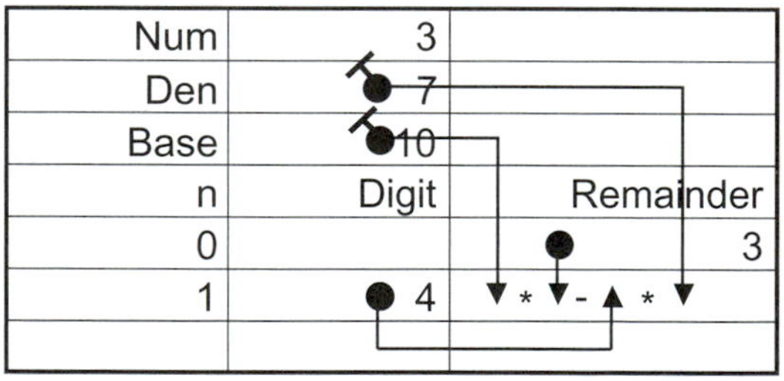

Figure 4.31

Next, we will use some of the spreadsheet's string functions to generate the formal decimal expression. This will look like a number, but it will really be a text, or string, expression.

Since all of our fractions are less than 1, we enter the string "0. as the first entry in the new column (Figure 4.32). In the next row, we convert the next digit into text with no decimals using the TEXT function. In general, the forms of the function that converts a number *num* into text with, say, 2 or 0 decimal places are

$$\text{TEXT}(num, \text{``0.00''}) \text{ or } \text{TEXT}(num, \text{``0''}).$$

We then append the new digit to the right of the previous expression using the text append (or concatenation) operation that is designated by &. As we copy this formula down its column, each new character is appended to the right end of the previous string (see Figure 4.32).

Num	3		
Den	7		
Base	10		
n	Digit	Remain	Decimal
0		3	● "0.
1	●4	2	▼ & TEXT(▲,0)

Figure 4.32

Now all that remains to complete the model is to copy these formulas as shown in Figure 4.33.

Num	3		
Den	7		
Base	10		
n	Digit	Remainder	Decimal
0		3	0.
1	4	2	0.4
2	2	6	0.42
3	8	4	0.428

Figure 4.33

This produces the output shown in Figure 4.34, with the second column showing the initial digits for the decimal expansion 0.428571428571428571 . . . .

Num	3		
Den	7		
Base	10		
n	Digit	Remainder	Decimal
0		3	0.
1	4	2	0.4
2	2	6	0.42
3	8	4	0.428
4	5	5	0.4285
5	7	1	0.42857
6	1	3	0.428571
7	4	2	0.4285714
8	2	6	0.42857142

Figure 4.34

From our output it is easy to see that the digits in the decimal expansion of 3/7 repeat in six-digit cycles beginning from the first term. We can use our model to discover other patterns that arise by trying various numerators and denominators. For example, we might try to discover which fractions have expansions that repeat immediately and find a pattern to the length of their repeated period. We might also find the fractions that have terminating expansions, those that repeat after a certain number of initial digits, and many more.

A nice feature of our model is that with it we can generate similar expansions for bases other than 10 just as easily. Strictly speaking, these are not "decimal" expansions, since decimal means base 10. However, for simplicity, we will refer to these as base b decimals. To generate a base b expansion with our model, with the original fraction still expressed in terms of base 10 integers for the numerator and the denominator, we need only enter the value of the parameter for the base. For instance, if the base is 4, then 3/7 becomes 0.123123 . . . (base 4). This expression means

$$\frac{3}{7} = \frac{1}{4^1} + \frac{2}{4^2} + \frac{3}{4^3} + \frac{1}{4^4} + \frac{2}{4^5} + \frac{3}{4^6} + \cdots$$

The spreadsheet output for this expansion is shown in Figure 4.35.

We can now repeat all of our previous investigations using a variety of different bases. However, note that if the base is larger than 10, then "digits" will include 11, 12, ... Perhaps the most important place in which using a base larger than 10 occurs is in computer science, where base 16 (or hexadecimal) notation frequently is

Num	3		
Den	7		
Base	4		
n	Digit	Remainder	Decimal
0		3	0.
1	1	5	0.1
2	2	6	0.12
3	3	3	0.123
4	1	5	0.1231
5	2	6	0.12312
6	3	3	0.123123
7	1	5	0.1231231
8	2	6	0.12312312

Figure 4.35

used. To generate expressions in this base, we create a list of symbols for the digits. The first ones (0, 1, . . . , 8, 9) are just the usual symbols. Typically one uses the letters A, B, C, D, E, and F for the next six characters, with A = 10, B = 11, . . . (Figure 4.36).

To use these with our model, we modify our string appending operation slightly by using the OFFSET function. If a is the next digit, then the function OFFSET($start,a,0$) begins from the cell location for $start$ (here, the first character in the list of hexadecimal symbols) and looks down a rows and over 0 columns (that is, in the same column). It then returns the character that is found there (refer to Figure 4.37).

Num	3			D	H
Den	14			0	0
Base	16			1	1
n	Digit	Remain	Hexadecimal	2	2
0		3	0.	3	3
1	3	6	& OFFSET(, ,0)	4	4
2	6	12		5	5
3	13	6		6	6
4	11	12		7	7
5	6	10		8	8
6	13	6		9	9
7	11	12		10	A
8	6	10		11	B
9	13	10		12	C
10	11	6		13	D
11	6	12		14	E
12	13	10		15	F

Figure 4.36

Hexadecimal
0.
0.3
0.36
0.36D
0.36DB
0.36DB6
0.36DB6D
0.36DB6DB
0.36DB6DB6
0.36DB6DB6D
0.36DB6DB6DB
0.36DB6DB6DB6
0.36DB6DB6DB6D

Figure 4.37

At this time, we observe that in traditional mathematical notation we can denote the recurrence relation for our process to obtain the (base 10) decimal representation for a fraction p/q (with $p < q$) as

$$r_0 = p \text{ and } a_{n+1} = \text{INT}\left(\frac{10r_n}{q}\right), r_{n+1} = 10r_n - a_{n+1}q \text{ for } n \geq 0$$

and for expansions in base b as

$$r_0 = p \text{ and } a_{n+1} = \text{INT}\left(\frac{br_n}{q}\right), r_{n+1} = br_n - a_{n+1}q \text{ for } n \geq 0$$

where $\text{INT}(x)$ represents the largest integer not exceeding the real number x.

Spreadsheets also allow us to illustrate decimal expansions in a variety of ways through the use of graphs. A column graph constructed from the first two columns of our original base 10 model is shown in Figure 4.38. The x-axis shows the counter values, n, while the y-axis shows the digits of the expansion.

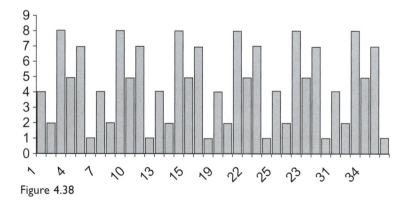

Figure 4.38

The next two graphs, showing the digits for the expansion of 3/7 in the bases $b = 10$ (Figure 4.39) and $b = 4$ (Figure 4.40), display the digits using b equally spaced points on a circle. Experimenting with different fractions and bases produces various designs, and you can investigate questions such as: For a given denominator, which numerators and bases produce either identical or similar graphs?

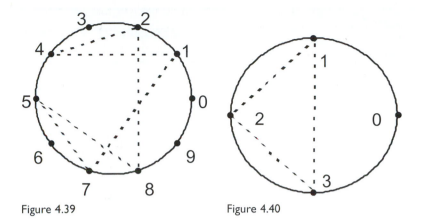

Figure 4.39   Figure 4.40

To construct this graph, we first use the techniques of the graphing sections to create an xy-graph of the circle, shown in the left (x, y) columns in Figure 4.41. In the central columns, we create the terms in the decimal expansion. Then we use two more columns of the model to find values $x = \cos(2\pi a_n/b)$ and $y = \sin(2\pi a_n/b)$ for each digit a_n and drag these columns into the graph of the circle to construct the dotted lines (see the sections on graphing in Chapter 5 and in the Appendix for details).

For example, if we are using base $b = 10$ and a digit is $a_n = 7$, then we need to plot a point that is 7/10 of the way around the circle, so we multiply the number of radians in a circle (that is, 2π) by 7/10 and then use the COS (cosine) and SIN (sine) functions. Alternatively, we can work in degrees and multiply 360 by the fraction a_n/b, and use the spreadsheet functions COS(RADIANS($360a_n/b$)) and SIN(RADIANS($360a_n/b$)) for the x- and y-coordinates.

With some spreadsheets, the integers shown in our graph can be included as interactive interior labels. In others they can be added as static symbols using a graph annotation mode.

We leave it as an exercise to interested readers to create this graph, or to design others to exhibit the patterns in decimal expansions creatively.

		circle						pattern	
deg	t	x_1	y_1	n	digit	rem		x_2	y_2
				0		3			
0	0.000	1.000	0.000	1	4	2		-0.809	0.588
1	0.017	1.000	0.017	2	2	6		0.309	0.951
2	0.035	0.999	0.035	3	8	4		0.309	-0.951
3	0.052	0.999	0.052	4	5	5		-1.000	0.000
4	0.070	0.998	0.070	5	7	1		0.309	-0.951
5	0.087	0.996	0.087	6	1	3		0.809	0.588
6	0.105	0.995	0.105	7	4	2		0.809	0.588
7	0.122	0.993	0.122	8	2	6		0.309	0.951
8	0.140	0.990	0.139	9	8	4		0.309	-0.951

Figure 4.41

1. Enter parameter values.
2. Generate a counter column.
3. Reproduce the numerator as the initial remainder, "0. as the initial decimal text.
4. Compute the first digit as =INT(base × rem/den)
5. Compute the next remainder as (base × remainder) − (den × digit)
6. Determine the next decimal form by converting the digit to text using =TEXT and concatenating onto the right of previous value.
7. Copy as indicated.

Num	3		
Den	7 } 1		
Base	10		
n	Digit	Remainder	Decimal
0	3	3	0.
1	4	2	0.4
2	2	6	0.42
3	8	4	0.428
4	5	5	0.4285
5	7	1	0.42857
6	1	3	0.428571
7	4	2	0.4285714
8	2	6	0.42857142

Exercises

1. Find the base 10, base 12, base 2, base 3, and base 6 expansions of the following rational fractions:

   a. $\dfrac{2}{5}$ b. $\dfrac{1}{3}$ c. $\dfrac{5}{6}$ d. $\dfrac{5}{17}$ e. $\dfrac{7}{12}$ f. $\dfrac{3}{23}$ g. $\dfrac{11}{72}$ h. $\dfrac{217}{240}$ i. $\dfrac{217}{241}$

2. Through spreadsheet experimentation and algebra analysis, discover patterns that describe which fractions a/b have base 10 decimal expansions that consist solely of a repeating component and that have an initial nonrepeating segment. Also, discover patterns for the length of the repeating component and the initial nonrepeating component. Then repeat the same process with other bases.

3. Design a spreadsheet model to write any rational fraction $\dfrac{p}{q}$ in the interval

   $0 < \dfrac{p}{q} < 1$ as the sum of distinct unit fractions, that is, fractions of the form $\dfrac{1}{k}$,

   where k is a positive integer. For example, $\dfrac{55}{111} = \dfrac{1}{3} + \dfrac{1}{7} + \dfrac{1}{52} + \dfrac{1}{13468}$. Some-

   times these are also called *Egyptian fractions*, since they were used in ancient Egypt. The key step in the process of writing a fraction as a sum of unit fractions

   is to find the smallest integer k so that $\dfrac{1}{k}$ is less than a fraction $\dfrac{a}{b}$ and then determine the next fraction that arises by subtracting. Use your algorithm to write the following fractions as the sum of distinct unit fractions. See Chapter 28 of Stewart (1952) for a further discussion of this topic, including a proof that every fraction can be represented as a sum of distinct unit fractions.

a. $\dfrac{3}{4}$ b. $\dfrac{6}{7}$ c. $\dfrac{5}{9}$ d. $\dfrac{19}{20}$ e. $\dfrac{37}{47}$

4. One ancient algorithm for multiplying positive integers is often called *Russian peasant multiplication*. We illustrate this algorithm through the display in Figure 4.42, where we show the computation of the product of 45×57. To carry out the algorithm, we repeatedly divide the first integer by 2, dropping any remainder, and double the second integer. We then mark and add those integers in the second column that correspond to odd integers in the first column. The sum of the marked integers is the desired product. Implement this algorithm in a spreadsheet. Note that there is a connection to the binary expansion of the first multiplicand.

45	57	x	57
22	114		
11	228	x	228
5	456	x	456
2	912		
1	1824	x	1824
	product		2565

Figure 4.42

Create a spreadsheet model to implement this algorithm. Then use it to perform the following multiplications.

a. 33×52 b. 45×47 c. 64×123 d. 31×52

5. [*] Create a spreadsheet graph to implement the diagram in Figure 4.43 for the base b positional expansion of a rational number (see Stewart, p. 345). The diagram in Figure 4.43 gives the base 3 expansion of $1/5 = 0.\overline{0121}_3$. The bar over the digits indicates periodically repeating digits.

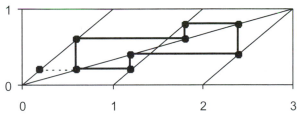

Figure 4.43

Use your model to create graphs for the following numbers.

a. $\dfrac{1}{5}$, base 8   b. $\dfrac{2}{3}$, base 7   c. $\dfrac{1}{4}$, base 9   d. $\dfrac{2}{3}$, base 5

4.3 Greatest Common Divisor

A divisor of a given integer is an integer that leaves no remainder when it is divided into the given integer. Thus, the divisors of 6 are 1, 2, 3, and 6, while the divisors of 5 are 1 and 5. The largest integer that is a divisor of two integers is called their greatest common divisor. This section presents an algorithm attributed to Euclid, a Greek mathematician of the third century B.C., which enables us to find the greatest common divisor of two integers. The greatest common divisor is used widely in doing mathematics that involves integers.

We call 6 a divisor of 30 because we can subdivide 30 into a whole number of 6s with no remainder left over. That is, we can write $30 = 5 \times 6$. This idea is illustrated in Figure 4.44. We also say that 6 divides 30.

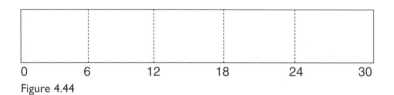

Figure 4.44

On the other hand, 7 is not a divisor of 30, since it is not possible to write 30 as a whole number of 7s. Equivalently, when we divide 30 by 7, we get a quotient of 4 and a nonzero remainder of 2. This is illustrated in Figure 4.45.

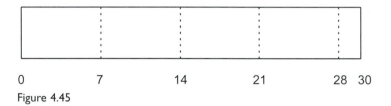

Figure 4.45

It is relatively easy to list all of the divisors of a small integer like 30. In this case, the divisors are 1, 2, 3, 5, 6, 10, 15, and 30. However, it is much more challenging to find all of the divisors of large integers such as 9,271,800 or 5,333,157.

> Frequently, in dealing with integers, we will want to find the largest integer that is a divisor of both of two given integers that we will call *large* and *small*, with *large* ≥ *small*. Such a divisor is called the *greatest common divisor* of the two numbers.

Sometimes we use the notation GCD*(large,small)* to denote the greatest common divisor of these two integers. For example, we can see that GCD(6,10) = 2, GCD(5,20) = 5, and GCD(8,3) = 1. To be consistent with GCD*(large, small)*, we should change these to GCD(10,6) and GCD(20,5) or else pick different names than *large and small.*

To try to find the greatest common divisor of *large and small,* we could try to list all of the divisors of each and then find the largest one that they share in common. This approach may be practical with small integers, but the work involved in such an approach would become excessive with much larger integers such as the two listed earlier.

Instead of embarking on such a tedious approach, we will employ a fundamental mathematical problem-solving technique in which we replace a possibly difficult or large problem with an equivalent problem that is in some sense easier or smaller. Then we successively repeat this process until we have reduced the problem that we need to solve to a simple one.

To do this in determining a greatest common divisor, we need to make another mathematical observation about common divisors. We will first present our approach in a visual format, illustrated in Figure 4.46 by using the integers 28 and 12. Later, we will justify the algorithm using formal mathematics.

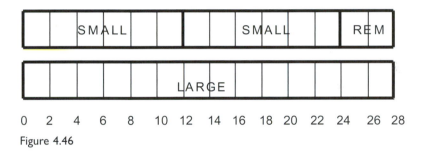

Figure 4.46

First, since *large* is greater than *small,* we can look at *large* as being the sum of one or more copies of *small,* with perhaps a positive integer, *remainder,* left over. For example, in Figure 4.46, we express 28 as 12 + 12 + 4, or as 28 = (2 × 12) + 4 (that is, as two copies of 12 with a remainder of 4). In general, we will have the relationship

$$large = (quotient \times small) + remainder.$$

From the diagram, we can observe that if an integer (here, 2) is a divisor of both *small* (12) and *remainder* (4), then clearly it must also divide *large* as well. Thus, any common divisor of *small* and *remainder* is also a common divisor of *small* and *large.*

Conversely, if an integer (here, 2) divides both *large* and *small,* then it must also divide *remainder* as well. For if we subtract successive blocks of *small* from *large,* then we simultaneously are subtracting whole number blocks of the given divisor at the same time. Thus, when the blocks of *small* are removed, a whole number of multiples of the divisor are removed, leaving a whole number of the units of the divisor remaining in *remainder.* Consequently, the given integer also divides *remainder.* Hence, a common divisor of *small* and *large* is a common divisor of *small* and *remainder.*

Since both pairs (*large,small*) and (*small,remainder*) have exactly the same common divisors, then the greatest common divisor of each pair will also be the same. Consequently, we will substitute the problem of finding the greatest common divisor of *small* and *remainder* for the problem of finding the greatest common divisor of *large* and *small*. As we will see shortly, the advantage of this swap is that we will now be working with a pair of smaller numbers. We can then repeat this process over and over and eventually obtain a problem whose answer will be obvious.

In our spreadsheet model, to find the greatest common divisor of two integers, we first enter the two integers *large* and *small* (Figure 4.47). We then determine the remainder, *remainder,* using the function MOD(*large,small*) that gives the remainder when *large* is divided by *small*.

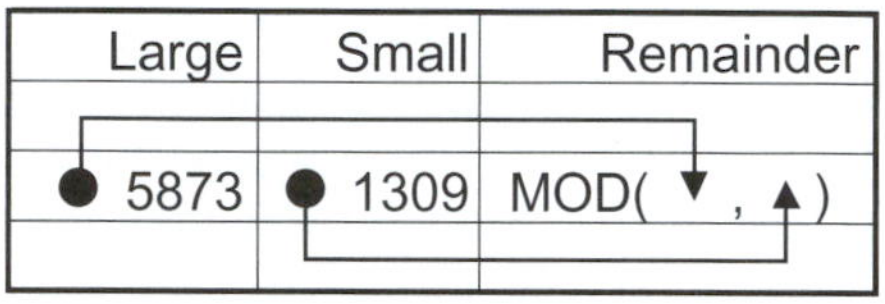

Figure 4.47

The *small* and *remainder* values in the current row now become the *large* and *small* values, respectively, for the next iteration of the algorithm (see Figure 4.48).

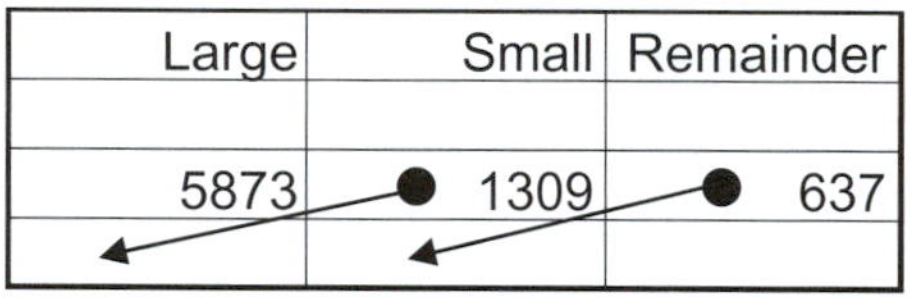

Figure 4.48

But now we simply need to repeat the process with these new values. To do this, we use the standard fill or copy process indicated in Figure 4.49.

Large	Small	Remainder
5873	1309	637
1309	637	35
637	35	7
35	7	0
7	0	#DIV/0!
0	#DIV/0!	#DIV/0!
#DIV/0!	#DIV/0!	#DIV/0!

Figure 4.49

In this figure, we see that the remainder eventually becomes 0. At that step, the smaller integer (here, 7) evenly divides the larger one and is therefore the greatest common divisor that we seek. Thus, GCD(5873,1309) = GCD(35, 7) = 7. We can also see that this value is the last nonzero remainder. We also need to observe that at the next step the smaller number is 0, so a "division by 0" error occurs in subsequent lines.

Although we can now simply look at our screen display and pick out the greatest common divisor, it is also possible to locate it by formulas. In Figure 4.50, the function MATCH(0,*remainder*,0) finds the location, *locate,* in the *remainder* column where 0 (this is the meaning of the first entry in the function) is found. The function INDEX(*remainder,locate* −1) then returns the value just above the location where 0 was found by the MATCH function.

Large	Small	Remainder	Locate	MATCH(0, ,0)		Locate	4
			GCD	INDEX(, -1)		GCD	7
5873	1309	637					
1309	637	35					
637	35	7					
35	7	0					
7	0	#DIV/0!					
0	#DIV/0!	#DIV/0!					
#DIV/0!	#DIV/0!	#DIV/0!					

Figure 4.50

We need to point out that the computations at the top of this model need to be adjusted for the case when the initial small number is a divisor of the large number. We leave finding a way to handle this special case to those who like challenges.

Once we have created this model, there are many ways to embellish it. In particular, we can rearrange the layout and set aside specific cells for the two numbers of choice, *large* and *small,* to be parameters. We can also use the model to find integers, *u* and *v,* for which

$$\text{GCD}(large,small) = (u \times large) + (v \times small).$$

In addition, we can find the least common multiple of the two numbers, or use the greatest common divisor to add fractions.

At the same time, note that for applications most spreadsheets have a built-in GCD function among their library functions.

Further Mathematics: Euclid's Algorithm

For those who are interested, we close this section with a more formal discussion of some of the points that we made in developing our algorithm, which is called Euclid's algorithm.

First, if *a* is an integer, then we say that an integer *d* divides *a*, or that *d* is a divisor of *a* if there is another integer *c* so that *a* = *cd*. We say that *d* is a common divisor of two integers large and small if it is a divisor of each of them. The largest such common divisor is called the *greatest common divisor* and is denoted by GCD(large, small).

Next, suppose that we wish to find the greatest common divisor of two integers, large and small. We can divide large by small to obtain large = (quotient · small) + remainder, where $0 \leq$ remainder $<$ small.

Thus, if d is a common divisor of small and remainder, then there are integers c_1 and c_2 so that small = $c_1 d$ and remainder = $c_2 d$. Since large = (quotient · small) + remainder, it follows that large = (quotient · $c_1 d$) + $c_2 d$ = (quotient · $c_1 + c_2$)d, so that d divides large. Conversely, if d divides both large and small, then there are integers c_3 and c_4 so that large = $c_3 d$ and small = $c_4 d$. Therefore, since remainder = large − small · quotient, we have remainder = $c_3 d$ − quotient · $c_4 d$ = (c_3 − quotient · c_4)d, and therefore d divides remainder. Since both pairs have exactly the same common divisors, they have the same greatest common divisor.

Finally, this algorithm can be written in symbols. Following from our work, here it is: Let a_0 = large, b_0 = small. Then if $a_n = q_n b_n + r_n$, where $0 \leq r_n < b_n$, then $a_{n+1} = b_n$, and $b_{n+1} = r_n$. If n is the first integer such that $r_{n+1} = 0$, then $\text{GCD}(q_0, b_0) = r_n$. This is quite a formal mouthful for what is a rather natural process to implement on a spreadsheet.

Construction Summary: Greatest Common Divisor

1. Enter parameter values, large and small.
2. Determine the initial remainder as MOD(large,small).
3. Reproduce the previous small as the new large.
4. Reproduce the previous rem as the new small. Copy as indicated.
5. Enter MATCH, INDEX functions.

	Large		Small		Rem	Locate	
						GCD	5
1	5873	1	1309	2	637		
3	1309	4	637		35		
	637		35		7		
	35		7		0		
	7		0		#DIV/0!		
	0		#DIV/0!		#DIV/0!		
	#DIV/0!		#DIV/0!		#DIV/0!		

Locate 5 { 4 , 7 }

Exercises

1. Use the greatest common divisor model to compute the greatest common divisor of the following pairs of integers.
   a. 56, 77   **b.** 79, 451   **c.** 393040, 1755675   **d.** 44814, 99910

2. Supplement your greatest common divisor model to compute additionally the *least common multiple* of a pair of positive integers. The least common multiple (LCM) is the smallest positive integer that is a multiple of the given integers. Try to find a relationship between the greatest common divisor and the least common

multiple of a pair of integers, and use this to compute the least common multiple for each pair of integers in Exercise 1. Hint: Observe that GCD(15,20) = 5, LCM(15,20) = 60, while GCD(12,28) = 4, LCM(12,28) = 84.

3. [*] Pairs of integers for which the greatest common divisor is 1 are called *relatively prime*. Create a spreadsheet model to find all of the integers less than a given integer n that are relatively prime to n.

4. [*] The number of integers that are less than integer n and relatively prime to n is denoted by $\phi(n)$, and the resulting function is called Euler's ϕ-function. Find $\phi(n)$ for each of the following values of n. Use your model from Exercise 3 rather than a function given in a number theory book.
   a. 7   **b.** 27   **c.** 29   **d.** 100   **e.** 6   **f.** 15   **g.** 800

5. *Excel* and many other spreadsheets have a built-in GCD function. You may first need to use an add-on command to ensure that it is available. Design a model that uses the GCD function together with a data table to evaluate $\phi(n)$ for $1 \le n \le 1000$.

4.4   Sieve of Eratosthenes

Some positive integers, or whole numbers, can be written as the product of other positive integers. For example, $6 = 2 \times 3$ and $60 = 6 \times 10$. However, for some integers, the only way to write them as a product is to use the integer itself and 1. For example, $2 = 1 \times 2$ and $5 = 1 \times 5$. We call the latter numbers *primes*. Primes are important because they can be considered as the building blocks of the integers, since every positive integer can be written as a unique product of primes. Thus, $60 = 2 \times 2 \times 3 \times 5$. Primes are also used in a variety of mathematical algorithms, including those for encrypting or coding messages. One ancient scheme for determining which positive integers are primes was developed by the Alexandrian mathematician Eratosthenes in the third century B.C. It is called a "sieve" because it systematically filters out the integers that are not primes, leaving only primes remaining.

> A *prime* number is a positive integer, p, that has no integer factors other than itself and 1.

Thus, the prime numbers include 2, 3, 5, ... On the other hand, 15 is not a prime number, since $15 = 3 \times 5$. The integer 1 is not considered to be a prime. One of the oldest traditional ways that has been used for generating prime numbers is a type of algorithm called a sieve. A sieve systematically examines all of the positive integers, eliminating those that are not prime.

The best-known algorithm of this variety is called the *sieve of Eratosthenes*. Its basic idea is quite simple. First, we list all of the positive integers from 2 on:

$$2, 3, 4, 5, 6, 7, 8, 9, \ldots$$

The smallest of these, 2, must be a prime. Next, we cross out all multiples of 2, other than 2, since these will have 2 as a factor and therefore will not be prime:

$$2, 3, x, 5, x, 7, x, 9, \ldots$$

The second smallest remaining integer, 3, is the next prime. We now cross out all of the remaining multiples of 3, other than 3, to obtain

$$2, 3, x, 5, x, 7, x, x, \ldots.$$

The next prime is now 5. By repeating this process often enough, we would eventually obtain any prime. Moreover, if p is the latest prime at some stage, not only is p prime, but after crossing out all of the multiples of p in the list, so are any of the remaining integers that do not exceed p^2.

However, one of the difficulties of implementing this sieve in a spreadsheet is that we would have to list all of the integers (at least through a certain size) and then make many repeated passes through the list. The resulting model would be too large for the computer's memory. In this section, we present two models to use the technique of the sieve of Eratosthenes to find the smaller primes, say up through 100 or 1000. The first of our examples can grow quite large, while the second one is made more compact by using some cells over and over.

1. Basic Model

Let us start the construction of our basic model to find the prime numbers up through 1000. We first generate an initial list consisting of the integers 2, 3, . . . , 1000 (or more if you prefer) down a column (Figure 4.51). We will apply the sieve to these integers. We determine the first prime, 2, in the top row by using the spreadsheet's minimum function, MIN, to find the smallest integer in the list (Figure 4.52).

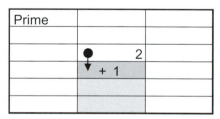

Figure 4.51   Figure 4.52

In the next column, we apply the sieve technique to filter out the multiples of the prime, p, that was just found. We test whether each integer, k, in the previous column is divisible by p. Since the function MOD(k,p) gives the remainder that results when k is divided by p, then k is divisible by p if MOD(k,p) = 0. If this is the case, then in the next column we indicate "crossing out" k by generating 9999 (or any number larger than those that we will be considering) in its place. Otherwise, we simply reproduce the number k itself. Thus, we enter the spreadsheet equivalent of the expression

IF MOD(k,p) = 0 THEN use 9999 ELSE copy k from the cell to the left.

Now all that we have to do is copy the minimum function across the first row and then copy the indicated IF formula into the remaining cells of the table (Figure 4.53). Before doing this, we must be careful to ensure that the reference to p always comes from the first row but still comes from the previous column. Notice that if a number, k, has already been eliminated, then 9999 results in either case. Thus, once an integer is eliminated, it remains eliminated. Also, observe that in order to find each new prime as the minimum in the current list, we must modify the sieve algorithm slightly by also crossing out each prime as it is generated. However, these primes remain for us to be seen in the first row.

prime	2				
	2	IF(MOD(▼ , ▼)=0 , 9999 , ▼)			
	3				
	4				
	5				

Figure 4.53

The display in Figure 4.54 shows some of the output that is produced. Although it appears that at this stage we have only found the first twelve primes, actually we can find many more by reading farther down the columns. For example, when we find that 7 is a prime, then any number that remains in the next iteration and is less than $7^2 = 49$ is a prime. In fact, if we start with a list of the first 1000 integers, then all of the numbers remaining in the last column will be prime since $37^2 > 1000$.

Prime	2	3	5	7	11	13	17	19	23	29	31	37
2	9999	9999	9999	9999	9999	9999	9999	9999	9999	9999	9999	9999
3	3	9999	9999	9999	9999	9999	9999	9999	9999	9999	9999	9999
4	9999	9999	9999	9999	9999	9999	9999	9999	9999	9999	9999	9999
5	5	5	9999	9999	9999	9999	9999	9999	9999	9999	9999	9999
6	9999	9999	9999	9999	9999	9999	9999	9999	9999	9999	9999	9999
7	7	7	7	9999	9999	9999	9999	9999	9999	9999	9999	9999
8	9999	9999	9999	9999	9999	9999	9999	9999	9999	9999	9999	9999
9	9	9999	9999	9999	9999	9999	9999	9999	9999	9999	9999	9999
10	9999	9999	9999	9999	9999	9999	9999	9999	9999	9999	9999	9999
11	11	11	11	11	9999	9999	9999	9999	9999	9999	9999	9999
12	9999	9999	9999	9999	9999	9999	9999	9999	9999	9999	9999	9999
13	13	13	13	13	13	9999	9999	9999	9999	9999	9999	9999
14	9999	9999	9999	9999	9999	9999	9999	9999	9999	9999	9999	9999
15	15	9999	9999	9999	9999	9999	9999	9999	9999	9999	9999	9999
16	9999	9999	9999	9999	9999	9999	9999	9999	9999	9999	9999	9999
17	17	17	17	17	17	17	9999	9999	9999	9999	9999	9999
18	9999	9999	9999	9999	9999	9999	9999	9999	9999	9999	9999	9999
19	19	19	19	19	19	19	19	9999	9999	9999	9999	9999

Figure 4.54

In some ways, this is a very good model. We have created it in essentially the same way in which would we proceed by hand. However, it has one evident drawback. If we were to use this approach to try to find the primes in the set of the first 10,000 integers and displayed several iterations, then the resulting model would be extremely large. We could improve the situation somewhat by eliminating all of the even integers from the start and, thereby, either cut the size of the model in half or double the number of integers that can be considered. However, we still end up using a very large model to generate a relatively small number of primes. Part of this is simply inherent in the inefficiency of the sieve of Eratosthenes. In any case, we next design a more compact way of carrying out our work.

2. Compact Version

The strategy that we will now pursue is to create a more compact model to repeatedly update a block of cells by copying the values from the next set of sieve computations back into them. We organize our model to determine all of the prime numbers among the first 100 integers displayed in a 10×10 block. However, our layout can be expanded to find the primes in the first 10,000 integers just as well by using a 100×100 block.

Our model uses two blocks to store the results of two successive iterations of the sieve. The top half contains the current list of candidates for prime numbers. We use it to locate the next prime. Multiples of that prime are then crossed out to form the bottom half. This time, the primes thus located will remain in the list. We next copy the values of the bottom half into the top half to repeat the process with the next prime. Again, we use 9999 or a suitable larger value to indicate that an integer has been "crossed out."

To build the model, we first enter the integers 1 through 100 in rows of 10 (Figure 4.55). We have entered 9999 in place of 1 to indicate that it has already been eliminated. Then we enter a formula that determines the next prime number, p, as the nth smallest integer. We also create a counter for the next iteration by adding 1 to the value of n above.

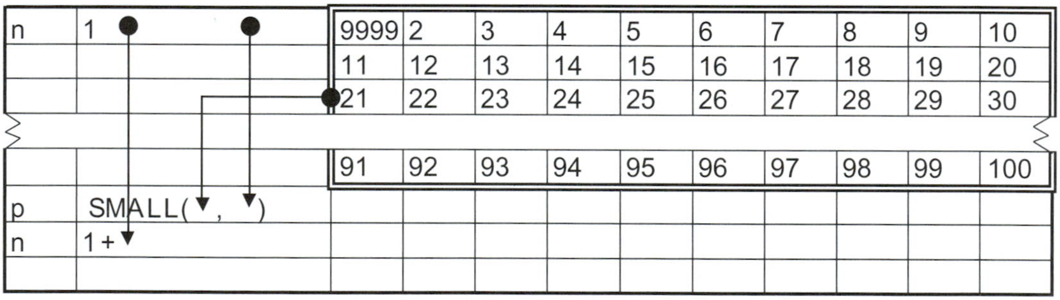

n	1	●	●	9999	2	3	4	5	6	7	8	9	10
				11	12	13	14	15	16	17	18	19	20
				21	22	23	24	25	26	27	28	29	30
				91	92	93	94	95	96	97	98	99	100
p	SMALL(▼ , ▼)												
n	1 + ▼												

Figure 4.55

Next, we determine whether each of the remaining entries is divisible by the current value of p and is larger than p. The last requirement ensures that the primes themselves will not be crossed out. We replace those integers that are to be eliminated as candidates by the number 9999 to indicate that they have been "crossed out." We then copy this expression throughout the bottom block of cells (Figure 4.56).

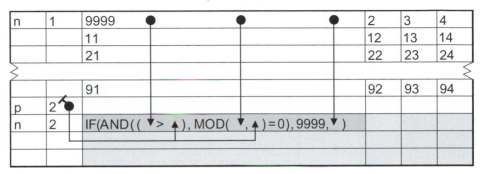

Figure 4.56

Figure 4.57 shows part of the initial output. Integers that have been "crossed out" are shown as 9999. We see that the multiples of 2, other than 2 itself, are "crossed out." Integers that are known to be prime because they are less than p^2, where p is the current prime found in the process, are highlighted by using the spreadsheet's conditional formatting command (see Appendix). The remaining integers are still candidates for being primes, but their status is yet to be determined by the sieve.

n	1	9999	2	3	4	5	6	7	8	9	10
		11	12	13	14	15	16	17	18	19	20
		21	22	23	24	25	26	27	28	29	30
		81	82	83	84	85	86	87	88	89	90
		91	92	93	94	95	96	97	98	99	100
p	2										
n	2	9999	2	3	9999	5	9999	7	9999	9	9999
		11	9999	13	9999	15	9999	17	9999	19	9999
		21	9999	23	9999	25	9999	27	9999	29	9999
		81	9999	83	9999	85	9999	87	9999	89	9999
		91	9999	93	9999	95	9999	97	9999	99	9999

Figure 4.57

Next we update our model by highlighting the cells in the bottom half of the model, copying them (including the value for n), and then pasting their values only into the cells of the top half of the model. This copying process can be recorded as a macro and then attached to a button, so that each time the button is clicked a new iteration will be performed.

The result of five iterations of the process is shown in Figure 4.58. By the time that we have established that 11 is the sixth largest prime, we know that all of the remaining numbers through 121 (and we only list those through 100) are prime. The conditional formatting command shows them by outlining their cells, as shown in the figure.

p	11									
n	6	9999	2	3	9999	5	9999	7	9999	9999	9999
		11	9999	13	9999	15	9999	17	9999	19	9999
		9999	9999	23	9999	9999	9999	9999	9999	29	9999
		31	9999	9999	9999	9999	9999	37	9999	9999	9999
		41	9999	43	9999	9999	9999	47	9999	9999	9999
		9999	9999	53	9999	9999	9999	9999	9999	59	9999
		61	9999	9999	9999	9999	9999	67	9999	9999	9999
		71	9999	73	9999	9999	9999	9999	9999	79	9999
		9999	9999	83	9999	9999	9999	9999	9999	89	9999
		9999	9999	9999	9999	9999	9999	97	9999	9999	9999

Figure 4.58

If you wish, you can now use this process with a 100×100 table to find the primes that are less than 10,000. See if you can predict how many iterations it will take to generate the complete list.

Construction Summary: Sieve of Eratosthenes

1. Generate an integer counter, n, starting with 2.
2. Compute the minimum value in the column as the initial prime.
3. Replace all remaining multiples of the last prime, p, through the formula =IF(MOD(n,p) = 0,9999,n).
4. Copy as indicated. Top row: right; Left column: down; Box 3: down and right.

Prime	2	2	3	5	7	11	13	17
	1	2	3	9999	9999	9999	9999	9999	9999
		3		3	9999	9999	9999	9999	9999
		4		9999	9999	9999	9999	9999	9999
		5		5	5	9999	9999	9999	9999
		6		9999	9999	9999	9999	9999	9999
		7		7	7	7	9999	9999	9999
		8		9999	9999	9999	9999	9999	9999

Exercises

1. Spreadsheets allow us to pursue alternate approaches for determining primes and the factorization of integers into a product of primes. A preliminary version of one such approach is outlined in Figure 4.59. A more advanced version of this model is described in Exercise 3. This model has been designed to investigate only odd integers, although we can generalize it if desired.

Num	Div	Rem	Factor	Type:	comp
				Prime Factors	
63	3	0	3	1	7
21	3	0	3	2	3
7	3	1	1	3	3
7	5	2	1	4	1
7	7	0	7	5	1
1	7	1	1	6	1
1	9	1	1	7	1
1	11	1	1	8	1
1	13	1	1	9	1
1	15	1	1	10	1

Figure 4.59

In the initial row of the display, we first enter the number to be examined. Here, it is 63. We then enter the initial odd integer (3) as a potential divisor and compute the corresponding remainder (0) using the MOD function. Next, in the fourth column, we use the IF function to reproduce the current potential divisor (3) as a prime factor if the remainder is 0, or a 1 otherwise.

Next, in the second row, we divide the previous number by the factor found in the row above. We then use an IF statement to repeat the previous divisor if it was a factor or to increment the potential divisor by 2 to obtain the next odd integer otherwise. In this preliminary model, simply incrementing an odd number by 2 may produce a composite number that of course will not generate a new prime factor.

We then copy these expressions through as many integers as we desire (limited by the spreadsheet's capabilities), say up to 1000 or 5000. Finally, in two additional columns (shown at the right in Figure 4.59), we use the spreadsheet's LARGE function to produce the largest prime factors that have been determined. At the top, we use another IF function to generate the strings "comp" or "prime" based upon the largest factor found. If the largest factor found is the initial number or 1, then the number is prime. Create this preliminary spreadsheet model and use it to test whether the following numbers are prime or composite.
   a. 141   b. 101   c. 12371   d. 89753   e. 89751

2. We now can use a data table to expand the model in Exercise 1 to display a list of integers, denoting whether they are prime or composite and providing their prime factorizations. To do this, we use a multiple-column data table, reading off the type

and the prime factors from the construction of Exercise 1. One possible display is provided in Figure 4.60. Here in the data table we have listed the odd integers starting with 3. However, we can use any odd integers, n, in the data table, provided the potential divisors used in the computation reach $y = g(x)$. In the dummy row at the top of the data table, we enter simple formulas to reproduce the type and the ten largest divisors, generating a blank ("") if a divisor is 1. The cell containing the test number (63 in Exercise 1) is the column input cell for the data table.

	comp		7	3	3							
3	prime		3									
5	prime		5									
7	prime		7									
9	comp		3									
11	prime		11	3								
13	prime		13									
15	comp		5	3								

Figure 4.60

Use your model to do each of the following:

a. Determine twin primes in the range $1001 \le n \le 1299$. *Twin primes* are primes that differ from each other by 2, for example, 5,7, and 10091, 10093.

b. Determine the number of primes less than 1000.

c. Find the prime factorizations of 6,023,943, 12,345, 665,511, 665,527.

3. In a more sophisticated and efficient approach to Exercise 1, instead of incrementing the potential divisors by 2 to obtain the next odd number (which may not be prime), we first can enter a list of prime numbers (say the first 200) into an array (shown at the right in Figure 4.61) and then use one of the spreadsheet's LOOKUP functions to generate the next prime as the next potential divisor rather than the next odd integer. Make this modification in your model. Typical output resulting from these modifications is shown in Figure 4.61 using 5733 as the number that is input.

				Type:	comp	Prime List	
Num	Div	Rem	Factor	Prime Factor		Prime	Next
5733	3	0	3	1	13	3	5
1911	3	0	3	2	7	5	7
637	3	1	1	3	7	7	11
637	5	2	1	4	3	11	13
637	7	0	7	5	3	13	17
91	7	0	7	6	1	17	19
13	7	6	1	7	1	19	23
13	11	2	1	8	1	23	29
13	13	0	13	9	1	29	31
1	13	1	1	10	1	31	37

Figure 4.61

4. [*] We can format the spreadsheet's display to show the factorization of a selected number as illustrated in Figure 4.62. Format your model to produce an effective user interface for this purpose.

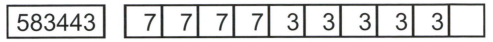

Figure 4.62

5. There is a clever way to create a visual display for the primes (see the Web site www.mathworld.wolfram.com/quadraticsieve.html). Create the graph of $x = y^2$ and draw lines connecting those points (k^2, k) with $|k| \geq 2$. The points of intersection of these lines "cross out" the composite integers, leaving the primes (see Figure 4.63). To see why this works, we can show algebraically that the lines through (k^2, k) and $(l^2, -l)$ intersect the x-axis at the point $x = kl$, which is a composite number. Moreover, any composite number $x = kl$ is on two of these lines, those determined by the pairs (k^2, k), $(l^2, -l)$ and (l^2, l), $(k^2, -k)$.

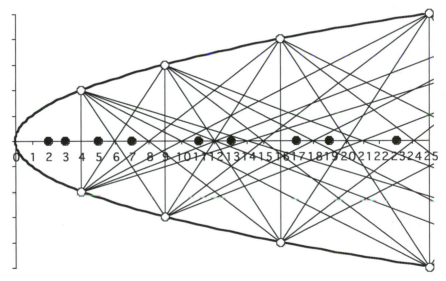

Figure 4.63

6. Design a spreadsheet model that uses a sieve approach to generate a list of the initial square-free integers: 2, 3, 5, 6, 7, 10, 11, 13, 14, 15, 17, 19, ... An integer n is square-free if it has no factors of the form k^2.

5

Functions and Graphing

CHAPTER OUTLINE

5.1   Graphing I: Basic Techniques

Graphs can be a great assistance to our studies of mathematical models, especially in enabling us to gain insights into the properties of functions. As we have found out in earlier examples, spreadsheets possess effective graphic facilities that also can help us to visualize both the output and the structure of our models. In this section we will use the xy-graph type to create graphs of functions of a single variable, $y = f(x)$. These are often needed in our mathematical investigations. The spreadsheet's xy-graph type plots a list of points by their (x, y)-coordinates, just as it is done in mathematics. Consecutively listed points are connected with line segments. By plotting points that are close to each other, the resulting output can produce very smooth curves. In using an xy-graph on a spreadsheet, we can choose to plot only markers at the points (x,y), only the connecting line segments, or both the lines and the markers. In addition to the material presented in this and the next section, further details for the process of creating an xy-graph are presented in the Appendix.

As an illustration of the manner in which the xy-graph plots points, we show three graphs that are generated from the following six points that lie on the curve $y = x^2$. We have purposely used unequal spaces between x-values to illustrate how the xy-graph works. To create the points, we first enter the x-values down the first column (Figure 5.1). We then use the exponentiation symbol ^ to create a formula in the second column that produces x^2 for the first value of x, and we then copy this expression down the second column.

The output is shown in Figure 5.2.

x	y
● 0.0 →	^2
0.4	
0.7	
1.2	
1.5	
2.0	

Figure 5.1

x	y
0.0	0.00
0.4	0.16
0.7	0.49
1.2	1.44
1.5	2.25
2.0	4.00

Figure 5.2

We next use the mouse to highlight the block of (x,y)-coordinates and then click on the Chart Wizard button (Figure 5.3).

Figure 5.3

In Step 1 of the Chart Wizard, we select the xy chart type and one of the suboptions. In Figure 5.4, Option 1 plots only the markers at the points; Option 2 plots only the lines connecting the points; and Option 3 plots both the markers and the lines. We will avoid using the middle two types, as they fit curves to the points through interpolation techniques that produce unwanted bumps and wiggles.

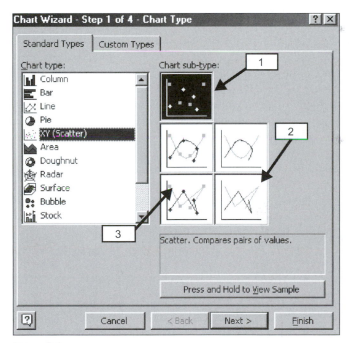

Figure 5.4

We can see the graphs that are produced in subsequent figures. The first graph (Figure 5.5) shows both markers and lines, the second one only markers (Figure 5.6), and the third one only lines (Figure 5.7). Of course, by using only six points lying fairly far apart from each other, the resulting graphs are only rough approximations of the curve $y = x^2$.

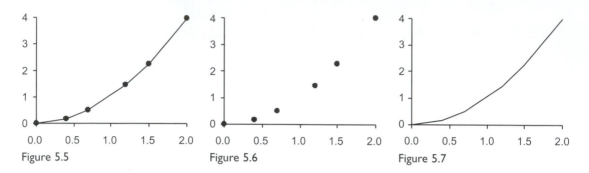

Figure 5.5   Figure 5.6   Figure 5.7

Spreadsheets provide two somewhat similar graph types, line and xy. The xy-graphs plot points by their (x, y)-coordinates, with connecting lines drawn to connect consecutively listed points. By contrast, line graphs plot points in equally spaced intervals going from left to right on the x-axis, regardless of the value of x. Thus, the x-values serve simply as labels and not as x-coordinates. Consequently, unless there is a good reason to use another graph type, we will usually use the xy graph type in drawing mathematical graphs.

1. Two Basic Examples

We next continue to illustrate fundamental graphing ideas by plotting the graph of the function $y = f(x) = 0.04x^4 - 0.4x^3 + x$ for $-2 \leq x \leq 2$. In this example, we have purposely kept the model simple, and we have chosen to delay introducing refinements in our approach until later examples. The model employs two columns to compute the (x, y)-components of points on the function's graph. In the first column a sequence of x-values is generated successively in step increments of size $dx = 0.02$. We enter the initial x-value, $x_0 = -2$, in the top cell of the column (Figure 5.8). Then in the cell below we enter a formula that adds 0.02 to the cell above. In future examples we will include x_0 and dx as separate parameters of our models. Next, in the top cell of the column of y-values, we enter the formula for $f(x)$. We enter the references to the x in the formula by using the cell at the left. Once the initial formulas are entered, we use a *copy* or *fill down* command to complete the table (Figure 5.9).

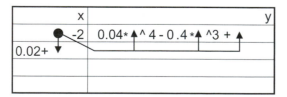

Figure 5.8   Figure 5.9

The resulting output is shown in Figure 5.10. Later we can set the format to show consistent numbers of decimal places in the display.

x	y
-2	1.84
-1.98	1.739738
-1.96	1.64213
-1.94	1.547141
1.98	-0.51018
2	-0.56

Figure 5.10

We now create an xy-graph for the function by first using the mouse to select the block of *xy*-values. We then click on the Chart Wizard button and choose the xy chart type. This defines the first column of values as the x-series and the second column as the y-series. The resulting graph is shown in Figure 5.11. In this case we have selected the option for plotting only the connecting line segments and not the markers.

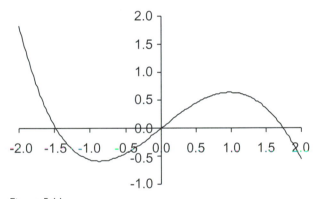

Figure 5.11

There are several ways in which we can improve this simple model. The easiest of these is to treat the initial value, x_0, the x step size, dx, and the coefficients, a_i, of a polynomial $f(x) = a_4x^4 + a_3x^3 + a_2x^2 + a_1x + a_0$ as parameters. We do this by entering their values in separate cells, typically located at the top of the model. The formulas for the *x*- and *y*-values then reference these cells. Consequently, we can modify and investigate the model simply by varying the values of the parameters. We now provide an outline of an improved model using the same function that we employed before.

First, we enter values for x_0, dx, and a_i, for $i = 4, 3, 2, 1, 0$ (Figure 5.12). Next, we use the first column to create the successive values of x in increments of size dx. Thus, the top cell contains a formula that reproduces the value of x_0, and the next one adds dx to the previous value of x (Figure 5.13). Notice that the value of dx refers to the same location anywhere it is used (so it is an absolute reference), while x is a relative reference. As before, we will complete this column by copying the formula for $x + dx$.

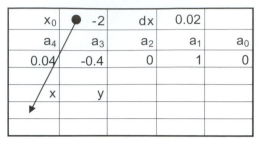

Figure 5.12   Figure 5.13

We now enter the formula for the first value of y (Figure 5.14). This time the values of the coefficients come from the parameter cells that are treated as absolute locations as the formula is copied down the column. We copy the formulas until we reach the value 2.0 for x (Figure 5.15). We can generalize the model even further by also including the exponents as parameters. Doing this is a nice exercise for you.

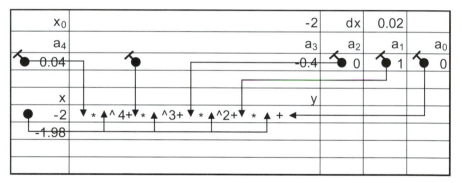

Figure 5.14

x	y
-2	1.84
-1.98	1.73974
-1.96	1.64213
-1.94	1.54714

Figure 5.15

Suppose that we now make a few changes in the parameters of the preceding model so that we get the display that is shown in Figure 5.16. In addition to changing the coefficients, setting x_0 to 0 plots points for an interval that is to the right of the previous one, while setting dx to 0.01 causes the points to be plotted over an interval that is half as wide as the one before. Thus, we generate coordinates of the graph

$$f(x) = -2x^4 + 3x^3 + 2.5x^2 - 3x + 0.5,\ 0 \leq x \leq 1.$$

x_0	0	dx	0.01	
a_4	a_3	a_2	a_1	a_0
-2	3	2.5	-3	0.5
x	y			
0	0.5			
0.01	0.4703			

Figure 5.16

The graph that is produced is shown in Figure 5.17.

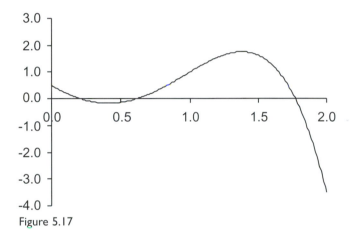

Figure 5.17

Construction Summary: Graphing Functions of One Variable

1. Enter initial value of x.
2. Add 0.02 to increment the previous value of x, copy down.
3. Enter the formula for $f(x)$ and copy down.

	x		y
1	-2.00	3	1.840
2	-1.98		1.740
	-1.96		1.642
	-1.94		1.547

2. Interactive Graphing

We next design a graphing model in a slightly different fashion in order to produce graphs that allow for even more user interaction. In fact, when such a model is coupled with the use of dialog boxes, scroll bars, buttons, and macros, it can make a spreadsheet look and act much like a graphing calculator—with the extra advantage of including the spreadsheet's many additional features.

In this design, rather than using an initial point of an x interval as a parameter as we did earlier, we select a central x-value, x_0, together with a step size for x of dx. We then use dx to create 100 points, x, in increments of size dx, lying on each side of x_0. These points are illustrated in Figure 5.18 and cover the interval

$$x_0 - 100dx \leqslant x \leqslant x_0 + 100dx.$$

In the spreadsheet, the points are labeled by a counter, n, that ranges from 0 to 200. This counter will be used with a scroll bar that must be linked to a cell that contains only positive integers.

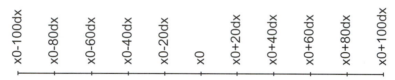

Figure 5.18

In addition to the graph of a function, we include a second y-series consisting of a single point to plot a separate point as a marker that can be used to trace points lying on the resulting curve for $y = f(x)$. The model allows us to move the marker along the curve to see how the resulting y-values change, to approximate the x-values of locations where the function's graph crosses the x-axis, and for other similar purposes. Our model allows us to select this point by entering a value of n, either directly or through a user-designed scroll bar. Observe that n is the number, or index, of the point to be selected.

We implement this approach in a model using the function

$$y = f(x) = 0.4x^3 + 0.1x^2 - x.$$

We first enter the identifying labels and the values of the parameters x_0 and dx (Figure 5.19). Here we set $x_0 = 0$ and $dx = 0.02$. Above the computations for the (x,y)-points of the function, we leave a blank row for later use. Then in the first column we generate a counter, n, in the usual manner. We then compute the first value of x (here -2.0) as $x_0 - 100dx$.

Center (x₀)	●	0	Step (dx)	● 0.02
	n	x	y	y₂
●	0	▼ -100 * ◄		
1+	▼			

Figure 5.19

Next, in the second column, we enter a formula to generate the second x-value by adding dx onto the first one (Figure 5.20). We need to ensure that dx is an absolute reference. We also calculate the first value of $f(x)$ in the third column. We then copy the formulas down their respective columns as indicated in Figure 5.21.

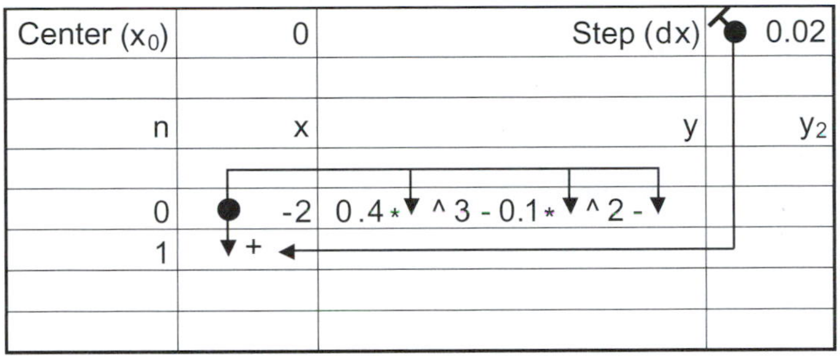

Center (x_0)		0			Step (dx)	0.02	
n		x			y		y_2
0		-2	0.4 * ^ 3 - 0.1 * ^ 2 -				
1	+						

Figure 5.20

Center (x_0)	0	Step (dx)	0.02
n	x	y	y_2
0	-2	-1.6	
1			

Figure 5.21

Now we create the coordinates for the trace point. To do this, we enter a value for the counter of a point, n_0, (here, $n = 132$) and a formula that looks up the coordinates of the corresponding point (Figure 5.22).

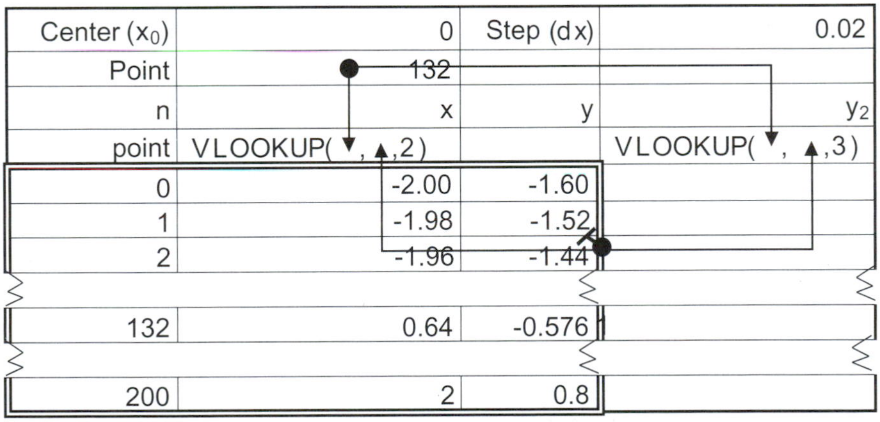

Center (x_0)		0	Step (dx)		0.02
Point		132			
n		x	y		y_2
point	VLOOKUP(,2)			VLOOKUP(,3)	
0		-2.00	-1.60		
1		-1.98	-1.52		
2		-1.96	-1.44		
132		0.64	-0.576		
200		2	0.8		

Figure 5.22

The first part of the output that is produced is shown in Figure 5.23.

Center (x_0)	0	Step (dx)	0.02
Point	132		
n	x	y	y_2
point	0.64		-0.5761
0	-2	-1.6	
1	-1.98	-1.517	
2	-1.96	-1.43597	
3	-1.94	-1.35691	

Figure 5.23

We next form an xy-graph by using the mouse to select, or highlight, the columns labeled *x*, *y*, and y_2, and choosing the xy chart type. A typical screen shot of a model as we are doing this is illustrated in Figure 5.24. As we create the xy-graph, we select the type that plots both markers and lines.

	A	B	C	D	E
1	Center (x_0)	0	Step (dx)	0.02	
2	Point	132			
3	n	x	y	y_2	
4	point	0.64		-0.5761	
5	0	-2	-1.6		
6	1	-1.98	-1.517		
7	2	-1.96	-1.43597		
8	3	-1.94	-1.35691		
9	4	-1.92	-1.2798		
10	5	-1.9	-1.2046		

Figure 5.24

This defines the column of *x*-values as the x-axis series and the two columns to its right as the first two y-series. The initial graph for the given values of our parameters is shown in Figure 5.25. Notice that by including the markers, we obtain a graph that is not as attractive as we might desire. However, there was a reason for doing this, and we will fix this soon. Also, in Step 3 of the Chart Wizard process we chose not to include a legend, and we removed the y-gridlines (for our book we also have reformatted the trace point so that it will show up better in the picture). We next make changes in the format of the graph by removing the gray background, thereby producing the graph shown in Figure 5.26. Details of the process involved are found in the Appendix.

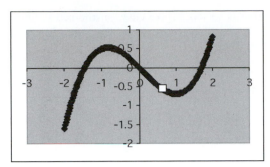

Figure 5.25

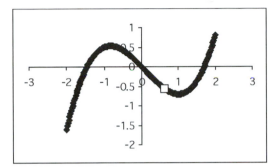

Figure 5.26

We now right click somewhere on the graph of the curve, and from the resulting menu we select the option Format Data Series (see Figure 5.27).

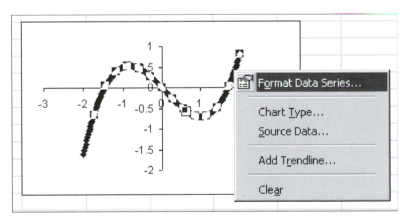

Figure 5.27

In the dialog box that is produced, we click on the tab Patterns, and then under Markers we select the option None to generate the graph shown in Figure 5.28.

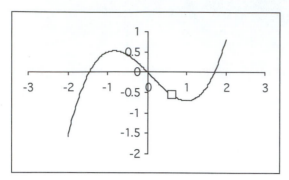

Figure 5.28

The marker for the auxiliary point still remains. This is the reason that we frequently start graphing by choosing to display both lines and markers. It is generally easier to remove unneeded markers than to add markers, especially if a graph series consists of a single point.

We now change the format on the marker that remains on the second y-series and add a label that supplies the (x,y)-coordinates of that point. Because the last feature is somewhat complex, the process for how to do this is provided in the Appendix. The graph that results is shown in Figure 5.29.

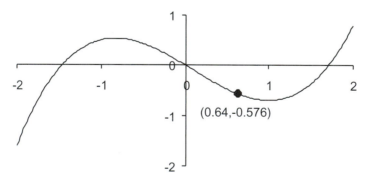

Figure 5.29

Next, we illustrate how our model can be used to examine a function's graph. First, we can vary the value of n, thereby causing the highlighted point to move. For example, we can vary n to search for the location of the right-most zero of the function, as shown in Figure 5.30 and Figure 5.31.

Center (x_0)	0	Step (dx)	0.02
Point	186		
n	x	y	y_2
point	1.72		0.0195
0	-2	-1.6	
1	-1.98	-1.517	
2	-1.96	-1.43597	
3	-1.94	-1.35691	

Figure 5.30

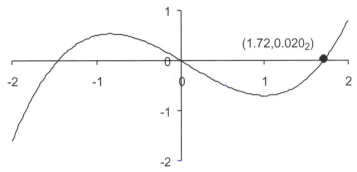

$(1.72, 0.020_2)$

Figure 5.31

This process is made much more effective by creating a scroll bar and connecting it to the cell that contains the value for n (see the Appendix), so that n can vary from 0 to 200. Then as we drag the slider on the scroll bar, the movement of the trace point becomes animated. We can also investigate the high and the low points on the curve in a similar fashion. In Figure 5.32, we see the values of the numerical display that gives us an approximation of a local minimum.

Center (x_0)	0	Step (dx)	0.02
Point	150		
n	x	y	y_2
point	1		-0.7
0	-2	-1.6	
1	-1.98	-1.517	
2	-1.96	-1.43597	
3	-1.94	-1.35691	

Figure 5.32

The illustrative screen shot (Figure 5.33) shows how the scroll bar approach can provide us with a much more effective means for carrying out an investigation. As we move the slider, we can home in on the local minimum.

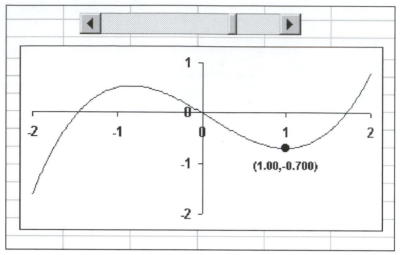

Figure 5.33

With our model we can also change x_0 to examine another portion of the curve, or zoom in on the region around x_0 by decreasing dx, or zoom out to see more of the curve by increasing dx. Entering the required commands as elementary macro commands into graph buttons and then clicking on them can mechanize these processes.

Another approach to modifying the basic model is to generate the auxiliary point directly from the formula for the function, as shown in Figure 5.34. This will allow us to locate the local maximum and minimum values of the function.

Center (x_0)	0	Step (dx)			0.02
n	x	y			y_2
point	1		0.4*▼^3-0.1*▼^2- ▼		
0	-2	-1.6			
1	-1.98	-1.517			

Figure 5.34

To approximate the local maximum that is at the left of the curve, we enter −1 as an estimate of the x-value of the point as illustrated in Figure 5.35. Next we select the spreadsheet's Solver command (see the Appendix) and choose the options that will set the y_2 cell to be a maximum by changing the x cell. The result provided in Figure 5.36 can be formatted to show more decimal places.

n	x	y	y_2
point	-1		0.5

Figure 5.35

n	x	y	y_2
point	-0.833		0.5324

Figure 5.36

The point on our graph moves to locate the local maximum (Figure 5.37). Of course, with just a little more work, both features can be built into the same model.

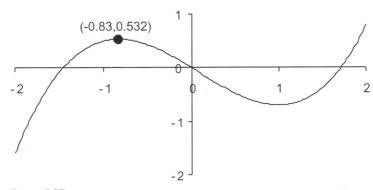

(-0.83,0.532)

Figure 5.37

Construction Summary: Graphing Functions with a Trace Point

1. Enter the x-value of the center of the domain, the x step size, and the number of a selected point.
2. Create a counter from 0 to 200.
3. Create the initial x-value at 100 steps to the left of the center.
4. Increment x by adding the step size to the previous x and copy.
5. Enter the formula for $f(x)$, copy.
6. Use =LOOKUP to find the selected point in the three left columns.

	Center(x_0)	[1]	{ 0	Step (dx)	[1]	0.02
	Point		{ 132			
	n		x	y		y_2
	point	[6]	0.64		[6]	-0.5761
[2]	0	[3]	-2.00	[5] -1.6000		
	1	[4]	-1.98	-1.5170		
	2		-1.96	-1.4360		
	3		-1.94	-1.3569		
	4		-1.92	-1.2798		
	5		-1.90	-1.2046		

3. Graphs of Several Functions

From time to time, we may need to create the graphs of two or more functions within the same display. There are several ways to do this on a spreadsheet. When all of the functions are defined over the same set of x-values, the easiest way to create a graph is to put the x- and y-values of the functions in adjacent columns if possible. Here we graph the functions $f(x) = 0.9 \sin(x)$ and $g(x) = 2 + f(x)$ over the interval $0 \leqslant x \leqslant 16$. After creating a column of the x-values, we compute values of $f(x)$ and $g(x)$, respectively, in the second and third columns (Figure 5.38). We then create the xy-graph after selecting all three columns for the graph (Figure 5.39). The first becomes the x-values, and the next two are the y-series.

x	f(x)	g(x)
0.00 ●	0.9 * SIN(▲) ●	2 + ▲
0.05		
0.10		

Figure 5.38

x	f(x)	g(x)
0.00	0.0000	2.0000
0.05	0.0450	2.0450
0.10	0.0899	2.0899

Figure 5.39

The graph produced is shown in Figure 5.40. Note that the graph of the function g is two units higher than the graph of the function f.

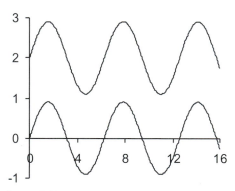

Figure 5.40

However, sometimes it is convenient (or even necessary) to arrange our work so that the columns for the y-values of two functions are not adjacent to each other. In the example shown in Figure 5.41, we have created the y-values for two functions in this manner, $f(x) = 0.9\sin(x)$ and $g(x) = |f(x)| + 2$, over the same interval as in the preceding example. We can notice that a column that has been used for an intermediate computation separates the columns of y-values. While different spreadsheets have varying ways of handling this situation, one possible approach that we can use with *Microsoft Excel* is to first create the graph of $y = f(x)$ (Figure 5.42) and then to select the column for $y = g(x)$ (Figure 5.43) and drag it into the spreadsheet graph as a new y-series. This process is examined in the Appendix.

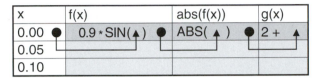

Figure 5.41

x	f(x)	abs(f(x))	g(x)
0.00	0.0000	0.0000	2.0000
0.05	0.0450	0.0450	2.0450
0.10	0.0899	0.0899	2.0899

Figure 5.42

x	f(x)	abs(f(x))	g(x)
0.00	0.0000	0.0000	2.0000
0.05	0.0450	0.0450	2.0450
0.10	0.0899	0.0899	2.0899

Figure 5.43

The final graph produced by this example is provided in Figure 5.44.

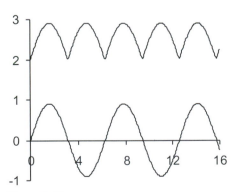

Figure 5.44

As a final example, we create the graphs of two curves that are defined on different sets of x-values. We do this with a line that is not the graph of a function, as we draw a vertical line from the x-axis to a high point on the curve at $x = 5\pi/2$. As in our last example, we first create the graph of the function, $f(x) = 0.9\sin(x)$, in the usual manner. However, this time we select both columns that provide the (x,y)-coordinates of the endpoints of the line and drag them into the existing graph. This process is described in Figure 5.45, Figure 5.46, and Figure 5.47.

x	f(x)	x	g(x)
0.00	0.0000	5*PI()/2	0
0.05	0.0450	5*PI()/2	0.9
0.10	0.0899		

Figure 5.45

x	f(x)	x	g(x)
0.00	0.000	7.854	0
0.05	0.045	7.854	0.9
0.10	0.090		

Figure 5.46

x	f(x)	x	g(x)
0.00	0.000	7.854	0
0.05	0.045	7.854	0.9
0.10	0.090		

Figure 5.47

The two stages in the construction of the graph appear in Figure 5.48 and Figure 5.49. Notice that in Figure 5.49 we also have reformatted the line so that it is dotted.

As a source for additional related spreadsheet graphing topics, see Arganbright (1993).

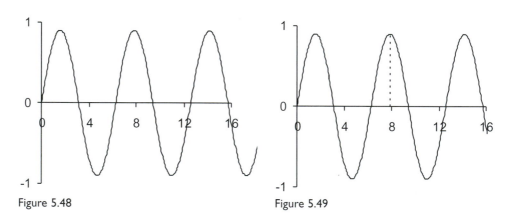

Figure 5.48   Figure 5.49

Construction Summary: Graphing Two Functions of One Variable

1. Enter the initial value of x.
2. Add the x-increment 0.05 to the previous value, copy down.
3. Enter the formula of $f(x)$, copy.
4. Enter the formula of $g(x)$, copy.

	x		f(x)		g(x)
1	0.00	3	0.0000	4	2.0000
2	0.05		0.0450		2.0450
	0.10		0.0899		2.0899
	0.15		0.1345		2.1345
	0.20		0.1788		2.1788

Exercises

1. Create a spreadsheet model to examine the effects of translations and stretches of a function by using the parameters a, b, c, d and graphing the functions $y = f(x)$ and $y = g(x) = cf(ax + b) + d$ on the same axes. Then link the parameter cells to scroll bars. Implement your model using the following functions and an appropriate set of x-values.

   a. $f(x) = \sin x$
   b. $f(x) = x^2$
   c. $f(x) = 1/x$
   d. $f(x) = |x|$
   e. $f(x) = \cos^2 x$
   f. $f(x) = 0.1x^3 - x$

2. Create a spreadsheet model that illustrates the composition of two functions. Show the graphs of $y = f(x)$, $y = g(x)$, and $y = (f \circ g)(x) = f(g(x))$ on the same axes. Also show $y = (g \circ f)(x) = g(f(x))$. Use the following pairs of functions with your model.

   a. $f(x) = x^2$, $g(x) = 2x$
   b. $f(x) = x)^2$, $g(x) = x - 2$
   c. $f(x) = |x|$, $g(x) = x - 2$
   d. $f(x) = x^2$, $g(x) = \sin(x)$

3. Create a spreadsheet model to produce graphs of a function $y = f(x)$ and its inverse relation $x = f(y)$ by simply interchanging the columns for x and y. The latter relation gives the graph of $y = f^{-1}(x)$ when f is a one-to-one function. Include the line $y = x$ in your graph.

   a. $f(x) = e^x$
   b. $f(x) = x^3 - x$
   c. $f(x) = x^2$
   d. $f(x) = \sin x$
   e. $f(x) = x^3 + 2x$
   f. $f(x) = \ln(x + 2)$

4. Create a graphic spreadsheet model that uses the solver to find the local maxima, minima, and zeroes of a function. Show these features in a graph of the function. Use your model to investigate the following functions.

   a. $f(x) = x^3 - x^2 + 1$, $-1 \le x \le 2$
   b. $f(x) = x \cos(x - 1)$, $-5 \le x \le 5$
   c. $f(x) = \cos(2x) - 3\sin x$, $-2\pi \le x \le 2\pi$
   d. $f(x) = xe^x$, $-4 \le x \le 1$

5. Design a spreadsheet model to create graphs of pairs of functions, using the solver to find points of intersection. Implement your model using the following functions.

   a. $f(x) = 4 - x^2$, $g(x) = 0.2x^3 - x + 1$ b. $f(x) = x + 1$, $g(x) = 0.2x^3 - x + 1$
   c. $f(x) = \cos x$, $g(x) = 0.2x^4 - 3x^2 + 2$ d. $f(x) = 0.2e^{-x} + 1$, $g(x) = x^2 - 2$

6. [*] [Calculus] Modify the animated model of this chapter to create a graph that incorporates a tangent line segment along with the moving point linked to a scroll bar (Figure 5.50). Also create a separate graph of the derivative function to show how it is related to the slope of the original function. Notice that the vector $[1, f'(x)]$ provides a segment in the correct direction. It can be scaled using its length to provide an appropriate tangent segment. Implement this using the functions of Exercise 3 and Exercise 4 from this section.

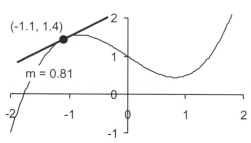

Figure 5.50

7. Create three-dimensional graphs for the following equations. In addition, include conditional formatting for the numerical output in the spreadsheet itself.

   a. $f(x, y) = ax^2 + by^2$ b. $f(x, y) = ax^2 - by^2$
   c. $f(x, y) = a + \sqrt{x^2 + y^2}$ d. $f(x, y) = a\sin x + b\sin y$

8. [Calculus and Theory] Newton's method is used to locate a zero of a differentiable function $y = f(x)$ as described in the section on Newton and secant methods later in this chapter. Create a model that illustrates the geometry of Newton's method. Then experiment with different functions to see that sometimes Newton's method converges to an unexpected zero, while at other times it diverges. Try the following situations.

   a. $f(x) = x^2 - 2$, $x_0 = 0.4$
   b. $f(x) = \cos x$, $x_0 = 0.30, 0.31, 0.36$

   c. $f(x) = \dfrac{4x - 3}{x - 1}$, $x_0 = -1.0, 0.0, 0.49, 0.5, 0.51, 0.7, 0.75, 0.99, 1.0$

   d. $f(x) = x^3 - 2x$, $x_0 = 0.1, 0.4, 0.5, 0.8$

9. [Calculus and Theory] Euler's method is used to approximate the solution of a differential equation initial value problem. We start with a differential equation $y' = f(x, y)$ and an initial condition $y(x_0) = y_0$. We then move a given step size, dx, from x_0 to a point $x_1 = x_0 + dx$, and approximate the new y-value by moving along the tangent line to obtain $y_1 = y_0 + f(x, y)dx$. We then repeat the process

to obtain successive approximations. To illustrate how the approximation behaves, take a differential equation that can be solved and compare the graph of the solution to Euler's approximation. The diagram in Figure 5.51 uses a step size of $dx = 0.1$ with the initial value problem $y' = x + y$, $y(0) = 1$ and its solution $y = 2e^x - x - 1$.

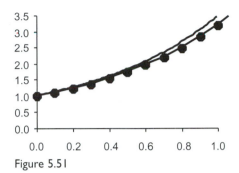

Figure 5.51

Use your model with the following initial value problems.

a. $y' = \cos(x)$, $y(0) = 0$, Solution: $y = \sin x$

b. $y' = \dfrac{1 - 2xy}{x^2}$, $y(1) = 2$, Solution: $y = \dfrac{1}{x} + \dfrac{1}{x^2}$

c. $y' = e^{-x} - y$, $y(0) = 3$, Solution: $y = xe^{-x} + 3e^{-x}$

5.2 Graphing II: Polar and Parametric Equations

1. Basic Construction

When we first encounter curves in the study of mathematics, they are usually constructed and interpreted graphically in terms of Cartesian, or rectangular coordinates, (x, y). Often the value of the y-coordinates of points on the curve are given by a function, $y = f(x)$. We examined the concepts of creating graphs of functions in the Cartesian coordinate system in the last section. In the Cartesian, or rectangular, coordinate system we locate points by their vertical and horizontal distances from the respective x- and y-axes.

However, this is not the only useful coordinate system. In this section, we consider curves that are defined in terms of polar coordinates, (r, t), where r is the distance of a point from the origin, and t is the angle that a line from the origin to the point makes with the positive x-axis. We can think of points as being located by a hand on a clock (see Figure 5.52). The length of the hand corresponds to the value of r, while the time corresponds to the value of t (with time measured counterclockwise from the positive x-axis).

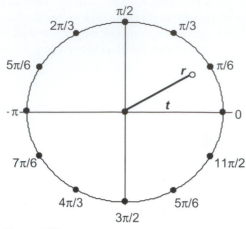

Figure 5.52

degrees	radians
0	0
30	$\pi/6$
45	$\pi/4$
60	$\pi/3$
90	$\pi/2$
120	$3\pi/2$
135	$3\pi/4$
150	$5\pi/6$
180	π
225	$5\pi/4$
270	$3\pi/2$
360	2π

Figure 5.53

When polar coordinates are used in a spreadsheet, the angle t is measured in radians, rather than in degrees. In using radian measurement a circle is divided into 2π units, while in degree measure the circle is divided into 360 degrees. A spreadsheet has a function, RADIANS, that converts degrees to radians. It also contains the inverse of this function, DEGREES, that converts radians to degrees. The table in Figure 5.53 gives equivalent measurements for some common angles. Thus,

$$\text{RADIANS}(360) = 2\pi, , \text{RADIANS}(180) = \pi, \text{RADIANS}(30) = \pi/6$$
$$\text{DEGREES}(2\pi) = 360, \text{DEGREES}(\pi) = 180, \text{DEGREES}(\pi/6) = 30$$

The ideas behind the rectangular and polar coordinate systems can be seen further in the graph in Figure 5.54. Here we have plotted a point with $r = 3$ and $t = 3\pi/y$ (the radian equivalent of 135 degrees). The corresponding (x, y)-coordinates are given by $x = r \cos(t)$ and $y = r \sin(t)$, or $(x, y) = (3/\sqrt{2}, 3/\sqrt{2} = (-2.12, 2.12)$. The (x, y)-coordinates can be seen from the dashed grid lines.

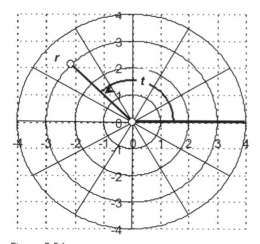

Figure 5.54

Although some spreadsheets provide a similar looking radar graph type, typically spreadsheets do not provide us with true polar graphs. Nonetheless, there is a way to create the graphs of polar functions on a spreadsheet by using the xy graph type. To do this, we convert from polar coordinate expressions into Cartesian coordinates through the equations

$$x = r \cos(t), \; y = r \sin(t).$$

The geometry is illustrated in the triangle shown in Figure 5.55.

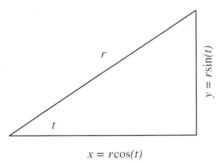

Figure 5.55

In Figure 5.56, we illustrate the approach in a model to graph polar functions of the form

$$r = a + b \cos(ct).$$

In our layout we generate a counter, n, to produce angles measured in degrees, usually in the range $0 \leq n \leq 360$, and subsequently compute their radian equivalent, t, where $0 \leq t \leq 2\pi$.

We first enter the parameter values of a, b, and c. Here we have chosen to graph the function defined by

$$r = 0 + 1.9 \cos(2t).$$

The first column of our model computes the values of the degree counter, n. The first entry is 0. The second value is incremented from the first by a degree step size entered as a parameter above. The second column generates the equivalent angle, t, in radians using the function RADIANS, that converts angles from degrees to radians. We also could have accomplished this by multiplying the degree measurements by the conversion factor, PI()/180.

		a	b	c	
		0	1.9	2	
step		1			
n (deg)	t (rad)	r	x	y	
0	RADIANS(▲)				
+					

Figure 5.56

In the third column we enter the function's formula as $a + b\cos(ct)$, where the value for t comes from the cell immediately to the left, while the others are absolute locations entered as parameters at the top of the model (Figure 5.57).

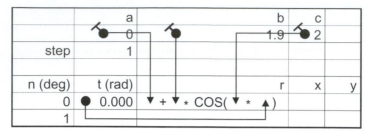

Figure 5.57

Now we compute x as $r\cos(t)$ and y as $r\sin(t)$ using the values for r and t that were computed in the same row (Figure 5.58).

		a	b	c	
		0	1.9	2	
step		1			
n (deg)	t (rad)		r	x	y
0	0.000		1.900	*COS(▲)	*SIN(▲)
1					

Figure 5.58

Finally, we complete the table as shown in Figure 5.59 by copying or filling down the formulas as usual.

		a	b	c	
		0	1.9	2	
step		1			
n (deg)	t (rad)	r	x	y	
0	0.0000	1.9000	1.9000	0.0000	
1	0.0175	1.8988	1.8986	0.0331	
2	0.0349	1.8954	1.8942	0.0661	
3	0.0524	1.8896	1.8870	0.0989	

Figure 5.59

A portion of the completed table is shown in Figure 5.60. From this display we create an xy-graph using the column of x values as the x-series, and the column of y values as the y-series.

		a	b	c	
	0	1.9	2	
step		1			
n (deg)	t (rad)	r	x	y
0	0.0000	1.9000	1.9000	0.0000
1	0.0175	1.8988	1.8986	0.0331
2	0.0349	1.8954	1.8942	0.0661
3	0.0524	1.8896	1.8870	0.0989
359	6.2657	1.8896	1.8986	-0.0331
360	6.2832	1.9000	1.9000	0.0000

Figure 5.60

The graph that is produced from the given parameters is shown in Figure 5.61. It is straightforward to experiment with the parameters to form other interesting graphs.

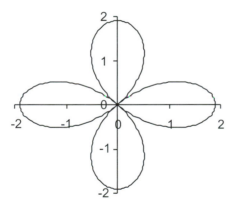

Figure 5.61

We can also use the same techniques to create graphs of parametric equations. For example, let us consider the parametric equations

$$x = \sin(2t), \ y = \sin(3t).$$

We show the formulas (Figure 5.62) and output (Figure 5.63) of a spreadsheet to create the graph. We create the graph by using the last two columns.

		a	b
		2	3
deg	t	x	y
0	RADIANS(→)	SIN(▼ * ▲)	SIN(▼ * ▲)
1			
2			
3			
4			

Figure 5.62

		a	b
		2.000	3.000
deg	t	x	y
0	0.000	0.000	0.000
1	0.017	0.035	0.052
2	0.035	0.070	0.105
3	0.052	0.105	0.156
4	0.070	0.139	0.208

Figure 5.63

Curves of this form are called *Lissajous figures*. The graph formed from our model is presented in Figure 5.64.

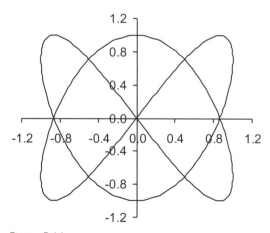

Figure 5.64

2. Incorporating Animation Effects

A spreadsheet allows us to build a variety of animation effects into our graphs. These effects can help us to visualize various mathematical aspects of our work. In this example, we add features that will help us to see how a polar curve is traced out by a point traveling along the curve and drawing the curve as the angle increases. The point is created as the endpoint of a vector of radius r drawn from the origin at an angle t with the positive x-axis, corresponding to the value $n = n_0$.

The output of the completed model is shown in Figure 5.65. The model first uses the x and y columns to plot the curve in one color. It then uses the y_2 column to plot as a heavier curve in a darker color only those points n in the range $0 \leq n \leq n_0$. Also, the two points in the y_3 column are used to draw the moving vector from the origin to the current point. Thus, as we increase the value of n_0 by one unit at a time, the number of points shown gradually increases to trace out the curve. We then enhance the operation of our model by incorporating a scroll bar to vary the value of n_0.

		a	b	c			
		0	1.9	2			
step		1			point, n_0		2
n (deg)	t (rad)	r	x	y	y_2		y_3
0	0.000	1.9000	1.9000	0.0000	0.0000		
1	0.0175	1.8988	1.8986	0.0331	0.0331		
2	0.0349	1.8954	1.8942	0.0661	0.0661		
3	0.0524	1.8896	1.8870	0.0989	#N/A		
4	0.0698	1.8815	1.8769	0.1312	#N/A		
360	6.2832	1.9000	1.9000	0.0000	#N/A		
			0				0.0000
			1.8942				0.0661

Figure 5.65

The graphic output of our model when $n_0 = 165$ is shown in Figure 5.66. Notice the vector that moves to trace out the curve.

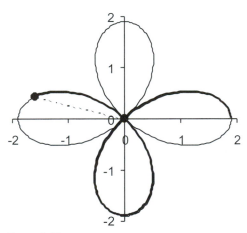

Figure 5.66

The formula that is used to create the second copy of the curve can be interpreted as

IF the value of n is less than or equal to n_0 ($n \leq n_0$)

THEN reproduce the y value ELSE generate the "not available" value

Points with the value #N/A will not be plotted. After we enter the expression for the first value of y_2, we then copy it down the y_2 column (Figure 5.67).

		a	b	c				
		0	1.9	2				
	step	1			point, n_0			2
n (deg)	t (rad)	r	x	y			y_2	y_3
0	0.000	1.900	1.900	0.000	IF(▼<= ▼, ▲,NA())			
1	0.017	1.899	1.899	0.033				
359	6.266	1.899	1.899	-0.033				
360	6.283	1.900	1.900	0.000				

Figure 5.67

The points that determine the vector from the origin to the moving point are entered next. We simply enter (0,0) for the origin and then use lookup functions to generate the coordinates of point n_0. Here the functions lookup the value n_0 in the indicated table (Figure 5.68).

		a	b		c		y_2			y_3
		0	1.9		2					
	step	1			point, n_0	2				
n (deg)	t (rad)	r		x	y	y_2				y_3
0	0.000	1.900		1.900	0.000	0.000				
1	0.017	1.899		1.899	0.033	0.033				
359	6.266	1.899		1.899	-0.033	#N/A				
360	6.283	1.900		1.900	0.000	#N/A				
				0						0
				VLOOKUP(▼, ▼,4)						VLOOKUP(▼, ▼,5)

Figure 5.68

Our animations can be made even more striking by eliminating the underlying full graph. To do this, we right click on the full curve and from the resulting menu we select Clear. Now as we trace out the curve we can observe how the construction develops. An illustration from a screen display that incorporates a scroll bar is shown in Figure 5.69.

For an extensive variety of related functions and ideas of topics that can be implemented with animated graphics in a spreadsheet, see Arganbright (1993).

	A	B	C	D	E	F
1	a,b =	2	2		n =	292
2	deg	t	r	x	y	Y
3	0	0.000	2.000	2.000	0.000	0.000
4	1	0.017	1.999	1.998	0.035	0.035
5	2	0.035	1.995	1.994	0.070	0.070
6	3	0.052	1.989	1.986	0.104	0.104
7	4	0.070	1.981	1.976	0.138	0.138
8	5	0.087	1.970	1.962	0.172	0.172
9	6	0.105	1.956	1.946	0.204	0.204
10	7	0.122	1.941	1.926	0.236	0.236
11	8	0.14	1.923	1.904	0.268	0.268
12	9	0.157	1.902	1.879	0.298	0.298
13	10	0.175	1.879	1.851	0.326	0.326
14	11	0.192	1.854	1.82	0.354	0.354
15	12	0.209	1.827	1.787	0.38	0.38
16	13	0.227	1.798	1.752	0.404	0.404

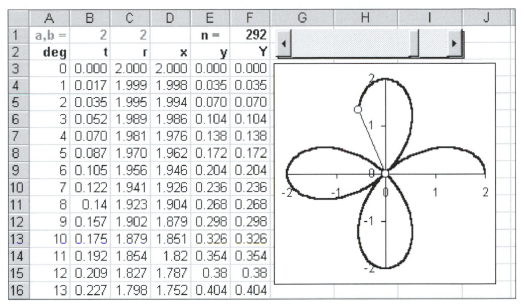

Figure 5.69

Construction Summary: Graphing Polar Equations

1. Enter the value of the parameters a, b, c and the degree step size.
2. Create a degree counter column.
3. Convert degrees to radians, t.
4. Enter a formula for r as $a+b\cos(ct)$.
5. Compute x as $r\cos(t)$.
6. Compute y as $r\sin(t)$.
7. Enter the number of degrees for the upper limit of moving display.
8. Use =IF to reproduce the y-values for degrees up to n_0.
9. Copy as indicated.

a,b,c	[1] 0	1.9	2		
step	1			n_0 [7]	4

n(deg)	t(rad)	r	x	y	y_2
[2] 0	[3] 0.000	[4] 1.900	[5] 1.900	[6] 0.000	[8] 0.000
1	0.017	1.899	1.899	0.033	0.033
2	0.035	1.895	1.894	0.066	0.066
3	0.052	1.890	1.887	0.099	0.099
4	0.070	1.882	1.877	0.131	0.131
5	0.087	1.871	1.864	0.163	#N/A
6	0.105	1.858	1.848	0.194	#N/A

Exercises

1. Create spreadsheet implementations for the following classical curves that are given in polar form. Include animation effects in your design. Some care may be needed in the interval selected for t. To generate the entire graph, you may need to select an interval other than $0 \leqslant t \leqslant 2\pi$.

   a. Cardioid: $r = 2a(1 + \cos t)$
   b. Limaçon: $r = b + 2a \cos t$
   c. Spiral of Archimedes: $r = at$
   d. Cissoid: $r = a \csc(bt) - \sin(ct)$
   e. Conchoid: $r = a \csc(bt) + c$
   f. Lemniscate: $r^2 = a^2 \cos(bt)$

2. Create spreadsheet models for the following classical curves that are given in parametric form. Include animation effects in your design. As in Exercise 1, you may need to adjust the interval chosen for t.

   a. Astroid: $y = \cos^3 t$, $y = \sin^3 t$
   b. Deltoid: $x = a(2\cos t + \cos 2t)$, $y = a(2\sin t - \sin 2t)$
   c. Nephroid: $x = a(3\cos t - \cos 3t)$, $y = a(3\sin t - \sin 3t)$
   d. Lissajous curves: $x = a\cos bt$, $y = c\sin dt$, where b and d are integers.
   e. Cycloid: $x = a(t - \sin t)$, $y = a(1 - \cos t)$
   f. Epicycloid: $y = na\cos t - a\cos nt$, $y = na\sin t - a\sin nt$

3. Use the ideas of this section to create an animated graph to show the creation of a function of the form $y = f(x)$. The illustration in Figure 5.70 uses $f(x) = \sin x$.

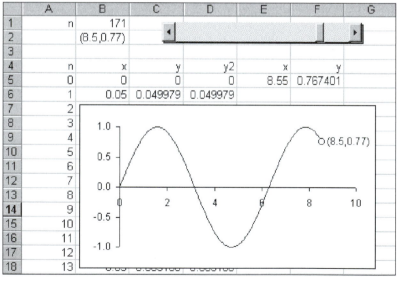

Figure 5.70

4. [*]The Archimedian spiral (see Exercise 1c) is characterized by the fact that after each complete pass around the center the distance from the center increases by the same amount. The Logarithmic spiral is characterized by the fact that after each complete pass around the center the distance from the center increases by the same factor. Draw both spirals in one graph and compare their behavior.

5.3 Bezier Curves

A Bezier curve is a fundamental tool for drawing smooth curves in the field of computer graphics. In this section, we develop a technique to create these curves that can be readily adjusted by varying a few control points.

Bezier curves are named after Pierre Bezier (1910–1999), a French applied mathematician and engineer who worked on automobile design problems and developed many of the mathematical concepts used in computer-aided design.

1. Basic Construction

To explain the process, we begin with four points, $P_{0,0} = (x_0, y_0)$, $P_{0,1} = (x_1, y_1)$, $P_{0,2} = (x_2, y_2)$, and $P_{0,3} = (x_3, y_3)$. We also select a real number, t, in the range $0 \leq t \leq 1$. We first illustrate our procedure in the graph in Figure 5.71 using $t = 0.3$. We begin by drawing line segments between the four initial points. Next, in the line segment from $P_{0,0}$ to $P_{0,.1}$, we create the point $P_{1,0}$ that lies $t = 0.3$ of the way between the endpoints. We then continue this process, creating similarly placed points $P_{1,1}$ and $P_{1,2}$ on the next two line segments.

We now repeat this process, drawing line segments connecting the points $P_{1,0}$, $P_{1,1}$, and $P_{1,2}$. On the segment between $P_{1,0}$ and $P_{1,1}$ we create the point $P_{2,0}$ that lies $t = 0.3$ of the way between the two endpoints. We create the point $P_{2,1}$ similarly on the second segment. Finally, we repeat this process yet again, drawing a line segment between $P_{2,0}$ and $P_{2,1}$, and creating the point $P_{3,0}$ that is $t = 0.3$ of the way between $P_{2,0}$ and $P_{2,1}$. This point, $P_{3,0}$, is one point on the curve that we wish to create.

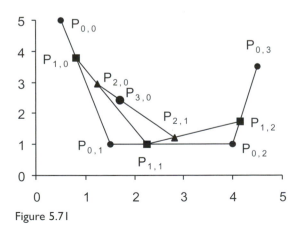

Figure 5.71

We now carry out this process for all other values of t in the interval $0 \le t \le 1$. The construction of the point $P_{3,0}$ for $t = 0.6$ is shown in Figure 5.72.

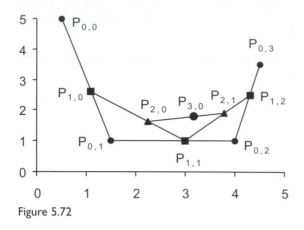

Figure 5.72

If we create points $P_{3,0}$ for all values of t in the interval $0 \le t \le 1$, then we generate a Bezier curve such as the one shown in Figure 5.73. The points $P_{0,0}$, $P_{0,1}$, $P_{0,2}$, and $P_{0,3}$ are called *control points*. We notice that when $t = 0$ we obtain the first point, $P_{0,0}$. When $t = 1$ we obtain the last point, $P_{0,3}$. The other control points generally do not lie on the Bezier curve. The process can easily be extended to more than four points.

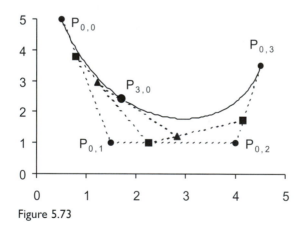

Figure 5.73

To construct our model, we observe that the point (x, y) that is a fraction, t, of the way between (x_0, y_0) and (x_1, y_1) is given by $x = x_0 + t(x_1 - x_0)$, $y = y_0 + t(y_1 - y_0)$. We use this observation repeatedly to create the (x, y) components of the point $P(t)$ that corresponds to t. The layout of our work to compute the coordinates of a point on the curve for a given value of t is shown in Figure 5.74. We use pairs of columns to list these (x, y)-coordinates at each stage in pairs of columns, or levels. Level 0 lists the initial control points $P_{0,i}$, Level 1 the points $P_{1,i}$, and so on. Notice that the level corresponds to the first subscript on a point. The coordinates for the point on the Bezier curve will be found in the right two columns, labeled Level 3.

t	0.3							
	Level 0		Level 1		Level 2		Level 3	
Ctrl	x	y	x	y	x	y	x	y
	0.5	5	P_{10}	P_{10}	P_{20}	P_{10}	P_{30}	P_{30}
	1.5	1	P_{11}	P_{11}	P_{21}	P_{10}		
	4	1.5	P_{12}	P_{12}				
	4.5	3.5						

Figure 5.74

To construct the first x-coordinate of Level 1, we enter the formula shown in Figure 5.75, $x = x_0 + t(x_1 - x_0)$. This references the appropriate two points located two columns to the left. Of course, the reference to the t value must be absolute. We then copy this into the cell to the right to generate $y = y_0 + t(y_1 - y_0)$ in the same way.

t	0.3			Level 1	
	Level 0				
Ctrl Pts	x	y		x	y
	0.5	5	▼ + ▼ * (▲ - ▼)		
	1.5	1			
	4	1.5			
	4.5	3.5			

Figure 5.75

In fact, this same process works for all of the remaining points. Each of the points is defined in terms of the coordinates of the points that are located two columns to the left. All that remains to do is to copy the generic formula into the remaining cells (Figure 5.76). Notice that this same construction can be used just as easily with any number of control points by extending the table in the natural way.

t	0.3							
	Level 0		Level 1		Level 2		Level 3	
Ctrl	x	y	x	y	x	y	x	y
	0.5	5	0.8	3.8	1.2	3	1.7	2.5
	1.5	1	2.3	1.2	2.8	1.4		
	4	1.5	4.2	2.1				
	4.5	3.5						

Figure 5.76

Using the output from the previous step, we next use the spreadsheet's data table command to generate a list of the points (x, y) on the curve corresponding to $t = 0.00$, 0.01, 0.02, . . . , 1.00. To do this we set up a data table with three columns shown at the right side of Figure 5.77, entering formulas in the top row that reproduce the final x- and y-coordinates. We then select these three columns and issue the data table command, using the cell for t as the column input location. Recall that the data table concept is discussed in the Appendix.

t	0.3									t	x	y
	Level 0		Level 1		Level 2		Level 3					
Ctrl	x	y	x	y	x	y	x	y				
	0.5	5	0.8	3.8	1.24	3.01	1.711	2.534		0.00		
	1.5	1	2.25	1.15	2.82	1.44				0.01		
	4	1.5	4.15	2.1						0.02		
	4.5	3.5								0.03		

Figure 5.77

The resulting display is shown in Figure 5.78. The last two columns on the right can be used to produce a graph of the curve. We can also supplement the graph by adding a scroll bar to vary the value of t and by incorporating the columns giving the other pairs of (x, y)-coordinates into the graph to show the construction lines. We can do the latter by dragging the array of the various (x, y)-components into the graph, as described in our graphing sections and the Appendix. Coupled together, these two additions can help us to better visualize the construction process.

t	0.3									t	x	y
	Level 0		Level 1		Level 2		Level 3				1.71	2.53
Ctrl	x	y	x	y	x	y	x	y				
	0.5	5	0.8	3.8	1.24	3.01	1.71	2.53		0	0.5	5
	1.5	1	2.25	1.15	2.82	1.44				0.01	0.53	4.88
	4	1.5	4.15	2.1						0.02	0.56	4.77
	4.5	3.5								0.03	0.59	4.65

Figure 5.78

It can be noted and proved that at the endpoints the Bezier curve is tangent to the left and right line segments that were described in the first stage of the construction process (see Figure 5.79).

After our model is completed, we can experiment by changing the control points to draw a variety of interesting curves. *Excel* even allows us to change the control points within the graph itself. For example, we can click on one of the control points in the graph to select the series. We then click again on the same point (not double-click) to select the particular point. This is illustrated in Figure 5.80. Then we hold down on the left mouse button and drag the point horizontally or vertically as shown in Figure 5.81. This not only moves the point and the resulting curve but updates values within the spreadsheet table itself.

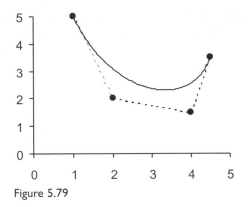

Figure 5.79

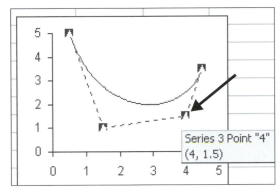

Figure 5.80

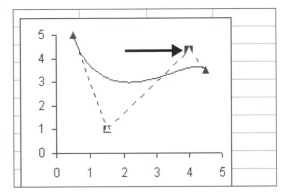

Figure 5.81

The Bezier curve model lends itself nicely to the use of a scroll bar. In the screen display in Figure 5.82, the scroll bar will trace out more of the curve as the slider is moved to the right. A similar construction can be used effectively with the lines and the points in the graph showing the construction of the curve.

While the construction of the Bezier curve is carried out through the use of control points, it is sometimes convenient to ensure that the curve passes through certain selected points and to be able to change points on the curve rather than to vary the control points directly. A discussion of this is included in the following section.

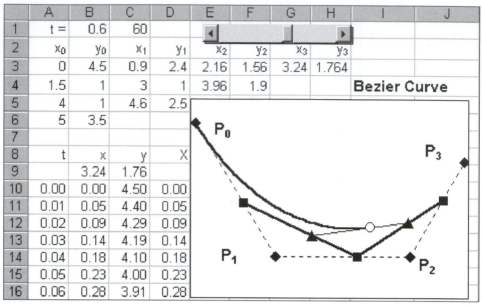

	A	B	C	D	E	F	G	H	I	J
1	t =	0.6	60		◀			▶		
2	x_0	y_0	x_1	y_1	x_2	y_2	x_3	y_3		
3	0	4.5	0.9	2.4	2.16	1.56	3.24	1.764		
4	1.5	1	3	1	3.96	1.9			Bezier Curve	
5	4	1	4.6	2.5						
6	5	3.5								
7										
8	t	x	y	X						
9		3.24	1.76							
10	0.00	0.00	4.50	0.00						
11	0.01	0.05	4.40	0.05						
12	0.02	0.09	4.29	0.09						
13	0.03	0.14	4.19	0.14						
14	0.04	0.18	4.10	0.18						
15	0.05	0.23	4.00	0.23						
16	0.06	0.28	3.91	0.28						

Figure 5.82

Construction Summary: Bezier Curves

1. Enter control point coordinates (x,y).
2. Enter an integer to link to a scroll bar.
3. Enter a formula for t by dividing the cell to the right by 100.
4. Compute the x-coordinate of the initial intermediate construction point using t.
5. Copy down and right as shown to generate intermediate points.

t =	3	0.3	2	30					
	Level 1		Level 2		Level 3		Level 4	
Ctrl	x	y	x	y	x	y	x	y
0.5	5	4	0.80	3.80	1.24	3.01	1.71	2.53
1.5	1		2.25	1.15	2.82	1.44		
4	1.5		4.15	2.10				
4.5	3.5							

Slider: Link scroll integer (Box 2) to a scroll bar that ranges from 0 to 100.
Graph: Create xy-graph from Level 1 columns; drag Level 2, Level 3, Level 4 columns into the graph.

2. An Alternate Approach

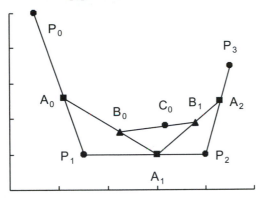

Figure 5.83

If we do some algebra, then we can generate formulas for the points on the Bezier curves. We refer to the drawing in Figure 5.83 in our construction.

Thus, suppose that A_0 and A_1 are the points that are a fraction t of the way on the line between the first and second control points, and $A_0 = (X_0, Y_0)$ and $A_1 = (X_1, Y_1)$. Then their coordinates are given by

$X_0 = x_0 + t(x_1 - x_0)$, $Y_0 = y_0 + t(y_1 - y_0)$ and $X_1 = x_1 + t(x_2 - x_1)$, $Y_1 = y_1 + t(y_2 - y_1)$.

Similarly, if $B_0 = (U_0, V_0)$ is the point the same fraction t of the way on the line between A_0 and A_1, then its coordinates can be determined similarly as

$$U_0 = X_0 + t(X_1 - X_0), \quad V_0 = Y_0 + t(Y_1 - Y_0).$$

By substituting the coordinates for X_0 and Y_0 into the latter expression and performing a little algebra, we obtain

$$U_0 = (1 - t)^2 x_0 + 2(1 - t)tx_1 + t^2 x_2, \quad V_0 = (1 - t)^2 y_0 + 2(1 - t)ty_1 + t^2 y_2.$$

If we do the same computations with A_1 and A_2 to find the corresponding point B_1, then we could continue in the same way to find that the coordinates of the point $C_0 = (X, Y)$ that is the same fraction of the way between the points B_0 and B_1 are given by

$$X = (1 - t)^3 x_0 + 3(1 - t)^2 tx_1 + 3(1 - t)t^2 x_2 + t^3 x_3,$$
$$Y = (1 - t)^3 y_0 + 3(1 - t)^2 ty_1 + 3(1 - t)t^2 y_2 + t^3 y_3.$$

From our discussion so far, the general pattern may now be clear. We can see that the points on the Bezier curve are given by a polynomial whose coefficients are the well-known binomial coefficients. Consequently, with $n + 1$ control points, (x_i, y_i), the formulas for the points on the associated Bezier curve would become

$$X = C(n,0)(1 - t)^n x_0 + C(n,1)(1 - t)^{n-1} tx_1 + \ldots$$
$$+ C(n,i)(1 - t)^{n-1} t^i x_i + \ldots + C(n,n)t^n x_n,$$
$$Y = C(n, 0)(1 - t)^n y_0 + C(n,1)(1 - t)^{n-1} ty_1 + \ldots$$
$$+ C(n,i)(1 - t)^{n-i} t^i y_i + \ldots + C(n,n)t^n y_n.$$

The explicit parametric functions for the (x,y)-coordinates allow us to do one more thing with our model. Suppose that we want to specify two points that the curve passes through in addition to the two endpoints. Let us call these points (a_1, b_1) and

(a_2, b_2). Then we must be able to select control points $P_1 = (x_1, y_1)$ and $P_2 = (x_2, y_2)$ and values of t, say t_1 and t_2, so that

$$a_1 = (1 - t_1)^3 x_0 + 3(1 - t_1)^2 t_1 x_1 + 3(1 - t_1)t_1^2 x_2 + t_1^3 x_3,$$
$$a_2 = (1 - t_2)^3 x_0 + 3(1 - t_2)^2 t_2 x_1 + 3(1 - t_2)t_2^2 x_2 + t_2^3 x_3.$$

Thus, we must solve for the values x_1 and x_2 in the system of equations

$$3(1 - t_1)^2 t_1 x_1 + 3(1 - t_1)t_1^2 x_2 = a_1 - (1 - t_1)^3 x_0 - t_1^3 x_3,$$
$$3(1 - t_2)^2 t_2 x_1 + 3(1 - t_2)t_2^2 x_2 = a_2 - (1 - t_2)^3 x_0 - t_2^3 x_3.$$

Written in matrix form, with

$$X = \begin{bmatrix} x_1 \\ x_2 \end{bmatrix}, \; A = \begin{bmatrix} 3(1 - t_1)^2 t_1 & 3(1 - t_1)t_1^2 \\ 3(1 - t_2)^2 t_2 & 3(1 - t_2)t_2^2 \end{bmatrix}, \; B_x = \begin{bmatrix} a_1 - (1 - t_1)^3 x_0 - t_1^3 x_3 \\ a_2 - (1 - t_2)^3 x_0 - t_2^3 x_3 \end{bmatrix},$$

this becomes the equation $AX = B_x$, whose solution can be found by using the inverse of A together with matrix multiplication, as $B_x = A^{-1}X$. Functions to carry out these matrix operations are available on *Excel* and are discussed in the Appendix. Alternatively, expressions for the operations can be entered directly. A similar result holds for the Y coordinates, B_x.

Although we will not list the details, here is an outline of what you can do. The full model is on the accompanying CD. First, in the block for the curve points, enter the four points through which the curve is to pass and the values of t to be used in the Bezier curve. The first and last of these points become the endpoints in the list of control points, with $t = 0$ and $t = 1$. Next, use the middle entries in the curve points to compute the components of the 2×2 array, A. If your spreadsheet provides formulas for the inverse of a matrix (MINVERSE in *Excel*), use it to compute A^{-1}; otherwise enter formulas from linear algebra to create this matrix. Next, create the arrays of constants in the equations as B_x and B_y. Finally, either use a built-in matrix multiplication function (MMULT in *Excel*), or provide equivalent formulas directly yourself, and compute $A^{-1}B_x$ and $A^{-1}B_y$. These give the two sought-for control points. Finally, enter formulas in the list of control points to read these off as shown in Figure 5.84 and Figure 5.85.

Curve Points				Control Points	
x	y	t		x	y
1	5	0			
2	4	0.3			
3	2.5	0.7			
4.5	3.5	1			
A				A^{-1}	
0.441	0.189			2.778	-1.190
0.189	0.441			-1.190	2.778
B_x	B_y			$A^{-1}B_x$	$A^{-1}B_y$
1.563	2.191			2.563	4.698
1.430	1.165			2.143	0.627

Figure 5.84

Curve Points				Control Points	
x	y	t		x	y
1	5	0		1	5
2	4	0.3		2.5635	4.6984
3	2.5	0.7		2.1429	0.627
4.5	3.5	1		4.5	3.5
A				A^{-1}	
0.441	0.189			2.7778	-1.190
0.189	0.441			-1.19	2.7778
B_x	B_y			$A^{-1}B_x$	$A^{-1}B_y$
1.5355	2.1915			2.5635	4.6984
1.4295	1.1655			2.1429	0.627

Figure 5.85

The resulting picture for our chosen points is shown in Figure 5.86, with the open circles being the two selected points for the curve to pass through. Notice that the control points have been adjusted to allow this to happen.

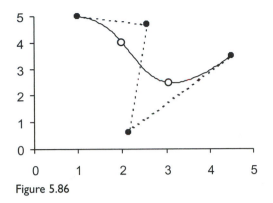

Figure 5.86

Construction Summary: Bezier Curves via Matrices

1. Enter 0 as the initial t value, add 0.01 to the cell above to increment t, copy down.
2. Enter a formula to reproduce the Bezier x-coordinate, copy to the right to get the y-coordinate.
3. Select double outlined block, use data table command with the cell for the scroll value of t (Box 3) as the column input cell.

	t		x	y		y_2
		6	1.71	2.53		
5	0.00	7	0.50	5.00	8	5.00
	0.01		0.53	4.88		4.88
	0.02		0.56	4.77		4.77
	0.03		0.59	4.65		4.65
	0.04		0.63	4.54		4.54

4. Use =IF formula to generate the y-coordinate of the points for which the left column t-value does not exceed the scroll value of t, copy.

Graph: Drag the columns of the (x,y)-coordinates into the graph as new series. Drag the last column into the graph as a y-series.

Exercises

1. Restructure the basic Bezier model to use a different number of control points. Create versions using the following sets of points. After you have completed a model, experiment with the placement of the control points by clicking twice on one of the control points in the chart and then dragging it to a new position.
   a. (0.5,2.0), (2.0,3.5), (2.6,1.0), (4.0,0.8), (3.2,4.0), (2.0,1.0)
   b. (0.5,2.5), (2.9,4.1), (0.6,1.7), (3.7,0.8), (1.7,4.1), (3.7,3.2), (1.5,1.0), (1.0,4.0)

2. Suppose that one of the control points is duplicated successively in the list of control points. Do you obtain the same curve by omitting one of the duplicate point listings? Experiment with one of your models.

3. Often rather than increasing the number of control points in a single Bezier curve, one frequently plots several cubic Bezier curves (each using four control points) and connects them systematically. For a curve to be continuous, the last control point of one of the successive curves must be the same as the first one of the succeeding one. However, with just this requirement the resulting curve may not be "smooth" (in calculus, this requires that the first and sometimes the second derivatives of the connecting curves agree at the endpoint). Construct a model to illustrate this preliminary approach. An illustration appears in Figure 5.87.

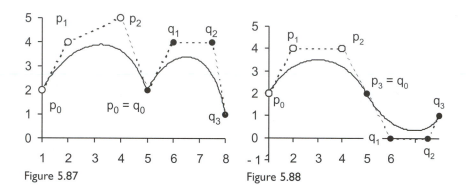

Figure 5.87   Figure 5.88

4. To guarantee that the slopes of the tangents at the connecting point of two Bezier curves are identical, the connecting point needs to be the midpoint of the control points immediately before and after it. Create a spreadsheet model using this idea. The output of one such curve is presented in Figure 5.88.

5. [*] The curve constructed in Exercise 4 typically will still fail to have the second derivative be equal at the junction point. Try to find a condition that ensures that the connection between the two curves is smooth, and use either the goal seek or the solver to modify one of the directional control points accordingly.

6. The concept of morphing gives us another graphic topic to pursue. To illustrate the idea, first consider two points in the plane, say (a,b) and (c,d). The vector between the points is given by $[c-a, d-b]$. Then for $0 \leq t \leq 1$ the vector $t[c-a, d-b]$ is t times as long. Thus, points that lie on the line segment connecting the two endpoints have the form

$$[a,b] + t[c-a, d-b] = (1-t)[a, b] + t[c, d], \ 0 \leq t \leq 1$$

as illustrated in Figure 5.89 with $t = 0.7$. Notice that $t = 0$ gives the point (a,b), while $t = 1$ gives the point (c,d).

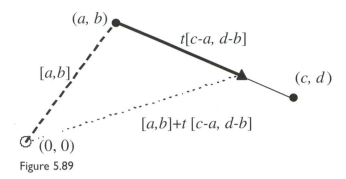

Figure 5.89

Now use the techniques for drawing polar curves to create two interesting closed curves. Then create a third curve whose points lie the portion *t* between the corresponding points on the two curves. Create a separate graph of the new curve and link the value of *t* to a scroll bar so that *t* varies from 0 to 1 in steps of size 0.01. Moving the slider will "continuously" transform one curve into the other. This process is called *morphing*. This is illustrated in Figure 5.90, Figure 5.91, and Figure 5.92 where a rabbit ($r = -2 + \cos 4t + \sin 3t$) is morphed into a butterfly: [$r = 1.5 + 1.75 \cos (4t - \pi) - 2.25 \sin(2t - \pi / 2)$].

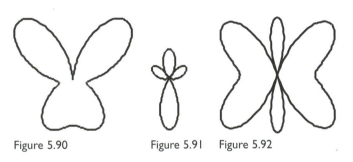

Figure 5.90   Figure 5.91   Figure 5.92

5.4 Bisection Algorithm

Frequently we want to find the solution of an equation of the form $f(x) = 0$. Such a solution, *r*, is also called a zero of the function $y = f(x)$. It satisfies the condition $f(r) = 0$. Sometimes we can find an exact solution of the equation directly by algebraic means. However, it is generally not possible to do so. In such cases it may be sufficient to obtain an approximation of the zero(s). Several methods are used for approximating zeros of continuous functions. In this section, we examine one of these, called the *bisection algorithm*. The graphs in Figure 5.93 need Figure 5.94 will help us to illustrate this algorithm. Initially we will assume that the function has only one zero on the interval that we are considering.

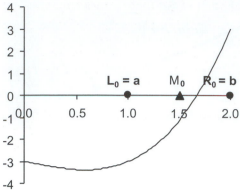

Figure 5.93

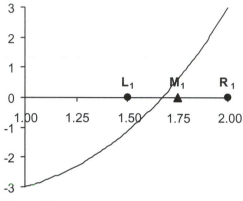

Figure 5.94

Our spreadsheet model for the bisection algorithm uses the function $f(x) = x^3 - x - 3$. If we first graph this function over the interval $0 \leq x \leq 2$, then we find that its graph looks like the one in Figure 5.93. With the bisection algorithm we first need to find two values of x, $x = a$ and $x = b$, that bracket a zero of the function. That is, a zero of the function is in the interval $a \leq x \leq b$.

Since the function f is continuous, from its graph we notice that the curve $y = f(x)$ must cross the x-axis at some point, r, between $x = 1$ and $x = 2$, since $f(1) < 0$ and $f(2) > 0$. We say that $f(1)$ and $f(2)$ have opposite signs. It follows that $x = 1$ and $x = 2$ bracket the zero, the point where the curve crosses the x-axis, so the zero is in the interval $1 \leq x \leq 2$.

Next we evaluate the function at the midpoint of that interval and find that $f(1.5) < 0$. Since we already know that $f(2) > 0$, it follows that $f(1.5)$ and $f(2)$ have opposite signs, so that 1.5 and 2 bracket the zero, and the zero is in the interval $1.5 \leq x \leq 2.0$.

We now repeat the process by evaluating the function at the midpoint of that interval. When we do this we find that $f(1.75) > 0$. Since $f(1.5) < 0$, it follows that $f(1.5)$ and $f(1.75)$ have opposite signs. Thus, 1.5 and 1.75 bracket the zero, so that the zero is in the interval $1.5 \leq x \leq 1.75$.

In short, we repeatedly find two points that bracket the zero and then cut the interval in half by using the midpoint and one of the endpoints to bracket the zero. We continue the process until the width of the interval is as small as we desire and then use the midpoint of that interval as an estimate of the zero.

In our spreadsheet model we begin by entering two values for x that bracket the zero, denoting these as the initial left and right endpoints, L_0 and R_0 (see Figure 5.95). We use the first column to number the iterations, beginning with 0 by using our standard counting technique. The next three columns calculate the left endpoint, the midpoint, and the right endpoint for each bracketing interval. The rightmost three columns evaluate the function at each of these points. We start by entering the formula for $f(L_0)$ as $x^3 - x - 3$.

n	L_n	M_n	R_n	$f(L_n)$	$f(M_n)$	$f(R_n)$
0	1		2	^3 - - 3		
1 +						

Figure 5.95

Now we compute the midpoint of the interval as $M_0 = (L_0 + R_0)/2$. In doing this, M_0 becomes the first approximation of the desired zero. We also copy the formula $f(L_0)$ into the two cells to the right to compute $f(M_0)$ and $f(R_0)$ (see Figure 5.96).

n	L_n	M_n	R_n	$f(L_n)$	$f(M_n)$	$f(R_n)$
0	1	(+)/2	2	-3		
1						

Figure 5.96

We now want to divide the interval in half and for the next iteration choose the half that contains the zero. To do this, we notice that $f(L_0) = -3$ and $f(M_0) = -1.125$ have the same signs, while $f(M_0) = -1.125$ and $f(R_0) = 3$ have opposite signs. Consequently, the zero actually lies between 1.5 and 2.0 (see Figure 5.97).

n	L_n	M_n	R_n	$f(L_n)$	$f(M_n)$	$f(R_n)$
0	1	1.5	2	-3	-1.125	3
1						

Figure 5.97

There is a standard way to test whether two real numbers have the same sign. If u and v are numbers with $uv < 0$, then u and v have opposite signs, since one must be positive and the other negative. Otherwise, either at least one of the numbers is 0 or they both have the same sign. In this example, since $f(L_0)f(M_0) > 0$, then $f(L_0)$ and $f(M_0)$ have the same sign.

In general, to obtain the algorithm's next approximation, we bisect the interval at the midpoint $x = M_0$. If $f(M_0) = 0$, then we have located a zero at M_0. Assuming that this is not the case, if $f(L_0)$ and $f(M_0)$ have opposite signs then the zero will be located in the left subinterval $L_0 < x < M_0$, and a new, smaller interval (of half the previous width) that brackets the zero will be $L_1 \leqslant x \leqslant R_1$, where $L_1 = L_0$ and $R_1 = M_0$. Otherwise, the zero lies in the right subinterval $L_1 \leqslant x \leqslant R_1$, where $L_1 = M_0$ and $R_1 = R_0$. Notice that in each case only one of the left or right endpoints changes in going from one iteration to the next.

We can express the test using an IF . . . THEN . . . ELSE construction:

IF $f(L_0)f(M_0) < 0$ THEN ($L_1 = L_0$ and $R_1 = M_0$) ELSE ($L_1 = M_0$ and $R_1 = R_0$).

We use this construction in carrying out the computations for the new endpoints in the L_n and R_n columns of the second row, as shown in Figure 5.98 and Figure 5.99.

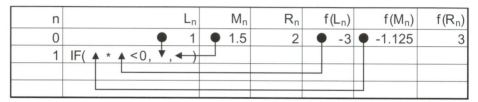

n				L_n	M_n	R_n	$f(L_n)$	$f(M_n)$	$f(R_n)$
0				1	1.5	2	-3	-1.125	3
1	IF(▲ * ▲ <0, ▼, ◄)								

Figure 5.98

n	L_n	M_n			R_n	$f(L_n)$	$f(M_n)$	$f(R_n)$
0	1	1.5			2	-3	-1.125	3
1	1.5	IF(▲ * ▲ < 0, ▼, ▼)						

Figure 5.99

The remainder of the algorithm simply consists of repeating this process. Thus, we can complete the table by using a copy or fill down command as shown in Figure 5.100.

n	L_n	M_n	R_n	$f(L_n)$	$f(M_n)$	$f(R_n)$
0	1	1.5	2	-3	-1.125	3
1	1.5	1.75	2	-1.125	0.6094	3
2	1.5	1.625	1.75	-1.125	-0.334	0.6094
3	1.625	1.6875	1.75	-0.334	0.1179	0.6094

Figure 5.100

We use the midpoints as successive approximations to the zero. In order to know when we have obtained a satisfactory approximation, an additional column gives us the maximum amount of error that can occur by using M_n as an approximation for the zero. Since the zero must lie either between L_n and M_n or between M_n and R_n (and the distance between each pair is the same), the maximum error that results in using M_n as the approximation for the zero is $R_n - M_n$. We use an additional column to compute this value at the maximum error and copy it down the column (Figure 5.101).

n	L_n	M_n	R_n	$f(L_n)$	$f(M_n)$	$f(R_n)$	Max Err
0	1	1.5	2	-3	-1.125	3	-
1	1.5	1.75	2	-1.125	0.60938	3	
2	1.5	1.625	1.75	-1.125	-0.33398	0.60938	
3	1.625	1.6875	1.75	-0.33398	0.11792	0.60938	

Figure 5.101

The output in Figure 5.102 shows the successive approximations of the zero as M_n in bold and displays the maximum size of the error at each stage in the rightmost column. Notice that while this algorithm always converges to a zero, the process tends to be slow. In particular, the value of the current zero to five decimal places is $r = 1.67170$. However, we will obtain this value only after 19 iterations.

n	L_n	M_n	R_n	$f(L_n)$	$f(M_n)$	$f(R_n)$	Max Err
0	1.00000	1.500000	2.00000	-3.00000	-1.12500	3.00000	0.50000000
1	1.50000	1.750000	2.00000	-1.12500	0.60938	3.00000	0.25000000
2	1.50000	1.625000	1.75000	-1.12500	-0.33398	0.60938	0.12500000
3	1.62500	1.687500	1.75000	-0.33398	0.11792	0.60938	0.06250000
4	1.62500	1.656250	1.68750	-0.33398	-0.11288	0.11792	0.03125000
18	1.67170	1.671701	1.67170	0.00000	0.00001	0.00003	0.00000191
19	1.67170	1.671700	1.67170	0.00000	0.00000	0.00001	0.00000095
20	1.67170	1.671700	1.67170	0.00000	0.00000	0.00000	0.00000048

Figure 5.102

A few observations should be made at this point. First, if the exact value of the zero is located in this process, the algorithm could stop. However, in order to simplify the description of the algorithm, we have not incorporated that situation in the model. Incorporating this feature is left as an exercise.

Also, to use the algorithm, one must first locate values L_0 and R_0 for which $f(L_0)$ and $f(R_0)$ have opposite signs. This may prove to be challenging to do for some functions. In addition, even if $f(L_0)$ and $f(R_0)$ have opposite signs, then all that we know is that there will be at least one zero in the interval $L_0 < x < R_0$. In fact, there could be three, five, or any other odd number of zeroes in the interval. Finally, even if $f(L_0)$ and $f(R_0)$ have the same sign, there may be zero, two, four, or any even number of zeroes in the interval $L_0 < x < R_0$. Thus, more analysis may have to be done to locate all of the solutions of an equation.

Construction Summary: Bisection Algorithm

1. Create a counter column.
2. Enter initial left, right estimates.
3. Compute midpoint, copy down.
4. Enter a formula to evaluate the function at the left endpoint, copy right and down.
5. Compute the maximum error as of right – mid.
6. Use =IF to compute new left point, copy.
7. Use =IF to compute new right point, copy.

n	Ln	Mn	Rn	f(Ln)	f(Mn)	f(Rn)	MaxErr
0	1.00	1.50	2.00	3.00	-1.13	3.00	0.50
1	1.50	1.75	2.00	-1.13	0.61	3.00	0.25
2	1.50	1.63	1.75	-1.13	-0.33	0.61	0.13
3	1.63	1.69	1.75	-0.33	0.12	0.61	0.06
4	1.63	1.66	1.69	-0.33	-0.11	0.12	0.03
5	1.66	1.67	1.69	-0.11	0.00	0.12	0.02
6	1.66	1.66	1.67	-0.11	-0.06	0.00	0.01
7	1.66	1.67	1.67	-0.06	-0.03	0.00	0.00
8	1.67	1.67	1.67	-0.03	-0.01	0.00	0.00
9	1.67	1.67	1.67	-0.01	-0.01	0.00	0.00
10	1.67	1.67	1.67	-0.01	0.00	0.00	0.00

Exercises

1. Use the bisection algorithm to locate the zeros of the following functions correct to 10^{-5}.
   a. $f(x) = x^2 - 2$
   b. $f(x) = \cos x$
   c. $f(x) = x^3 - x - 1$
   d. $f(x) = x^3 + \cos x$
   e. $f(x) = x - 1 - 0.5 \sin x$
   f. $f(x) = x^2 - \ln x - 2$

2. Use the bisection algorithm to find the points of intersection of the following pairs of functions correct to 10^{-5}.
   a. $f(x) = 5 - x^2$, $g(x) = 1 + 0.2e^{-x}$
   b. $f(x) = 4 - 3x^2$, $g(x) = x^3 - 2x$
   c. $f(x) = \sin(x - 1)$, $g(x) = x^2 - 3$
   d. $f(x) = \ln(x + 3)$, $g(x) = x^2 + x - 3$

3. [*] In the bisection algorithm we find real x-values a and b that bracket a zero of a function $y = f(x)$ and then divide the interval $a \leqslant x \leqslant b$ into two equal subintervals, locating the subinterval that contains the zero. However, using a spread-

sheet we can just as well divide the interval into 10 equal subintervals and employ a similar technique. This allows us to locate one additional decimal place of the zero with each iteration. Design a spreadsheet model to carry this out, either using a format similar to that presented in the text or by using a condensed technique that is updated by a copy routine that can be incorporated into a macro. Also design illustrative graphs, as provided in Figure 5.103 and Figure 5.104 for the function $f(x) = x^2 - 2$. Notice the 10 subintervals in each step.

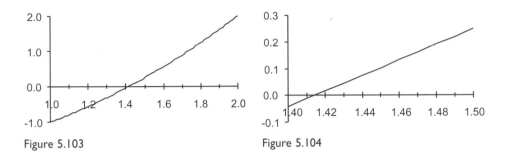

Figure 5.103   Figure 5.104

4. Recall that a real number a is called a *fixed point* of a function $y = g(x)$ if $g(a) = a$. For example, $a = 0$ and $a = 1$ are fixed points of the function $g(x) = x^2$, since $g(0) = 0$ and $g(1) = 1$. A fixed point can be interpreted as a point at which the curve $y = g(x)$ intersects the line $y = x$. For certain functions we can locate a fixed point by starting with an initial approximation x_0 and then computing the next approximation as $x_1 = g(x_0)$. We then repeat the process until it converges to a point, which will be a fixed point, or until we see that the process diverges. It can be shown that the process will converge on an interval in which the function intersects the line $y = x$ and the first derivative satisfies $|g'(x)| < 1$. The process can be illustrated as shown in Figure 5.105 for $g(x) = 5^{-x}$ starting with $x_0 = 0.3$. The fixed point is approximately $a = 0.469622$. Notice that the points shown are $(x_0, 0)$, $(x_0, g(x_0))$, $(g(x_0), g(x_0))$, $(g(x_0), g(g(x_0)))$, . . .

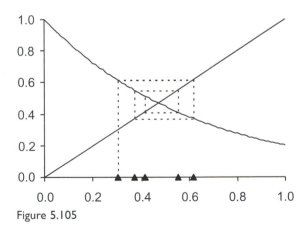

Figure 5.105

Design a spreadsheet model for the fixed-point algorithm. Also, create a graph similar to the one shown in Figure 5.105. Implement the model using the following functions. Notice that the algorithm is not always able to locate all of the zeroes.

    a. $g(x) = \cos x$ **b.** $g(x) = x^2$ **c.** $g(x) = e^{-x}$ **d.** $g(x) = \dfrac{1}{x}$

5. We often can find a zero of a function $y = f(x)$ by finding a fixed point of a related function $y = g(x)$ by solving for x in the function g. We illustrate several possible functions $y = g(x)$ using the function $f(x) = x^3 - x^2 - 2$. For example, if x is a zero of the function, then $x^3 - x^2 - 2 - 0$, so that $x^3 = x^2 + 2$, $x^2 = \dfrac{x^2 + 2}{x}$, and $x = \sqrt{\dfrac{x^2 + 2}{x}}$. Thus, x is a fixed point of $g(x) = \sqrt{\dfrac{x^2 + 2}{x}}$. Show that the fixed points of each of the following functions $y = g(x)$ provide a zero of the function $f(x) - x^3 - x^2 - 2$. Then use your fixed-point spreadsheet model to observe the rate of convergence or divergence of this algorithm using each of the functions.

    a. $g(x) = \sqrt{\dfrac{x^2 + 2}{x}}$ **b.** $g(x) = \sqrt{\dfrac{2}{x - 1}}$ **c.** $g(x) = \sqrt[3]{x^2 + 2}$

    d. $g(x) = \sqrt{x^3 - 2}$ **e.** $g(x) = \dfrac{x^2 + 2}{x^2}$

6. Find the zeroes of the following functions $y = f(x)$ by first determining an appropriate function $y = g(x)$ for the fixed-point algorithm and then using the algorithm to locate them if possible.
    a. $f(x) = x^3 - 2x - 1$ **b.** $f(x) = x^3 + x - \tan x$
    c. $f(x) = x - e^{-x}$ **d.** $f(x) = x^2 + 10 \cos x$

7. Modify our model for the bisection algorithm to recognize that one of the midpoints in the process is in fact a zero.

5.5 The Newton and Secant Methods

In the previous section we saw that the bisection algorithm gives us a way to locate a zero of a continuous function. However, that method generally converges to the zero rather slowly. Moreover, we need to be able to find x-values that bracket the zero before we begin. Often we can improve upon that method when we want to find a zero of a function $y = f(x)$ whose derivative is known. In this setting Newton's method, which is sometimes called the Newton-Raphson method, provides us with an algorithm that often converges very rapidly. Although this particular algorithm uses a concept from calculus, at the end of the section we present a variant, called the secant method that can be used without reference to calculus.

I. Newton's Method

Newton's method has a nice geometric interpretation that can be seen in the graph in Figure 5.106. To visualize the algorithm, let x_0 be an approximation of the desired zero. We first construct the line tangent to the curve $y = f(x)$ at the point $(x_0, f(x_0))$. The line's equation $y = f(x_0) + f'(x_0)(x - x_0)$ uses the derivative of the function f. We then find the next approximation as the point x_1 at which the tangent line intersects the x-axis. Since this occurs when $y = 0$, the next approximation is obtained by solving the resulting equation $0 = f(x_0) + f'(x_0)(x_1 - x_0)$ for x_1 to obtain

$$x_1 = x_0 - f(x_0)/f'(x_0).$$

In most cases (but not all!), x_1 provides a much better approximation of the zero than x_0. We then repeat the process by finding the next approximation, x_2, and subsequent ones in the same manner.

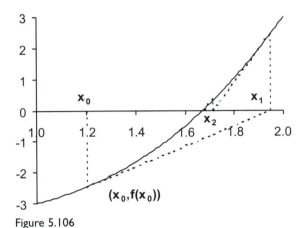

Figure 5.106

We now create a spreadsheet model for Newton's method using the function $f(x) = x^3 - x - 3$. We recall that its derivative is $f'(x) = 3x^2 - 1$. As usual, we generate an iteration counter, n, down the first column (see Figure 5.107). We then enter a value for an initial estimate of the zero, x_0, at the top of the x_n column. Here we have used $x_0 = 1.2$ as our initial estimate of the zero. Next, we compute $y = f(x) = x^3 - x - 3$ and $y' = f'(x) = 3x^2 - 1$ for the initial value, $x = x_0$.

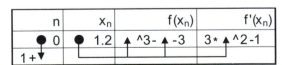

Figure 5.107

Now we enter the key step. The new approximation, x_1, is formed from the last one by subtracting the quotient $f(x_0)/f'(x_0)$ from the previous approximation, x_0 (see Figure 5.108).

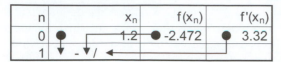

Figure 5.108

We now complete the process by using a fill down or copy command (Figure 5.109). We can copy the expressions until we obtain the accuracy that we desire, or until divergence is indicated.

n	x_n	$f(x_n)$	$f'(x_n)$
0	1.2	-2.472	3.32
1	1.9446	2.4086	10.344
2	1.7117	0.3037	7.7901
3	1.6727	0.0077	7.3943

Figure 5.109

The display in Figure 5.110 indicates the output that is produced from the model. Here our model gives us the desired zero to within at least 0.000001 by Step 6.

n	x_n	$f(x_n)$	$f'(x_n)$
0	1.200000	-2.472	3.32
1	1.944578	2.408621	10.34415
2	1.711730	0.303671	7.790057
3	1.672748	0.007744	7.394257
4	1.671701	5.5E-06	7.383749
5	1.671700	2.79E-12	7.383741
6	1.671700	0	7.383741

Figure 5.110

Newton's method generally produces rapid convergence to the expected zero. However, our spreadsheet model also gives us an excellent means to illustrate some of the unexpected results that Newton's method may produce. For example, the algorithm may diverge and miss the zero, as shown in Figure 5.111, where

$$f(x) = \arctan(x), \quad f'(x) = \frac{1}{1 + x^2}, \quad \text{and} \quad x_0 = 1.395.$$

In addition, the method may result in an infinite cycle as in Figure 5.112 where

$$f(x) = -(1 - x)^{\frac{1}{2}}, \ x \geq 1, \ f(x) = (x - 1)^{\frac{1}{2}}, \ x < 1, \text{ and } x_0 = 1.5.$$

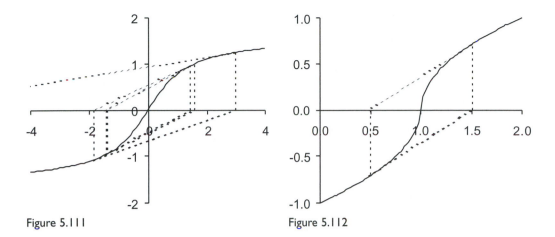

Figure 5.111   Figure 5.112

Newton's method also can be quite sensitive to the initial approximation, x_0, especially when x_0 is near a point where the derivative is 0. The examples in Figure 5.113 and Figure 5.114 show the results of using Newton's method for $f(x) = \cos(x)$ and $f'(x) = -\sin(x)$. Starting with $x_0 = 0.39$, the algorithm converges to $-3\pi/2$, while for $x_0 = 0.40$, which lies only 0.01 unit to the right of the previous value, it converges to $3\pi/2$. Moreover, if $x_0 = 0.41$, then the nearest zero, $\pi/2$, is obtained.

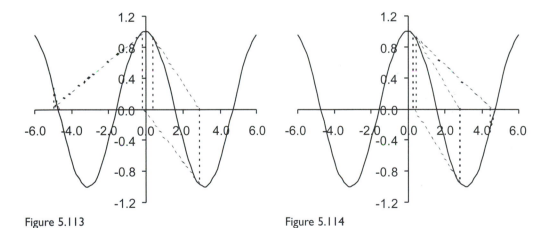

Figure 5.113   Figure 5.114

In summarizing our look at Newton's method, it should be observed that we have used the recurrence relation:

$$x_{n+1} = x_n - f(x_n)/f'(x_n).$$

Construction Summary: Newton's Method

1. Generate a counter column.
2. Enter an initial estimate of the zero, x_0.
3. Enter the formula for $f(x_0)$.
4. Enter the formula for the derivative, $f'(x_0)$.
5. Compute the next x, x_1, as $x_0 - f(x_0)/f'(x_0)$.
6. Copy as indicated.

	n		x_n		$f(x_n)$		$f'(x_n)$
1	0	2	1.2000	3	-2.4720	4	3.3200
	1	5	1.9446		2.4086		10.3442
	2		1.7117		0.3037		7.7901
	3		1.6727		0.0077		7.3943
	4		1.6717		0.0000		7.3837
	5		1.6717		0.0000		7.3837
	6		1.6717		0.0000		7.3837
	7		1.6717		0.0000		7.3837
	8		1.6717		0.0000		7.3837

2. The Secant Method

One difficulty with Newton's method is that to use it we need to know the derivative of a function. A very similar algorithm, the secant method, avoids this difficulty (see Figure 5.115). With this algorithm, instead of starting with a single estimate, x_0, and using the derivative at each point as we do with Newton's method, we start with two estimates, x_0 and x_1, and then use the slope of the line determined by the previous two points (that is, a secant line) instead of using the derivative (that is, a tangent line).

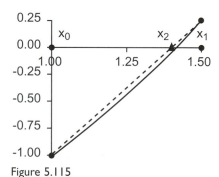

Figure 5.115

We next create a model for the secant method using the function $f(x) = x^2 - 2$. We again use the first column as an iteration counter (see Figure 5.116). In the top two cells of the second column we enter two different estimates of the zero, here 1.0 and 1.5. We then enter the formula of $f(x)$ for the first value and copy it down for the second value.

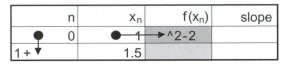

Figure 5.116

Next we compute the slope of the line segment for these two points (Figure 5.117).

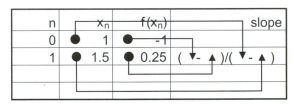

Figure 5.117

Now we use this slope to compute the next value of x just as we did in Newton's method, as $x_{new} = x_{last} - f(x_{last})/slope$. Here x_{last} is the most recent approximation and *slope* is the slope of the line determined by the last two approximations (Figure 5.118).

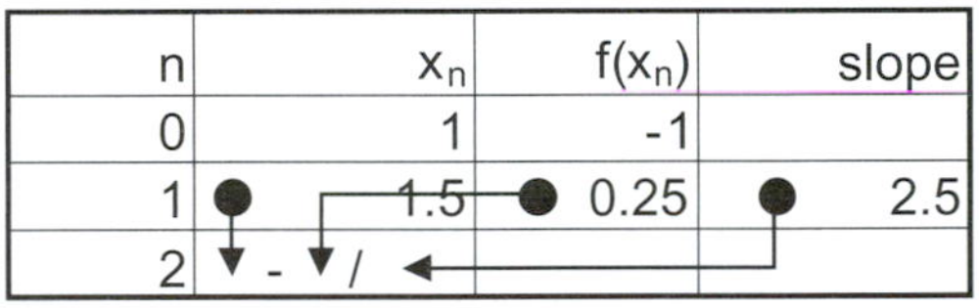

n	x_n	$f(x_n)$	slope
0	1	-1	
1	1.5	0.25	2.5
2	- /		

Figure 5.118

We then copy the expression in each column down to complete the model (Figure 5.119).

n	x_n	$f(x_n)$	slope
0	1	-1	
1	1.5	0.25	2.5
2	1.4	-0.04	2.9
3	1.413793	-0.0012	2.813793
4	1.414216	6E-06	2.828009

Figure 5.119

The output of our model is shown in Figure 5.120. Notice that we have found the expected zero, $\sqrt{2}$. Also note that as we near the zero the x-values of successive approximations become essentially equal, generating a division by zero error when we try to find the slope of the line generated by two such points.

n	x_n	$f(x_n)$	slope
0	1	-1	
1	1.5	0.25	2.5
2	1.4	-0.04	2.9
3	1.4137931	-0.001189	2.8137931
4	1.4142157	6.007E-06	2.8280088
5	1.4142136	-8.93E-10	2.8284292
6	1.4142136	0	2.8284302
7	1.4142136	0	#DIV/0!

Figure 5.120

While the secant method also contains the same drawback regarding convergence as mentioned earlier for Newton's method, it does provide us a nice method for locating zeroes without the need to employ calculus. Its recurrence relation is given by

$$x_{n+1} = x_n - \frac{f(x_n)}{\left[\dfrac{f(x_n) - f(x_{n-1})}{x_n - x_{n-1}}\right]}, n \geqslant 1$$

Construction Summary: Secant Method

1. Generate a counter column.
2. Enter two initial estimates of the zero, x_0 and x_1.
3. Enter $f(x_0)$ and copy down.
4. Enter the formula for the slope of the secant line as $(f(x_1) - f(x_0))/(x_1 - x_0)$.
5. Compute x_2 as $x_1 - f(x_1)/\text{slope}(x_0, x_1)$.
6. Copy as indicated.

n	x_n	$f(x_n)$	slope
0	1.0000	-1.0000	
1	1.5000	2.5000	2.5000
2	1.4000	-0.0400	2.9000
3	1.4138	-0.0012	2.8138
4	1.4142	0.0000	2.8280
5	1.4142	0.0000	2.8284
6	1.4142	0.0000	2.8284
7	1.4142	0.0000	#DIV/0!
8	#DIV/0!	#DIV/0!	#DIV/0!

Exercises

1. Use the secant method to approximate the zeroes of the following functions.
   a. $f(x) = x^2 - 2$
   b. $f(x) = \cos x$
   c. $f(x) = x^3 - x - 1$
   d. $f(x) = x^3 + \cos x$
   e. $f(x) = x - 1 - 0.5 \sin x$
   f. $f(x) = x^2 - \ln x - 2$

2. [Calculus] Use Newton's method to approximate the zeroes of the functions of Exercise 1.

3. [*]Design a spreadsheet model for another algorithm, the method of *false position*, to approximate the zeroes of a continuous function. As illustrated in Figure 5.121, we first locate two real numbers a_1 and b_1 for which the function has opposite signs, so that the numbers bracket the sought for zero. That is, the zero lies in the interval $[a_1, b_1]$. We then connect the points $[a_1, f(a_1)]$ and $[b_1, f(b_1)]$ with a line segment and determine the point p_1 where the line intersects the x-axis. We then use the techniques of the bisection algorithm to select as the next subinterval $[a_2, b_2]$ either $[a_1, p_1]$ or $[p_1, b_1]$, depending upon which contains the zero. We then repeat the process until the desired level of accuracy is reached. The second iteration is shown in Figure 5.122.

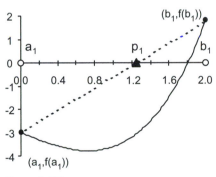

Figure 5.121   Figure 5.122

4. [*] Use the method of false position to approximate the zeroes of the functions in Exercise 1.

5. Implement the following elementary method for approximating the square root of a positive real number, r. First choose an initial estimate, a, of the square root. If a is a square root of r, then $r/a = a$. Otherwise, compute the average of a and r/a as the next approximation of the square root. Continue this process until the desired accuracy is obtained. Try to express this method as a special case of Newton's method.

6. Try to find similar algorithms to compute the cube and nth roots of a positive real number r and implement it into a spreadsheet model.

7. The technique of synthetic division is sometimes used in finding the zeroes of a polynomial. Create a spreadsheet model for this process. Test the real numbers 2 and -3

as possible zeroes of the polynomial $p(x) = x^3 + 3x^2 - 4x - 12$. Locate all of the zeroes of the function. The layout of the synthetic division process is presented in Figure 5.123 and Figure 5.124, here testing to find if 2 is a zero of the function. To carry out the algorithm, we first copy down the leading coefficient, multiply it by 2, write the product beneath the second coefficient, and add. We then repeat the process. The number under the constant coefficient is $f(c)$. In this case $f(2) = 0$, so that 2 is a zero of the polynomial function. In addition, the quotient that results when $f(x)$ is divided by $x - 2$ is $x^2 + 5x + 6$.

```
2 | 1  3  -4  -12        2 | 1   3   -4  -12
                                 2   10   12
  _____           _____
                             1   5    6    0
```

Figure 5.123   Figure 5.124

8. Use your synthetic division model to determine the zeroes of the following polynomials.
   a. $f(x) = x^3 - 4x^2 - 4x + 16$ b. $f(x) = x^4 - 8x^3 + 17x^2 - 16x + 30$
   c. $f(x) = x^3 - 9x^2 + 4x + 15$ d. $f(x) = 8x^4 - 54x^3 + 135x^2 - 148x + 40$

5.6 Fitting a Curve to Data

In virtually every realm of human activity, whether it be commercial, academic, medicine, or sports, people gather, process, and interpret vast amounts of data. In this process it is often desirable to be able to construct a function that describes the data under consideration and then to use this function in applications. In this section, we look at one such technique, polynomial interpolation, in some detail and then briefly consider another possible approach.

1. Polynomial Interpolation

Suppose we are driving down a highway, and at 1:45 P.M. we notice that we are at Milepost 183. Later, at 2:05 P.M., we are at Milepost 207. From these two observations, can we tell where we were at 2:00 P.M.? Of course we cannot do this from only the information that is given. However, we certainly can make a reasonable guess. For if we assume that we have traveled at a constant velocity, then we could construct a straight-line segment between these two data points in a graph. In doing this, for the sake of simplicity we will measure time in minutes and measure the time from noon. In this example we measure distance in miles, although it is also easy to adjust the example for kilometers.

Under our assumption, the graph of our travel would be that given in the graph in Figure 5.125. We make this assumption, even though we may have been driving in a more variable fashion, driving increasingly faster at first and then slowing down just before 2:05, so that our actual function is given by the curve in Figure 5.126. In this case, the line only approximates our true driving record.

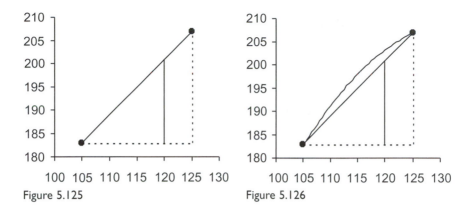

Figure 5.125   Figure 5.126

The traditional way of estimating our distance at 2:00 P.M. is called linear interpolation. Since 2:00 is three-fourths of the time under consideration (15 minutes out of 20), we estimate that we will have gone three-fourths of the total distance traveled during that time, $\frac{3}{4} \times 24$ miles, or 18 miles. Thus, we estimate that at 2:00 P.M. we were at Milepost $183 + 18$, or 201.

Looking at this from another viewpoint, we assume we know the nature of our driving (constant velocity) and then find a function, f, that has this property (a linear function) and also passes through the known data. There are a variety of ways in which we can come up with the equation. If we determine the velocity as the slope of the line, or $24/20 = 1.2$ miles/minute, then the equation of the line is $f(x) = 183 + 1.2(x - 105)$. We could then use that equation to approximate our location at 120 minutes (for 2:00 P.M.) as $f(x) = 183 + 1.2(120 - 105) = 201$.

Many times linear interpolation is quite satisfactory, especially between relatively short periods, because the changes between two points in naturally occurring, continuous phenomena is "essentially" linear in the short run. However, other times this is definitely not the case.

For example, suppose that a ball is thrown from a cliff and we would like to determine an equation for its height at any given time. We assume that we are somehow able to measure the height precisely at two times. Unfortunately, we do not know what the initial velocity of the ball was or the height from which it was thrown. We also do not remember the gravitational constant that we vaguely recall having been told in the distant past. But we do remember that the height of such an object can be described by a parabola and that a function for the height will have the form $f(t) = at^2 + bt + c$. This is sometimes called a second-degree polynomial function.

Can we determine its height at any time from knowing its height at two different times? Unfortunately, the answer is "No," for unlike the line that we used for linear interpolation where two points were enough to determine it, many parabolas (in fact, an infinite number of them) pass through two given points. Thus, if we know the data given in Figure 5.127 from our measurements then unfortunately we can find many possible fits, some of which are most unlikely. Three possibilities are shown in Figure 5.128. Further, it would not make sense to use linear interpolation here, as shown in Figure 5.129, because this would assume that the ball always continued to go down (or up) at a constant velocity!

time (sec.)	height (feet)
1	116
2	100

Figure 5.127

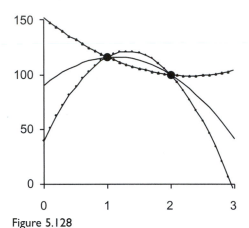

Figure 5.128

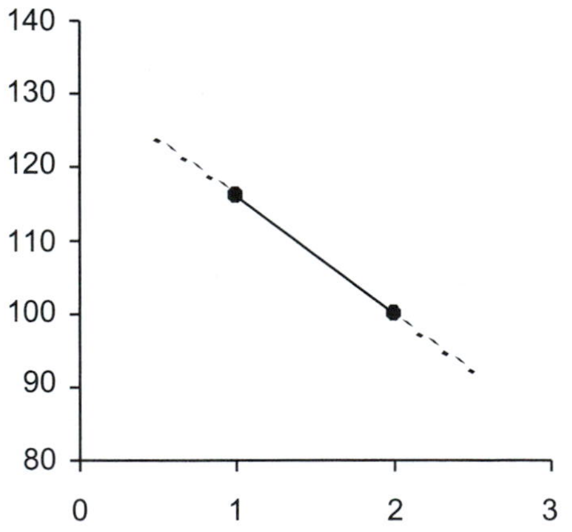

Figure 5.129

time (sec.)	height (feet)
1	116
2	100
3	52

Figure 5.130

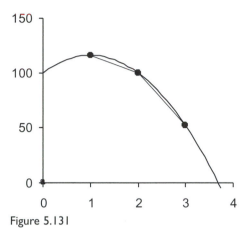

Figure 5.131

However, if we know the height of the ball at three different times, then the situation is quite different. For in this case it can be shown mathematically that there is one, and only one, parabola that fits the given data. With the three data points given in Figure 5.130, we obtain the curve in Figure 5.131, and from it and its equation we can find the remaining details of the initial velocity, the initial height, the time until it hits the ground, and so on. The parabola that fits our given data is shown in Figure 5.131.

The next question that we might consider is, How do we determine the equation? Actually, there are many ways that lend themselves to a spreadsheet solution. We present two of them here. In both of these, we illustrate the techniques through the use of four points, to which we can fit a cubic equation of the form

$$f(x) = a_0 + a_1x + a_2x^2 + a_3x^3.$$

However, the methods work just as well with any number of points.

Our first model uses the spreadsheet's Solver. To start off, at the top of our model we initially enter 0s (or a better approximation if we have one) for the values of the coefficients (a_0, a_1, a_2, a_3) that we seek to discover (Figure 5.132). Then in the first two columns we enter the x- and y-coordinates of the four points through which we

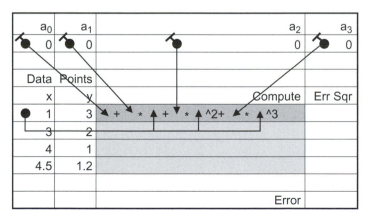

Figure 5.132

want our curve to pass. Next, we enter the formula for the cubic polynomial, using the yet-to-be-determined coefficients as absolute references, and copy it down the third column.

Next we need to find a way of forcing all of the computed values of the function to be equal to the y-coordinates of the given data points. We use the Solver to do this. In the fourth column we find the square of the differences between the actual and computed y-values. Ultimately, we want each of these cells to be 0. The square of the differences ensures that each of the results is non-negative. Because of that fact, when we add all of the error terms the sum resulting cannot become zero by having negative terms cancel out positive ones. The only way it can be zero is if all terms are zero. After we enter the first expression, we copy it down and then sum up these error terms (Figure 5.133).

a_0	a_1	a_2	a_3
0	0	0	0
Data Points			
x	y	Comp	Err Sqr
1	3	0	(-)^2
3	2	0	
4	1	0	
4.5	1.2	0	
		Error	SUM( )

Figure 5.133

We now use the Solver to set the sum to zero by varying the coefficients. We observe that the only way the sum can become 0 is for all of the individual squared errors to become zero, or if the data points and the computed points are the same! Initially we see the display in Figure 5.134. We then issue the Solver command and obtain the display in Figure 5.135. Notice that the errors are not quite zero but have been made sufficiently close to 0. The appropriate coefficients then appear at the top.

a_0	a_1	a_2	a_3
0	0	0	0
Data Points			
x	y	Comp	Err Sqr
1	3	0	9
3	2	0	4
4	1	0	1
4.5	1.2	0	1.44
		Error	15.44

Figure 5.134

a_0	a_1	a_2	a_3		
-0.805	6.1905	-2.702	0.3167		

Data Points				Curve	
x	y	Comp	Err Sqr	x	y
1	3	3.0001	2E-08	0	-0.805
3	2	2.0011	1E-06	0.1	-0.213
4	1	0.9976	6E-06	0.2	0.3273
4.5	1.2	1.2013	2E-06	0.3	0.8173
				0.4	1.259
		Error =	9E-06	0.5	1.6542

Figure 5.135

In the section of our model shown in Figure 5.135, we use the coefficients to plot the (x,y)-values of the curve in an xy-graph. Then we drag the block of the data values into the graph as a new series, plotting only markers but not the connecting lines. To obtain the solution using the Solver, in the dialog box that appears after we issue the Solver command we opt to minimize the value of error cell by changing the values of the coefficients. Details of this process are discussed in the Appendix.

The initial data points are show in Figure 5.136, with the interpolating polynomial in Figure 5.137. Note that *interpolation* means "reading between the points." In general, we can use our polynomial to approximate the values in the range $1 \leq x \leq 4.5$, but it is unwise to use the formula to estimate values outside of this range (this is called extrapolation) because the curve often does not continue in a manner that is suggested by the data. This is certainly the situation in this example. Exceptions to this rule can be made if we know more about the function, as in the case of the thrown object in the earlier example where we know the function's graph must be a parabola.

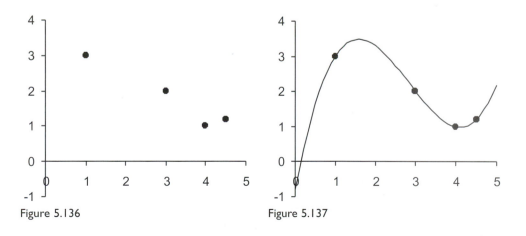

Figure 5.136

Figure 5.137

Construction Summary: Fitting a Curve to Data

1. Enter initial estimates for the coefficients.
2. Enter coordinates of data points.
3. Compute function value from the a_i for the given x, copy.
4. Compute the square of difference between the two columns to the left, copy.
5. Compute the sum of the column.
6. Enter initial x-value for curve.
7. Add an x-step to the previous value of x, copy.
8. Compute the y-values from the a_i for the corresponding x-values in the column to the left, copy.

a_0	a_1	a_2	a_3		
0	0	0	0		

Data Points				Curve	
x	y	Comp	Err Sqr	x	y
1	3	3	0	4	9	6	0	8	0
3	2	0	4	7	0.1	0
4	1	0	1	0.2	0
4.5	1.2	0	1.44	0.3	0
				0.4	0
2		Error =	5	15.4	0.5	0

Solver: Use the solver to minimize the value of the error cell by changing the coefficients a_i.

2. A Matrix Approach to Interpolation

One of the drawbacks with our solver approach is that if we change any of the data, then we must reuse the solver to create the resulting interpolation polynomial. This is because the solver solution is not automatically updated. In the study of numerical analysis, a number of interesting computational techniques for interpolation can be implemented on a spreadsheet. However, here we use a simpler approach that uses matrices. Suppose that we want to find a cubic polynomial, $f(x) = a_0 + a_1x + a_2x^2 + a_3x^3$, that fits the four points given in Figure 5.138.

x	y
1.0	3.0
2.0	4.0
3.5	2.0
4.5	3.0

Figure 5.138

If we substitute into the equation, then we obtain a system of four linear equations in four unknowns:

$$a_0 + 1.0a_1 + 1.0^2a_2 + 1.0^3a_3 = 3.0$$
$$a_0 + 2.0a_1 + 2.0^2a_2 + 2.0^3a_3 = 4.0$$
$$a_0 + 3.5a_1 + 3.5^2{+}a_2 + 3.5^3a_3 = 2.0$$
$$a_0 + 4.5a_1 + 4.5^2{+}a_2 + 4.5^3a_3 = 3.0$$

If we expand these in matrix form, then we have the matrix equation $CA = B$, whose solution is $A = C^{-1}B$, where the matrices are given by

$$C = \begin{bmatrix} 1.0 & 1.0 & 1.0^2 & 1.0^3 \\ 1.0 & 2.0 & 2.0^2 & 2.0^3 \\ 1.0 & 3.5 & 3.5^2 & 3.5^3 \\ 1.0 & 4.5 & 4.5^2 & 4.5^3 \end{bmatrix}, \quad A = \begin{bmatrix} a_0 \\ a_1 \\ a_2 \\ a_3 \end{bmatrix}, \quad B = \begin{bmatrix} 3.0 \\ 4.0 \\ 2.0 \\ 3.0 \end{bmatrix}.$$

We create our model by entering the data points and the equations to generate the values of the coefficient matrix (Figure 5.139). In each row we first enter the number 1 in the first column and then multiply this value by the x-coordinate in the next column, treating the reference as absolute for the column component only. We then copy that expression across to complete the row, and copy the row to complete the matrix (Figure 5.140).

Data Points		Matrix Coefficients			
x	y	1	x	x^2	x^3
1	3				
2	4				
3.5	2				
4.5	3				

Figure 5.139

Data Points		Matrix Coefficients			
x	y	1	x	x^2	x^3
1	3	1	1	1	1
2	4	1	2	4	8
3.5	2	1	3.5	12.3	42.9
4.5	3	1	4.5	20.3	91.1

Figure 5.140

The resulting output is shown in Figure 5.141.

Data Points		Matrix Coefficients			
x	y	1	x	x^2	x^3
1	3	1	1	1	1
2	4	1	2	4	8
3.5	2	1	3.5	12.25	42.88
4.5	3	1	4.5	20.25	91.13

Figure 5.141

Next we generate the polynomial coefficients as previously described using the spreadsheet's matrix multiplication and inverse in a single statement to compute $A = C^{-1}B$ (see Figure 5.142.) See the Appendix for details of matrix operations. The vector A gives us the coefficients of the interpolation polynomial.

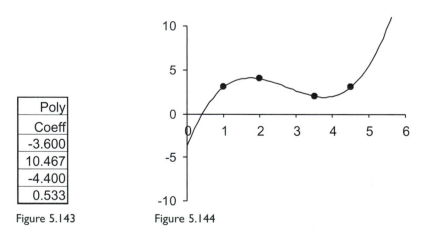

Data Points		Matrix Coefficients				Poly
x	y	1	x	x^2	x^3	Coeff
1	3	1	1	1	1	MMULT(MINVERSE(▲), ▲)
2	4	1	2	4	8	
3.5	2	1	3.5	12.25	42.88	
4.5	3	1	4.5	20.25	91.13	

Figure 5.142

The resulting coefficients are shown in Figure 5.143.

We can now plot the graph of the resulting function together with the interpolating points using the same technique as presented in the previous example. The resulting graph is shown in Figure 5.144. This time our model and graph are updated automatically if we alter one of the data points.

Poly
Coeff
-3.600
10.467
-4.400
0.533

Figure 5.143

Figure 5.144

As discussed in the Appendix, in *Excel* it is possible to vary one of the data points directly in the graph. To do this, we click on one of the data points with the mouse and then click again to select the point. At that time we hold down the left mouse button and drag the point either horizontally or vertically.

3. A Glance at the Least-Squares Approximation

If we look at the data in the preceding example before we fit an interpolation polynomial to it, it looks as if the data well may come from a linear relationship between the x- and y-values. However, a straight line cannot pass through all four data points. As a result, another approach is often used in the field of statistics. Here we try to find that straight line that best fits the data. We define that line as the one that minimizes the sum of the squares of the differences in the y-coordinates of the data set and the y-coordinates of the corresponding points on the line. Thus, if we let d_i represent this distance for a typical point, then we want to minimize $d_1^2 + d_2^2 + d_3^2 + d_4^2$ as illustrated in Figure 5.145.

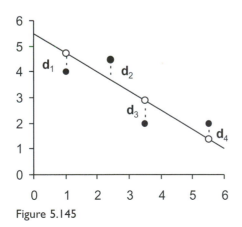

Figure 5.145

Our model uses the Solver (see Appendix), and our approach is similar to the one we used for interpolation (see Figure 5.146). We first enter the data points in the first two columns and leave room for the coefficients of the line $y = a_0 + a_1x$ at the top of the layout. Initially, the values of these coefficients are set to 0 or to another appropriate value. We use these coefficients to compute $f(x)$ for the data values in the third column. Finally, we square the differences between the given data and the line's computed value (that is, the errors in the approximation of the data by the line at the data points) and find their sum (Figure 5.147).

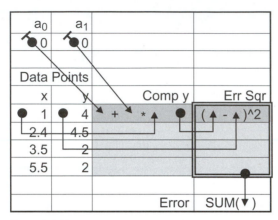

Figure 5.146

a_0	a_1		
0	0		
Data Points			
x	y	Comp y	Err Sqr
1	4	0	16
2.4	4.5	0	20.25
3.5	2	0	4
5.5	2	0	4
		Error	44.25

Figure 5.147

Figure 5.148 shows what the screen view will look like initially.

	A	B	C	D
1	a_0	a_1		
2	0	0		
3				
4	Data Points			
5	x	y	Comp y	Err Sqr
6	1	4	0	16
7	2.4	4.5	0	20.25
8	3.5	2	0	4
9	5.5	2	0	4
10				
11			Error	44.25

Figure 5.148

We now select the menu options Tools, Solver. In the ensuing solver dialog box we choose the option to minimize the error term in the sum cell by changing the values of a_1 and a_2 (Figure 5.149).

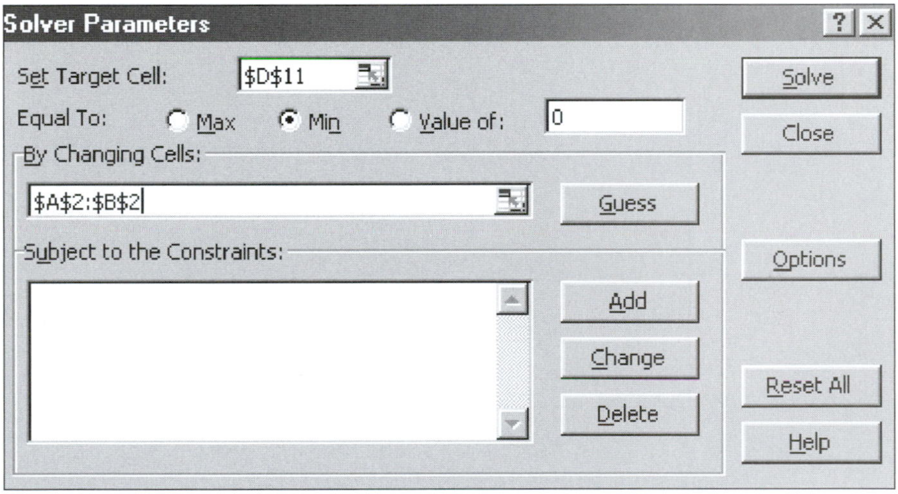

Figure 5.149

The solver now varies the values of a_0 and a_1 so as to minimize the error. The resulting output is shown in Figure 5.150.

a_0	a_1		
4.8297	-0.55		
Data Points			
x	y	Comp y	Err Sqr
1	4	4.2798	0.0783
2.4	4.5	3.5099	0.9802
3.5	2	2.905	0.8191
5.5	2	1.8052	0.0379
		Error	1.9155

Figure 5.150

The line that is produced from the coefficients is called the *least-squares regression line*. Thus, for our data, the equation of the least-squares line is $f(x) = 4.8297 - 0.55x$. To approximate the value of a point with $x = 2$, we would find $f(2) \approx 3.7297$. Again, it is generally unwise to use this function to predict values outside of the data range of $1 \leqslant x \leqslant 5.5$.

We can use our model to create an xy-graph from the first three columns, using the first column for the x-values and the others for two y-series, plotting markers only for the first series and lines for the second series. The graph that is produced is shown in Figure 5.151.

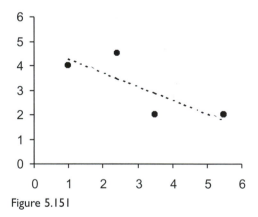

Figure 5.151

In elementary statistics classes and texts, you will usually see a derivation of the equation for the least-squares regression line, $y = ax + b$. If there are n data points (x, y), then

$$a = \frac{n \sum xy - \sum x \sum y}{n \sum x^2 - \left(\sum x \right)^2}, \quad b = \bar{y} - a\bar{x},$$

where $\bar{x}$ and $\bar{y}$ are the means of the data values

$$\bar{x} = \frac{\sum x}{n} \quad \text{and} \quad \bar{y} = \frac{\sum y}{n}.$$

Although we do not present the details here, implementing these formulas is a good application for a spreadsheet. A typical layout and the output for the computations are shown in Figure 5.152. Readers are invited to supply the formulas using the preceding expressions.

		mean x	3.1	slope a	-0.5499
		mean y	3.125	intcpt b	4.82971
Count	x	y	x^2	xy	
1	1	4	1	4	
2	2.4	4.5	5.76	10.8	
3	3.5	2	12.25	7	
4	5.5	2	30.25	11	
Sums	12.4	12.5	49.26	32.8	

Figure 5.152

We also can create a column of the points on the regression line for the x-values in the data by using the equation $y = ax + b$ (see Figure 5.153).

		mean x	3.1	slope a	-0.550	
		mean y	3.125	intcpt b	4.830	
Count	x	y	x^2	xy	y(reg)
1	1	4	1	4	
2	2.4	4.5	5.76	10.8	
3	3.5	2	12.25	7	
4	5.5	2	30.25	11	
Sums	12.4	12.5	49.26	32.8	

Figure 5.153

If the x-values are listed in order, then it is also possible to illustrate the regression line in a graph. We first create the graph by using the x- and y-values and plotting only the markers (see Figure 5.154).

		mean x	3.1	slope a	-0.5499	
		mean y	3.125	intcpt b	4.82971	
Count	x	y	x^2	xy	y(reg)	
1	1	4	1	4	4.280	
2	2.4	4.5	5.76	10.8	3.510	
3	3.5	2	12.25	7	2.905	
4	5.5	2	30.25	11	1.805	
Sums	12.4	12.5	49.26	32.8		

Figure 5.154

We then drag the column of computed y-values of the regression line into the graph, plotting both lines and markers as shown in Figure 5.155.

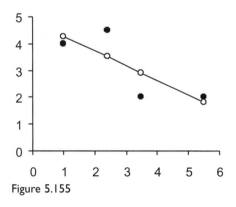

Figure 5.155

Finally, note that spreadsheets provide us with two other ways to create regression lines. First, *Excel* has built-in functions that compute the slope and the y-intercept of the least-squares regression line, as SLOPE(y-array,x-array) and INTERCEPT(y-array,x-array).

Second, suppose that we simply create an xy-graph to plot the data points from the previous example. We can obtain the regression line by right-clicking on the series and choosing the option Add Trendline from the menu box that is generated (Figure 5.156).

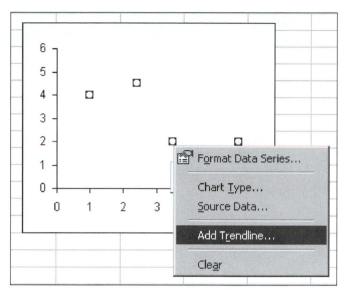

Figure 5.156

From the dialog box that appears, we choose the option Linear as the type of trend-line to create. This generates the graph that is shown in Figure 5.157. We can then right-click on the line and choose the option Format Trendline. From the ensuing dia-log box, we can choose several options to enhance our graph, including one to display the equation of the line within the graph, as illustrated in Figure 5.158. We should note, however, that the desired equation is not generated in the spreadsheet itself.

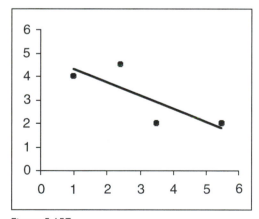

Figure 5.157

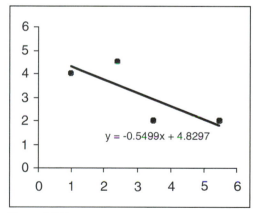

Figure 5.158

Construction Summary: Least-Squares Regression

1. Create a counter column.
2. Enter the (x,y) data points.
3. Compute the square of the x-coordinates.
4. Compute the product xy.
5. Sum the columns.
6. Compute means of x- and y-values.
7. Compute the slope and the y-intercept of the least-squares line.
8. Compute the y-coordinates on the least-squares line for the given x.

	mean x	[6]	3.1	slope a	[7]	-0.550	
	mean y		3.13	intcp b		4.830	
			[2]				
Cnt	x		y	x^2		xy	y(reg)
[1] 1	1		4	[3] 1.00 [4]		4.0	[8] 4.280
2	2.4		4.5	5.76		10.8	3.510
3	3.5		2	12.25		7.0	2.905
4	5.5		2	30.25		11.0	1.805
Sum	[5] 12.4		12.5	49.26		32.8	

Exercises

1. Use the techniques in this section to find the polynomials that fit the following data sets.

   a. (1,2), (3,6), (4,3), (5,5)
   b. (0,3), (2,3), (4,2.5), (5,3), (6,3)
   c. (0,0), (1,1), (2,3), (3,6), (4,10)
   d. (0,0), (1,1), (2,2), (3,3), (4,2), (5,1), (6,0)

2. Create a spreadsheet model to compute the coefficients of an interpolation polynomial. Use these to compute the value of $f(x)$ for one value of x. Then use a data table to compute the values of $f(x)$ for a range of values of x.

3. Use the least-squares algorithm of this section with the data values for a function $f(x)$ listed in Figure 5.159 to predict other values of $f(x)$ for the indicated values of x.

| x | 0.5 | 1.3 | 3.2 | 3.2 | 4.1 | 5.7 |
| y | 4.2 | 6.1 | 7.3 | 9.1 | 9.8 | 11.5 |

Figure 5.159

   a. 1.0   b. 1.3   c. 2.0   d. 3.3   e. 5.0

4. We can obtain practice at visualizing fitting a line to data by plotting points using an xy-graph as shown in Figure 5.160. We then use a straightedge to estimate the best-fitting line and then estimate the slope and y-intercept of the line that we

have chosen. We can compare our answer to one determined using methods that are presented in this section. Create a spreadsheet model to do this using the following points: (0.5,4.4), (1.5,3.7), (2.0,2.4), (3.3,2.1), (4.7,0.5).

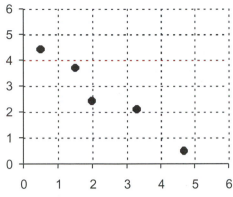

Figure 5.160

5. In *Excel* we can find the least-squares line by right-clicking on the data and choosing the options Add Trendline, Linear. We display its equation choosing the Series option of the same command and selecting Options and ticking the box to display the equation. The result is shown in Figure 5.161. Use this approach to modify your model from Exercise 4.

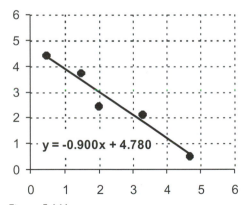

y = -0.900x + 4.780

Figure 5.161

6. Your spreadsheet often supplies other types of curves to fit to data. In *Excel* this is found under the same Trendline command. Experiment with these types using the following data sets.

   a. (0.5,0.6), (1.5,0.7), (2.0,0.9), (3.1,1.2), (3.6,1.6)
   b. (0.5,1.5), (1.5,1.3), (3.0,1.1), (4.3,0.4), (5.7,1.0)
   c. (1.0,2.0), (1.5,3.4), (2.0,3.7), (3.1,4.0), (4.8,5.5)

    d. (0.5,0.4), (1.5,0.5), (2.0,0.7), (3.1,1.0), (3.6,1.3)
    e. (0.5,3.0), (1.5,2.1), (2.0,2.0), (3.1,2.5), (3.6,5.0)
    f. (1.0,0.3), (1.5,0.5), (2.0,0.9), (3.1,1.8), (3.6,2.3)
    g. (0.8,3.6), (1.4,4.5), (2.3,8.4), (3.5,21.4), (4.1,33.4)
    h. (0.5,5.4), (1.5,3.0), (3.0,2.0), (4.3,1.4), (5.7,1.3)

7. Use a least-squares approach together with the Solver command to design models to compute the best fits for a set of data using the following types of curves. The letters should be treated as parameters in your models. You can vary these manually to experiment and then use the solver to find the best fit. Experiment with the data sets in Exercise 6.

    a. Quadratic: $f(x) = a_2x^2 + a_1x + a_0$
    b. Cubic: $f(x) = a_3x^3 + a_2x^2 + a_1x + a_0$
    c. Exponential: $f(x) = ce^{ax}$
    d. Translated exponential: $f(x) = ce^{ax} + b$
    e. Power: $f(x) = ax^n$
    f. Translated power: $f(x) = ax^n + b$
    g. Reciprocal: $f(x) = \dfrac{a}{x}$
    h. Translated reciprocal: $f(x) = \dfrac{a}{x} + b$
    i. Translated logarithmic: $f(x) = a\ln(x) + b$

8. Use spreadsheet matrix operations (see Appendix) to solve the following systems of equations.

    a. $\begin{aligned} 3x_1 + x_2 &= 5 \\ 4x_1 - 2x_2 &= 8 \end{aligned}$
    b. $\begin{aligned} 2x_1 - 3x_2 + 4x_3 &= 5 \\ 3x_1 + 2x_2 + 5x_3 &= 8 \\ -x_1 - 2x_2 + 2x_3 &= 3 \end{aligned}$
    c. $\begin{aligned} 2x_1 - 3x_2 + 4x_3 - x_4 &= 5 \\ 3x_1 + 2x_2 + 5x_3 + x_4 &= 8 \\ x_1 - 2x_2 + 2x_3 + 2x_4 &= 3 \\ 5x_1 + 5x_2 - 2x_3 - 3x_4 &= 1 \end{aligned}$

5.7 The Binomial Coefficients

Binomial coefficients appear in many different places in mathematics, from the coefficients in certain polynomials to the number of ways of choosing subcommittees of a given size from a group of people of a given (larger) size and the computation of probabilities for a wide class of problems. Many readers will be familiar with the process of creating them as Pascal's triangle, which is named after the French mathematician and philosopher Blaise Pascal (1623–1662), but whose use can be traced back into antiquity in various civilizations. The method of constructing these numbers provides us with a valuable tool to use in implementing recurrence relations in mathematics.

In this section we look at the polynomials that arise as powers of $1 + x$:

$$1 + x = 1 + x$$
$$(1 + x)^2 = 1 + 2x + x^2$$
$$(1 + x)^3 = 1 + 3x + 3x^2 + x^3$$
$$(1 + x)^4 = 1 + 4x + 6x^2 + 4x^3 + x^4$$

We want to find a method to calculate the coefficients of the powers of x in these polynomials. To find these coefficients, we set up a spreadsheet (see Figure 5.162).

Degree	x^0	x^1	x^2	x^3	x^4
1	1	1	0	0	0

Figure 5.162

The second row tells us that for the polynomial of Degree 1, x^0 has coefficient 1, $x^1 = x$ has coefficient 1, and all powers of x of higher order have coefficient 0. Therefore the polynomial represented by the row labeled Degree 1 is $1 + x$.

Two polynomials are added by adding the coefficients for the same power of x for both polynomials. Now let us use a shorthand notation and let $(1,1)$ be a representation of $1 + x$ and let $(0,1,1)$ be a representation of $x + x^2$. For another example, let $(1,0,4,7)$ be a representation of $1 + 4x^2 + 7x^3$. We will call these the row representations of polynomials. In this shorthand notation, we see that multiplying a polynomial by x shifts the coefficient row to the right by one element and adds a leading 0. Thus, when $(1,0,4,7)$ is shorthand for $1 + 4x^2 + 7x^3$, $(0,1,0,4,7)$ is shorthand for $x(1 + 4x^2 + 7x^3) = (x + 4x^3 + 7x^4)$.

Multiplying by $1 + 4x^2 + 7x^3$ by $1 + x$, or computing

$$(1 + x)(1 + 4x^2 + 7x^3) = (1 + 4x^2 + 7x^3) + x(1 + 4x^2 + 7x^3)$$

translates into adding the polynomials represented by $(1,0,4,7)$ and $(0,1,0,4,7)$. Adding the coefficients at the same positions produces $(1 + 0,0 + 1,4 + 0,7 + 4,7)$, or $(1,1,4,11,7)$ to produce the product $1 + x + 4x^2 + 11x^3 + 7x^4$.

Multiplying a polynomial by $(1 + x)$ therefore is equivalent to adding the original row representation of the polynomial and the row representation of the polynomial shifted to the right by one position.

The first element of the product polynomial therefore is equal to the first element of the original polynomial. The second element (corresponding to x) is the sum of the second element of the original row representation and the second element of the shifted row representation, which is the first element of the original row representation. The third element, corresponding to x^2, is the sum of the third element of the original row representation and the second element of the original row representation, and so on. Let us translate this into a spreadsheet model to compute the powers of $1 + x$. We have just learned that the first element of the spreadsheet row for polynomial of Degree 2 is equal to the element above. In the description shown in Figure 5.163, we have also incorporated the formula for the degree of the polynomial under consideration, which is one more than the degree of the polynomial in the row above.

Degree	x^0	x^1	x^2	x^3	x^4
1	1	1	0	0	0
+1					

Figure 5.163

The second element in this row is the sum of the element above and the element above and to the left (see Figure 5.164).

Degree	x^0	x^1	x^2	x^3	x^4
1	● 1	● 1	0	0	0
2	1 ► + ▼				

Figure 5.164

The same is true for all the elements to the right of this element; therefore we can copy the formula as shown in Figure 5.165.

Degree	x^0	x^1	x^2	x^3	x^4
1	1	1	0	0	0
2	1	2	1	0	0

Figure 5.165

For all the polynomials of higher degree, the construction method is the same: add a copy of the row above and a shifted copy of the row above. As a consequence, the formulas for the coefficients can be copied down and we can compute the coefficients of the polynomial $(1 + x)^n$ for any n (see Figure 5.166).

Degree	x^0	x^1	x^2	x^3	x^4
1	1	1	0	0	0
2	1	2	1	0	0
3	1	3	3	1	0
4	1	4	6	4	1

Figure 5.166

The numbers in this table are known as the polynomial coefficients, and they are usually written as $\binom{n}{k}$. It can be proved that they also can be calculated as

$$\binom{n}{k} = \frac{n(n-1)(n-2)\cdots(n-k+1)}{1 \cdot 2 \cdot 3 \cdot \ldots \cdot k} = \frac{n!}{k!\,(n-k)!}$$

where $n!$ is defined as $n! = 1 \cdot 2 \cdot 3 \ldots n$, the product of all integers from 1 up to n. The resulting output is also called Pascal's triangle.

If $B(n, k)$ is used to give the entry in the row for Degree n and the column for the coefficient of x^k of our table, we can rewrite the construction method of our spreadsheet as

$$B(1, 0) = 1$$
$$B(1, 1) = 1$$
$$B(1, k) = 0 \qquad\qquad\qquad\text{for } k > 1$$
$$B(n, 0) = B(n - 1, 0) \qquad\quad\text{for } n > 1$$
$$B(n, k) = B(n - 1, k) + B(n - 1, k - 1) \text{ for } n > 1 \text{ and } k > 0$$

These equations are just another way of expressing the structure of our spreadsheet table for the binomial coefficients.

Construction Summary: The Binomial Coefficients

1. Generate a counter column.
2. Enter coefficients of $1 + x$.
3. Reproduce the initial coefficient, copy.
4. Compute the sum of the cell above and to the left and the cell above.
5. Copy appropriately.

Degree	x^0	x^1	x^2	x^3	x^4
1	1	1	0	0	0
2	1	2	1	0	0
3	1	3	3	1	0
4	1	4	6	4	1

Exercises

1. Modify the binomial spreadsheet to get the coefficients of $(2 + 3x)^n$.

2. Modify the binomial spreadsheet to obtain the coefficients of $(2 + 3x)^n$. Generalize your model for $(a + bx)^n$.

3. Try to modify the spreadsheet in a way to get the coefficients of $(1 + x + x^2)^n$. (Hint: Multiplying a polynomial by x^2 shifts the row representation to the right by 2.)

4. [*] Design a spreadsheet to multiply any two polynomials of degree up to 10. (Hint: the function SUMPRODUCT might be useful.)

5. [*] By looking at the structure of the spreadsheet, can you see why the sum of the coefficients of $(1 + x)^n$ is 2^n? Can you give a similar formula for the sum of the coefficients of $(1 + x + x^2)^n$?

6. Modify the binomial model to create the expansion of the expression $(p + q)^n$, where $p + q = 1$, or $q = 1 - p$, and $0 \leqslant p \leqslant 1$. The entry in Column k of Row n is then the probability of obtaining k successes in n repetitions of a binomial experiment with probability of success p for $k = 0, 1, 2, \ldots, n$.

7. Modify the binomial model so that the computations are carried out modulo 2. Then use conditional formatting to color the 1 cells red and the 0 cells white. Observe the effect that is created. You can increase the visual effect by using font colors the same as the cell color, reduce the column width and the percent of zoom, and generate many rows and columns to obtain output like that shown in Figure 5.167.

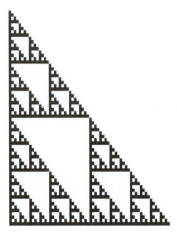

Figure 5.167

8. Suppose that a jar contains m red balls and n green balls, for a total of $N = m + n$ balls. The balls are well mixed and drawn one at a time at random without replacement. Create a *Excel* model to find the probability that after k balls are drawn exactly i of them are red. Hint: Construct a table in which the first column counts the number, k, of balls drawn and the top row counts the number, i, of red balls that are obtained in the process. In order to have i red balls at Step k, either (a) we have $i - 1$ red balls at Step $k - 1$ and draw a red ball on the kth draw, or (b) we already have i red balls on draw $k - 1$ and draw a green ball on the kth draw. As one application of this result, suppose that in Lotto $m = 6$ winning numbers (red balls) are selected from $N = 45$ possible numbers (call the remaining $n = N - m = 39$ numbers green balls) and you have selected $k = 6$ numbers. What is the probability that exactly i (for $i = 0,1,2, \ldots ,6$) of your selections are among the winning numbers?

Creative Mathematics in Everyday Settings

6.1   A Bicycle Investigation

We will study the following question: Somebody riding a bicycle makes a full turn of the pedals once per second. At what speed is this person moving?

To be able to calculate this, we need to know a few things. If the bicycle were a unicycle, one revolution of the legs would correspond to one revolution of the wheel, and therefore the distance per second would be the perimeter of the wheel. So we need to know the perimeter of a bicycle wheel. The size of bicycle wheels is usually given by the diameter in inches, and typical values are 24, 26, 27 and 28 inches.

Since we are not studying a unicycle, we need some additional knowledge. Bicycles have gears. The sprocket wheel with the pedals attached has two or three gear rims with 35 to 55 sprockets. The back wheel usually has five or six gear rims with 14 to 28 sprockets.

We will now study our problem for a bicycle with two gear rims with 42 and 52 sprockets on the pedaling gear and five gear rims with 14, 17, 20, 24, and 28 sprockets on the wheel. We will work with 28-inch wheels (see Figure 6.1). To start, we calculate the perimeter of the wheel (multiplying the diameter with PI(), the function returning the value π of PI, 3.14159) and also the number of inches per mile (12 inches to a foot, 3 feet to a yard, 1760 yards to a mile).

wheel			
diameter			
● 28			
wheel perimeter		inches	
inches		per mile	
▼ *PI()		12 * 3 * 1760	

Figure 6.1

Next, we enter a value for revolutions per second and calculate revolutions per hour from this number (Figure 6.2).

wheel			
diameter			
28			
wheel perimeter		inches	
inches		per mile	
87.96459		63360	
revolutions	per sec	● 1	
	per hour	▼ * 3600	

Figure 6.2

Now we need to take care of the gears. Dividing the number of sprockets on the pedal wheel by the number of sprockets on the back wheel yields the number of revolutions of the back wheel per full turn of the pedals (Figure 6.3).

wheel			
diameter			
28			
wheel perimeter		inches	
inches		per mile	
87.96459		63360	
revolutions	per sec	1	
	per hour	3600	
sprockets		turning	
front gear	back gear	ratio	
● 52	● 14	▲ / ▼	
52	17		

Figure 6.3

Multiplying the turning ratio by the number of revolutions per hour, then by the perimeter of the wheel, and then dividing by inches per mile finally gives us speed in miles per hour (Figure 6.4). We use absolute references to make the formula "copyable."

Copying down gives the speed for all combinations of sprocket wheels (Figure 6.5).

wheel			
diameter			
28			
wheel perimeter		inches	
inches		per mile	
87.96459		63360	
revolutions	per sec	1	
	per hour	3600	
sprockets		turning	speed
front gear	back gear	ratio	
52	14	3.714286	*▼ * ▼ /▼
52	17		

Figure 6.4

wheel			
diameter			
28			
wheel perimeter		inches	
inches		per mile	
87.96459		63360	
revolutions	per sec	1	
	per hour	3600	
sprockets		turning	speed
front gear	back gear	ratio	
52	14	3.714286	18.56396
52	17	3.058824	15.28796
52	20	2.6	12.99477
52	24	2.166667	10.82897
52	28	1.857143	9.281978
42	14	3	14.99396
42	17	2.470588	12.34797
42	20	2.1	10.49578
42	24	1.75	8.74648
42	28	1.5	7.496982

Figure 6.5

Experienced bicyclers know that one should not use a combination like a 52 sprocket wheel and a 28 sprocket back wheel because the chain then gets very skewed, resulting in mechanical problems. So the "usable" sprocket combinations are 52 with 20 or lower, and 42 with 20 or higher. This reduces the number of gears from 10 to 6, and looking at our spreadsheet we see that we do not lose too much because the unpractical combinations lie somewhere in between practical combinations.

Construction Summary: Bicycle Investigation

1. Enter wheel diameter in inches.
2. Compute wheel perimeter.
3. Compute inches per mile.
4. Enter revolutions per second.
5. Compute revolutions per hour.
6. Enter number of sprockets (front).
7. Enter number of sprockets (rear).
8. Compute ratio front/rear.
9. Compute speed.
10. Copy as indicated.

wheel diameter (in.)		1	28		
wheel perimeter (in.)		2	87.965		
inches per mile		3	63360		
revolutions	per second	4	1		
	per hour	5	3600		
sprockets			turning		
front gear	back gear		ratio		speed
6 52	7 14	8	3.714	9	18.564
52	17		3.059		15.288
52	20		2.600		12.995

6.2 Ski Racing

When watching downhill ski races on television, one sometimes can hear the reporter say something like the following: The second skier took 5/100 of a second longer than the first one. If they had started at the same time, the second one would have gone through the finish 5 feet behind the first one.

We will try to find a reasonable mathematical model for this kind of statement. A typical downhill course is about 2 miles long, and top-class skiers need about 2 minutes to finish the race. We want to find out how to express a time difference of 1/100 second as a distance between two skiers.

We start with simplification: we assume that the competitors go down the race-course with constant speed. Later we will discuss how much our results should change because this assumption is not fulfilled in reality.

Now let us assume that the winner of a race on a 2-mile-long course takes exactly 2 minutes, and the second-place finisher takes 1/100 second more. In our spreadsheet, we also calculate the distance in feet (Figure 6.6).

distance (mi)		2	
distance (ft)		* 1760 * 3	
		winner	2nd
time (sec)		120	120.01

Figure 6.6

We can calculate the speed of both skiers by dividing the distance by the respective times (Figure 6.7).

distance (mi)		2	
distance (ft)		10560	
		winner	2nd
time (sec)		120	120.01
		/	

Figure 6.7

Since we used an absolute reference to the distance, we can copy the formula to the right and get the speed for the second racer (Figure 6.8).

distance (mi)		2	
distance (ft)		10560	
		winner	2nd
time (sec)		120	120.01
		88	87.99267

Figure 6.8

To see how far the second racer had come when the winner passed the finish line, we multiply the time of the winner with the speed of the second racer (Figure 6.9).

distance (mi)		2	
distance (ft)		10560	
		winner	2nd
time (sec)		120	120.01
		88	87.99267
distance 2nd			*

Figure 6.9

Subtracting this distance from the total length of the racecourse gives us the (virtual) distance between the winner and the second racer (Figure 6.10).

distance (mi)		2	
distance (ft)		10560	
		winner	2nd
time (sec)		120	120.01
		88	87.99267
distance 2nd			10559.12
difference			-

Figure 6.10

This finishes our spreadsheet and we see that the distance between the two ski racers is slightly more than 1 foot (Figure 6.11).

distance (mi)	2	
distance (ft)	10560	
	winner	2nd
time (sec)	120	120.01
speed (ft/sec)	88	87.99267
distance 2nd		10559.12
difference		0.87993

Figure 6.11

We can now change the times of the winner and the second-place finisher and see that the distance in feet is slightly less than the time difference in hundredths of a second. Two things are worth considering further.

Since time is measured with a precision of 1/100 second, 120 seconds might be any number between 119.9951 seconds and 120.0049 seconds. Entering these two numbers gives distances of 1.4 feet and 0.4 foot respectively. So we see that with this kind of calculation it does not make any sense to give the distance with a precision of more than 0.5 foot. Some television reporters have given the distance in inches. We have just seen that this is "pretended precision."

Furthermore, the basic assumption was that the racers maintained constant speeds along the course, which is not true. Usually the finish of a race is the fastest segment, so the racers do have a higher speed than the average speed at this stage. The racers can be up to 40 percent faster in the finish, and therefore we can conclude that the distance between two racers, had they started simultaneously, would be slightly larger (when expressed in feet) than their time difference in hundredths of a second.

More precise statements are not very sensible.

Construction Summary: Ski Racing

1. Enter the race distance in miles.
2. Compute the distance in feet.
3. Enter the time of the winner and the second-place finisher.
4. Compute the speed, copy right.
5. Find the approximate distance of the second-place finisher when the winner crosses the finish line.
6. Compute the distance between the first two finishers.

			winner		2nd
distance (mi)	1	2			
distance (ft)	2	10560			
			winner		2nd
time (sec)	3	120			120.01
speed (ft/sec)	4	88			87.99267
distance (2nd)				5	10559.12
difference				6	0.87993

6.3 The Golden Ratio

The *golden ratio* arises from an attempt to find an aesthetically pleasing rectangle. If the two adjacent sides of a rectangle are almost equal, the rectangle is perceived as a square, and if the two sides are very different, it is perceived as a "stick." In both cases it is not perceived as a "typical" rectangle.

We will construct a rectangle that most people consider typical and well designed. We start with a rectangle with a long side of 1.5 and a short side of 1 and put a square on top of it (Figure 6.12).

Figure 6.12

We want the larger rectangle (the combination of the small rectangle and the square) to have the same side ratio as the small rectangle. To check if this is the case, we turn it by 90 degrees and put the smaller rectangle on top of it (Figure 6.13).

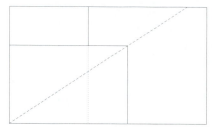

Figure 6.13

If the ratios of the sides were equal for the smaller and the larger rectangle, the extension of the diagonal of the smaller rectangle would pass through the corner of the larger rectangle. This is not true, and therefore we will try to modify the ratio of sides until we can achieve this goal.

So let us set up a spreadsheet model containing the size of the small rectangle (Figure 6.14).

	large rectangle	small rectangle
long edge	1.5	
short edge	1	

Figure 6.14

The short edge of the large rectangle is equal to the long edge of the small rectangle (Figure 6.15) and the long edge of the large rectangle is the sum of the long edge and the short edge of the small rectangle (Figure 6.16).

	small rectangle	large rectangle
long edge	●—1.5	
short edge	1	

Figure 6.15

	small rectangle	large rectangle
long edge	●—1.5 ►	+
short edge	●——1	1.5

Figure 6.16

Now we calculate the side ratio for the small rectangle, as shown in Figure 6.17. We then copy our results to get the ratio for the large rectangle (Figure 6.18).

	small rectangle	large rectangle
long edge	● 1.5	2.5
short edge	● 1	1.5
side ratio	↓ /	↓

Figure 6.17

	small rectangle	large rectangle
long edge	1.5	2.5
short edge	1	1.5
side ratio	1.5	1.6666667

Figure 6.18

We want these two ratios to be equal, so we calculate the difference (see Figure 6.19).

	small rectangle	large rectangle
long edge	1.5	2.5
short edge	1	1.5
side ratio	● 1.5	1.666667
difference		- ↓

Figure 6.19

This difference equals 0 when the two ratios are equal. This is not the case at the moment (see Figure 6.20).

	small rectangle	large rectangle
long edge	1.5	2.5
short edge	1	1.5
side ratio	1.5	1.6666667
difference		-0.166667

Figure 6.20

We can use trial and error to find a better solution. We may change the long edge of the small rectangle and try to find a value that produces a difference of 0 or almost 0. For example, using 1.6 instead of 1.5 produces a difference of –0.0025.

Modern spreadsheet programs like *Microsoft Excel* do offer us a more convenient way of finding the value that we need. They have a "goal seek" procedure to accomplish this task.

In *Excel*, the goal seek procedure is invoked by selecting the Goal Seek menu item from the Tools menu and then indicating the cell that should have a given value (in our case the cell with the difference in ratios), the target value itself (it has to be entered as a number into the dialog box, it cannot be given by a cell reference), and finally the cell that may be varied (in our case the cell with the longer edge of the small rectangle). When this is done, *Excel* does a few iterations and finds the approximate solution shown in Figure 6.21.

	small rectangle	large rectangle
long edge	1.617992	2.6179916
short edge	1	1.6179916
side ratio	1.617992	1.6180502
difference		-5.86E-05

Figure 6.21

The ratio of the side lengths is called the *golden ratio,* and we just numerically derived that its value is approximately 1.618.

We can also find this value analytically: When the short edge of the small rectangle is 1 and the long edge of the small rectangle is x, the short edge of the large rectangle is x and the long edge of the large rectangle is $x + 1$. The side ratios being equal is expressed by

$$\frac{x}{1} = \frac{x+1}{x}$$

or equivalently by

$$x^2 = x + 1.$$

This quadratic equation has two solutions, $\dfrac{-\sqrt{5}+1}{2}$ and $\dfrac{\sqrt{5}+1}{2}$. Since only

$\dfrac{\sqrt{5}+1}{2}$ is positive, that has to be the size of the long edge. Its numerical value is

approximately 1.61803. This shows that our initial approximate solution was quite good.

Using the ratio of sides that we just derived, we see that the graph containing both rectangles has the desired property: the extension of the diagonal of the smaller rectangle passes through the corner of the larger rectangle (see Figure 6.22).

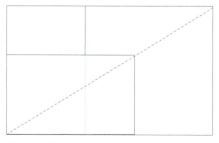

Figure 6.22

Another way of finding the golden ratio is to start with a square and put a second square on top of it, producing a 1 by 2 rectangle (Figure 6.23).

Now, add a square to the right, again producing a rectangle, this time with sides 2 by 3 (see Figure 6.24).

Add a square on top, again producing a rectangle (Figure 6.25) and now add a square to the left (Figure 6.26).

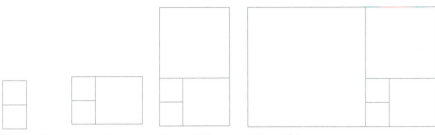

Figure 6.23   Figure 6.24   Figure 6.25   Figure 6.26

Each of the rectangles formed by the union of squares has the properties that its shorter side is equal to the longer side of the previous rectangle and its longer side is equal to the sum of the shorter side and the longer side of the previous rectangle. So the side lengths have the property that the first two lengths are 1 and all the following lengths are the sum of their two predecessors. We already know this sequence. It is the sequence of Fibonacci numbers.

We may perceive that the side ratios of the rectangles that we get are close to the golden ratio. To check this, we create a sheet with the Fibonacci numbers and calculate the ratio for each pair of consecutive numbers. First, we create the formula for the numbers (Figure 6.27).

Figure 6.27

Then, we create the formula for the ratios (Figure 6.28).

Figure 6.28

Finally, we copy the formulas down (Figure 6.29).

1	
1	
2	2
3	1.5
5	1.666667
8	1.6
13	1.625
21	1.615385
34	1.619048
55	1.617647
89	1.618182
144	1.617978
233	1.618056
377	1.618026
610	1.618037
987	1.618033
1597	1.618034

Figure 6.29

We see that the ratio of sides very quickly approaches the golden ratio. We can gain insight into this fact with some algebra.

Let F_n denote the nth Fibonacci number. We know that we have $F_{n+1} = F_n + F_{n-1}$. Then $r_n = \dfrac{F_n}{F_{n-1}}$ is the ratio of sides for the nth rectangle in our sequence. Using the Fibonacci equation $F_{n+1} = F_n + F_{n-1}$ we have

$$r_{n+1} = \frac{F_{n+1}}{F_n} = \frac{F_n + F_{n-1}}{F_n} = 1 + \frac{1}{r_n}.$$

If two consecutive rectangles in our sequence had exactly the same ratio of sides, $r = r_{n+1} = r_n$, then this ratio would satisfy the equation $r = 1 + \dfrac{1}{r}$.

Solving $r = 1 + \dfrac{1}{r}$ is equivalent to solving $r^2 = r + 1$, and we already know that the solution that is larger than 1 is $\dfrac{\sqrt{5} + 1}{2}$, the golden ratio. It is a special property of the sequence r_n that it converges to the golden ratio.

Construction Summary: Golden Ratio (Direct), Compute Ratio Directly

1. Enter the dimensions of the small rectangle.
2. Reproduce the long edge of the small rectangle as the short edge of the large rectangle.
3. Compute the sum of the sides of the small rectangle as the long edge of the large rectangle.
4. Compute the ratio of sides, copy.
5. Compute the difference of the side ratios.
6. Use Goal Seek to make difference 0.

	small rectangle		large rectangle	
long edge	1	1.5	3	2.5
short edge	1	1	2	1.5
side ratio	4	1.5	4	1.6667
difference			5	-0.1667

Construction Summary: Golden Ratio (Iteration), Compute Ratio by Iteration

1. Enter 1 as the first two Fibonacci numbers.
2. Compute the next Fibonacci number as the sum of the two cells above.
3. Find the quotient of the previous and current Fibonacci numbers.
4. Copy as indicated.

1	
1	
2	2
3	1.5
5	1.666667
8	1.6
13	1.625
21	1.615385

6.4 Paper Formats

American standard letter paper is 8.5 by 11 inches. There is no special geometric property associated with this size. Things are different for European paper sizes. Standard European paper sizes are named A0, A1, A2, and so on, and each size is produced by folding the previous size in half, as shown in Figure 6.30.

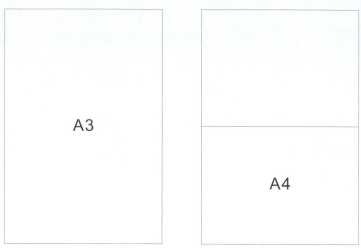

Figure 6.30

The ratio of the shorter edge to the longer edge is not arbitrary. It is chosen in such a way that the "folded-in-half" rectangle has the same ratio of sides as the full rectangle.

Graphically, that has a consequence that is very easy to see: First, turn the large rectangle by 90 degrees and put the smaller rectangle on top of it. Next, draw the diagonal of the smaller rectangle and extend it to the boundary of the larger rectangle (see Figure 6.31).

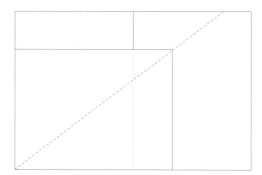

Figure 6.31

If the smaller rectangle and the larger rectangle had the same side ratio, the diagonal would pass through the corner of the larger rectangle. So let us for the moment assume that the shorter length of the larger rectangle is 1 and the length of the longer edge is 1.5. Then the "half-size" paper has longer edge 1 and shorter edge 0.75. These are the lengths in the graph in Figure 6.31.

The problem is similar to the golden ratio problem. In both cases, we want to have a smaller rectangle that has the same ratio of sides as a larger rectangle. In the paper format problem, the smaller rectangle is produced by folding the larger rectangle in half, or equivalently, the larger rectangle is produced by joining two copies of the smaller rectangle. In the case of the golden ratio problem, the larger rectangle is produced by adding a square to the smaller rectangle.

Let us set up a spreadsheet model for the paper format problem now. First, we enter the size of the larger rectangle (Figure 6.32).

	full size	half size
long edge	1.5	
short edge	1	

Figure 6.32

The long edge of the smaller rectangle is equal to the short edge of the larger rectangle (see Figure 6.33), and the short edge of the smaller rectangle is equal to half the length of the long edge of the large rectangle (Figure 6.34).

	full size	half size
long edge	1.5	
short edge	● 1	

Figure 6.33

	full size	half size
long edge	● 1.5	1
short edge	1	/ 2

Figure 6.34

So we have the numbers shown in Figure 6.35.

Now we calculate the ratio of the long edge to the short edge for both rectangles by entering the formula shown in Figure 6.36.

	full size	half size
long edge	1.5	1
short edge	1	0.75

Figure 6.35

	full size	half size
long edge	● 1.5	1
short edge	● 1	0.75
ratio	/	

Figure 6.36

Copying it gives the result shown in Figure 6.37.

Our goal is to have these two ratios be equal by changing the size of the long edge of the full-size rectangle. Trying 1.4 for this size produces the values shown in Figure 6.38.

	full size	half size
long edge	1.5	1
short edge	1	0.75
ratio	1.5	1.333333

Figure 6.37

	full size	half size
long edge	1.4	1
short edge	1	0.7
ratio	1.4	1.428571

Figure 6.38

The two ratios are close together now, but we really want them to be equal. Therefore, we compute the difference of the two ratios, as shown in Figure 6.39.

	full size	half size
long edge	1.5	1
short edge	1	0.7
ratio	1.5	1.333333
difference		0.166667

Figure 6.39

Now we want to change the value of the long edge of the full size in such a way that the difference becomes 0. This can be done using the goal seek procedure that we already used in the golden ratio problem. So, we select the Goal Seek menu item from the Tools menu and then indicate the cell that should have a given value (in our case the cell with the ratio difference), the target value itself (it has to be entered as a number into the dialog box, but it cannot be given by a cell reference), and finally the cell that may be varied (in our case the cell with the full-size long edge). When this is done, *Excel* does a few iterations and finds the approximate solution shown in Figure 6.40.

	full size	half size
long edge	1.414263	1
short edge	1	0.707132
ratio	1.414263	1.414164
difference		9.96E-05

Figure 6.40

The solution seems familiar. It looks like the solution might be $\sqrt{2}$. We also see that the solution is not exact, but it is off only by less than 1/1000, which is close enough for practical purposes.

We can derive this fact mathematically. If x is the full-size long edge, the property of equal proportions in the large and the small rectangles can be expressed as

$$\frac{x}{1} = \frac{1}{\frac{x}{2}}$$

or, equivalently

$$x^2 = 2,$$

and therefore the length of the longer edge has to be $\sqrt{2}$ times the length of the shorter edge. This ratio is preserved by folding the paper in half.

Figure 6.41 is our picture modified so the rectangles have a ratio of the sides of √2.

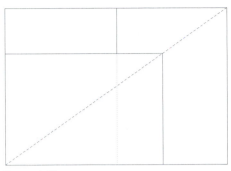

Figure 6.41

We see that the diagonal of the smaller rectangle now also passes through the larger rectangle's corner.

This still does not give us the size of the paper. We know the ratio of the long edge and the short edge, but we have no rule for how to choose one of them. An additional condition tells us how to calculate the numbers: Paper size A0 is exactly 1 square meter of paper. So we can set up the spreadsheet model shown in Figure 6.42 and calculate the area of the rectangle (Figure 6.43), yielding the values shown in Figure 6.44.

paper size	long edge	short edge	area
A0	sqrt(2)* ← ●	1	

Figure 6.42

paper size	long edge	short edge	area
A0	1.404214	● ——1 ►	* ↓

Figure 6.43

paper size	long edge	short edge	area
A0	1.414214	1	1.414214

Figure 6.44

Using Goal Seek again with a target value of 1 for the area by modifying the short-edge length, we get the values shown in Figure 6.45.

paper size	long edge	short edge	area
A0	1.189363	0.8410066	1.000262

Figure 6.45

Now we can calculate the smaller paper formats by the formulas shown in Figure 6.46 (using the "folding-in-half" method).

paper size	long edge	short edge	area
A0	1.189363	0.8410066	1.000262
A1	/ 2		

Figure 6.46

Copying down gives us all of the paper sizes, with lengths in meters (Figure 6.47).

paper size	long edge	short edge	area
A0	1.189363	0.8410066	1.000262
A1	0.841007	0.5946815	
A2	0.594682	0.4205033	
A3	0.420503	0.2973408	
A4	0.297341	0.2102517	
A5	0.210252	0.1486704	
A6	0.14867	0.1051258	

Figure 6.47

Adding two more columns, we can express the sizes in inches (multiplying the size in meters with 100 and dividing by 2.54) by first creating a formula (Figure 6.48) and then copying the formula to the places where it is needed (Figure 6.49), yielding the final table (Figure 6.50).

We see that European A4 paper is slightly narrower and higher than U.S. letter paper.

paper size	long edge	short edge	area	long edge	short edge
A0	1.189363	0.8410066	1.000262	* 100 / 2.54	
A1	0.841007	0.5946815			
A2	0.594682	0.4205033			
A3	0.420503	0.2973408			
A4	0.297341	0.2102517			
A5	0.210252	0.1486704			
A6	0.14867	0.1051258			

Figure 6.48

paper size	long edge	short edge	area	long edge	short edge
A0	1.189363	0.8410066	1.000262	46.82532	33.1105
A1	0.841007	0.5946815		33.1105	23.41266
A2	0.594682	0.4205033		23.41266	16.55525
A3	0.420503	0.2973408		16.55525	11.70633
A4	0.297341	0.2102517		11.70633	8.277624
A5	0.210252	0.1486704		8.277624	5.853164
A6	0.14867	0.1051258		5.853164	4.138812

Figure 6.49

paper size	long edge	short edge	area	long edge	short edge
A0	1.189363	0.8410066	1.000262	46.82532	33.1105
A1	0.841007	0.5946815		33.1105	23.41266
A2	0.594682	0.4205033		23.41266	16.55525
A3	0.420503	0.2973408		16.55525	11.70633
A4	0.297341	0.2102517		11.70633	8.277624
A5	0.210252	0.1486704		8.277624	5.853164
A6	0.14867	0.1051258		5.853164	4.138812

Figure 6.50

Construction Summary: Paper Formats

1. Enter 1 as estimate of short edge.
2. Multiply short edge by $\sqrt{2}$.
3. Compute area, then use Goal Seek to set area to 1 by changing the short edge.
4. Reproduce old short edge as new long edge, copy.
5. Compute half of old long edge as new short edge, copy.
6. Multiply edge by 100/2.54, copy right and down.

size	long ed	shrt ed	area	long ed	shrt ed
A0	[2] 1.19	[1] 0.84	[3] 1.00	[6] 46.83	33.11
A1	[4] 0.84	[5] 0.59		33.11	23.41
A2	0.59	0.42		23.41	16.56
A3	0.42	0.30		16.56	11.71
A4	0.30	0.21		11.71	8.28
A5	0.21	0.15		8.28	5.85
A6	0.15	0.11		5.85	4.14

6.5 Game of Life

We will now study a system of cells living on a rectangular grid and evolving through time. Each cell may be either alive or dead. Cells can survive or die from one point in time to the next one, and dead cells may come alive at the next point in time. Being dead or alive in the next generation depends upon the number of living neighboring cells. Each cell has eight neighbors. In the graph in Figure 6.51, Cell 1 has as neighbors Cell 2 to Cell 9.

25	10	11	12	13
24	9	2	3	14
23	8	1	4	15
22	7	6	5	16
21	20	19	18	17

Figure 6.51

The following simple rules govern whether the state of a cell is dead or alive in the next generation:

- A live cell with two or three living neighbors stays alive.
- A dead cell with exactly three living neighbors comes alive.
- For any other conditions, the cell is dead.

The English mathematician John Horton Conway developed the resulting process, called the *Game of Life* (see Sigmund (1993) and Eigen et al (1993)).

The cell system starts with an arbitrary initial configuration. Next, the number of living neighbors is computed for each cell. From these results the next state of all the cells is computed. Then this process is repeated over and over again.

Let us illustrate the step from one point in time to the next one with a simple example. Coloring dead cells white and live cells black, we see that the state in Figure 6.52 evolves into the one in Figure 6.53.

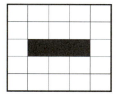

Figure 6.52

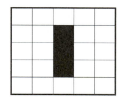

Figure 6.53

We will now set up a worksheet computing the states of one generation of cells from the previous generation. We start with an 8 by 8 grid containing an arbitrary initial configuration (see Figure 6.54). In our spreadsheet, we will represent live cells by 1 and dead cells by 0 (or by an empty spreadsheet cell). We need to provide additional empty rows and columns around the grid in order to be able to set up a straightforward formula for computing the number of live neighbors.

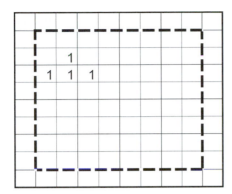

Figure 6.54

To compute the number of live neighbors, we reserve a second 8 by 8 grid to the right of the original grid (Figure 6.55). In this grid (marked by the dashed border), we compute the number of live neighbors of the upper left cell of the original grid by computing the sum of the nine-cell neighborhood of the cell (including the cell itself) and subtracting the value of the cell itself.

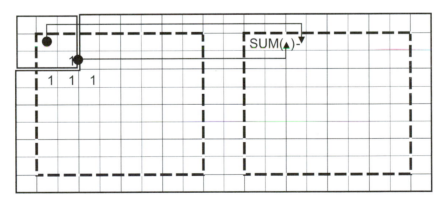

Figure 6.55

Since all of the references in this formula are relative, we can copy the formula to the rest of the grid that computes the number of live neighbors (Figure 6.56).

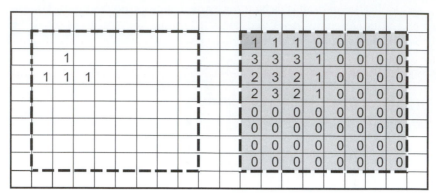

Figure 6.56

From the original grid and the grid showing the number of live neighbors, we now can compute the next grid. The formula that we need is somewhat complex, and it is created using nested IF functions (see Figure 6.57).

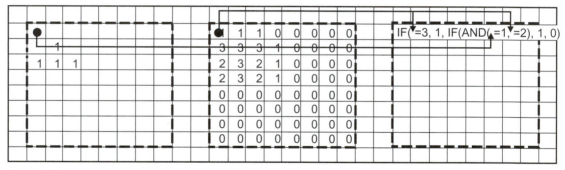

Figure 6.57

This function expresses the rules in a slightly rephrased form:

- A cell with three living neighbors stays or becomes alive.
- A live cell with exactly two living neighbors stays alive.
- Otherwise, the cell is dead.

Copying this formula creates the grid for the next generation of cells (Figure 6.58).

The easiest way to compute one more generation is to copy the values of the newly computed generation into the grid of the original generation. In other words, we copy the values from the rightmost of our three 8 by 8 grids to the leftmost of the three grids. Just copying everything, however, does not work since that would copy not only the values but also the formulas. Therefore, we first select the rightmost grid and copy the range. In the next step, we select the upper left cell of the leftmost 8 by

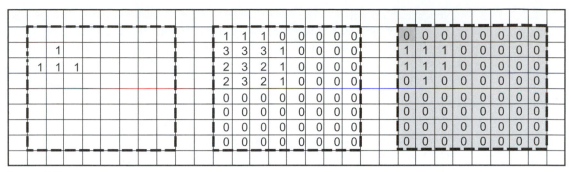

Figure 6.58

8 grid and then select the options Edit, Paste Special from the menu bar. Then in the dialog box that pops up, we mark the Paste, Values option button and click OK.

Now our worksheet is updated. The left grid now contains the second generation of our iterations, while the rightmost grid contains the third generation (Figure 6.59).

0	0	0	0	0	0	0	0		2	3	2	1	0	0	0	0		0	1	0	0	0	0	0	0
1	1	1	0	0	0	0	0		3	5	3	2	0	0	0	0		1	0	1	0	0	0	0	0
1	1	1	0	0	0	0	0		4	6	4	2	0	0	0	0		0	0	0	0	0	0	0	0
0	1	0	0	0	0	0	0		3	3	3	1	0	0	0	0		1	1	1	0	0	0	0	0
0	0	0	0	0	0	0	0		1	1	1	0	0	0	0	0		0	0	0	0	0	0	0	0
0	0	0	0	0	0	0	0		0	0	0	0	0	0	0	0		0	0	0	0	0	0	0	0
0	0	0	0	0	0	0	0		0	0	0	0	0	0	0	0		0	0	0	0	0	0	0	0
0	0	0	0	0	0	0	0		0	0	0	0	0	0	0	0		0	0	0	0	0	0	0	0

Figure 6.59

Repeating this Edit, Copy and Edit, Paste Special, Values process several times iterates the generations further. Doing this manually, however, is quite tedious. Therefore, we will use the macro recorder to automate this process. Details on how to use the macro recorder are given in the Appendix. We will just briefly describe the necessary steps here.

We start by selecting the options Tools, Macro, Record New Macro from the menus. Next, a dialog box appears prompting us for the name of the macro that we will define. We may choose the name freely. A reasonable suggestion for the name is "CopyGen" for copy generation. After entering the name and clicking OK, a toolbar with a button for switching off the macro recorder appears (Figure 6.60).

Figure 6.60

Now we again perform the whole series of operations needed to copy the values of the new generation into the grid of the old generation. To make the flickering selection indicator disappear, we press the Escape key after the paste operation is performed. To remove the selection color background from the target range of our paste operation, we select the cell to the left and above the leftmost 8 by 8 grid.

When that is finished, we click the leftmost button on the toolbar that just appeared. This button stops the recording of the macro. Alternatively, we can choose the menu options Tools, Macro, Stop Recording.

At this time we create a new toolbar by using the menu sequence Tools, Customize, Toolbars, New and name it "Life." In the Customize dialog box, from the Commands menu and the Macros item in the left list, we choose Custom Menu Item and drag that onto the newly created Life toolbar. Right-clicking on this new menu item brings up the context menu for this menu item, allowing us to change the name to "Step 1" and to attach the macro "CopyGen" to the button via the Assign Macro option. We then close the dialog box.

Now just clicking on the button will iterate one generation in our implementation of the Game of Life.

Displaying the numbers 0 and 1 to indicate if a cell is alive contains the full information about the system. However, we can improve the visual appearance of our worksheet by using one more spreadsheet concept, conditional formatting, a topic that is discussed in the Appendix.

To do so, we select the leftmost 8 by 8 grid (the one containing the "initial" generation). Then we select Format, Conditional Formatting from the menus. In the dialog box that appears we set the background pattern to black if the cell value equals 1. That gives us completely black worksheet cells for cells containing a value of 1. The cells containing 0 still display a black 0 on white background. If we want to change that also, we set conditional formatting to use a white font color if the cell value equals 0.

Now, we can enter any initial configuration in the leftmost grid and perform iterations to see what patterns emerge. Our grid is not very large. Using a larger grid allows us to study more complex initial configurations and also to investigate moving configurations of cells.

The Game of Life is the best-known example of cellular automata. General cellular automata use the same basic mechanism as the one in the Game of Life. Each cell can have a state from a given set of states (alive or dead in our case). The new state of a cell is calculated from the state of a cell and the state of the neighboring cells. This process is repeated many times. The rules for computing the new state of a cell from its previous state and the previous state of its neighbors can describe cellular automata, and different sets of rules can generate vastly different behavior.

1.

1								
	1							
1	1	1						

2.

1	1	1	0	0	0	0	0
3	3	3	1	0	0	0	0
2	3	2	1	0	0	0	0
2	3	2	1	0	0	0	0
0	0	0	0	0	0	0	0
0	0	0	0	0	0	0	0
0	0	0	0	0	0	0	0
0	0	0	0	0	0	0	0

3.

0	0	0	0	0	0	0	0
1	1	1	0	0	0	0	0
1	1	1	0	0	0	0	0
0	1	0	0	0	0	0	0
0	0	0	0	0	0	0	0
0	0	0	0	0	0	0	0
0	0	0	0	0	0	0	0
0	0	0	0	0	0	0	0

1. Enter 1 to designate the initial live cells.
2. Enter the formula to find the number of neighboring live cells. Copy throughout.
3. Enter the IF formula to generate a 1 (live) or 0 (dead). Copy throughout.
4. Copy the values of the rightmost block into the leftmost block.

Exercises

1. Modify the worksheet so that the basic grid for the Game of Life is 100 by 100 instead of 8 by 8. Since worksheets usually are only 256 columns wide, you cannot put the three grids needed in our calculation side by side. Therefore, arrange the three grids vertically instead of horizontally. Use your model to find and investigate interesting starting conditions. Many of these can be found in books discussing automata and the Game of Life.

2. Modify the rules for how the next state of a cell is calculated. A general rule could be:
   - If the cell is alive and the number of live neighbors is between a and b, the cell stays alive.
   - If the cell is alive and the number of live neighboring cells is outside of the range from a to b, the cell dies.

- If the cell is dead and the number of live neighbors is between c and d, the cell becomes alive.
- If the cell is dead and the number of live neighboring cells is outside of the range from c to d, the cell remains dead.
- Using the values 2, 3, 3, 3 for a, b, c, and d turns these more general rules into the original rules used in this section. Put the values for a, b, c, and d into cells of the worksheet and design a formula to implement the rules just listed.

3. Using the spreadsheet set up in the previous exercise, try to find "interesting" rules. Rules are interesting when they produce complex or unexpected behavior over time.

4. Use *Excel*'s random number function to implement the following game on a 6 by 6 grid. Design your model to incorporate a recorded macro linked to a button. Start with cells that are alternately 0 (white) and 1 (black). Simulate rolling a pair of dice to obtain the coordinates (m,n), $1 \le m \le 6$, $1 \le n \le 6$, of a point on the grid. Next, simulate flipping a coin to determine if the color of that point becomes white or black. Then repeat the process indefinitely. The starting display is shown in Figure 6.61, with a typical display after a number of future iterations shown in Figure 6.62. For further discussion of this game, see Eigen and Winkler (1993), *Laws of the Game*, page 30. In that reference you will also find a wide variety of other games that can be implemented on a spreadsheet.

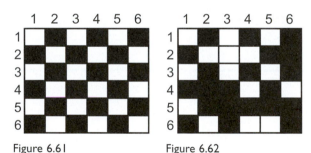

Figure 6.61   Figure 6.62

6.6 Problems to Solve by Yourself

Let us have some fun now. After reading and working through all the projects in this section, you might be interested in solving some more interesting problems yourself.

Here is a list of such problems, some of them related to the problems of this section. Do not try to solve the problems with paper and pencil before you use a spreadsheet to compute the numerical solutions; go right ahead and develop your models directly as a spreadsheet.

Exercises

1. A teacher asks her class to write down an integer between 101 and 200, inclusive. If there are 20 students in the class, what is the probability that at least two students will have written the same number? If the teacher wants to be 95 percent certain that a duplicate will occur, from how many numbers should she have her students choose?

2. A basketball player makes free throws with probability p. Assuming that the probability does not change from shot to shot, what is the most likely outcome of a 1-and-1 free throw situation? In this situation the player gets a second shot only if she makes the first one. Create an xy-graph showing the probability of each possible outcome (that is, making zero, one, or two shots) for $p = 0.00$, $0.01, 0.02, \ldots 1.00$.

3. In baseball a player's batting average is computed by dividing the number of hits by the number of at bats. Is it possible for Sluggo to have a better batting average against both left- and right-handed pitchers than Wimpy but for Wimpy to have a better overall batting average than Sluggo (see Reinhardt (1981))? Create a spreadsheet to do the computations. Examine the data in Figure 6.63. Can you design a graph to illustrate the seeming paradox?

Batter	Against RHP		Against LHP	
	AB	Hits	AB	Hits
Wimpy	100	30	200	80
Sluggo	300	95	50	21

Figure 6.63

4. Suppose that a corporation decides to find all of the drug users among its 10,000 employees, although management acknowledges that there are relatively few of them. To do this, it uses a test that is known to be 95 percent accurate. Design a spreadsheet model to compute the number and percentage of "false positives" that will result if, in fact, the company has only 1 percent drug users. Design the test so that the accuracy is a parameter, and vary that value through a scroll bar. Try to design an insightful graph to accompany the model.

5. Refer to the previous exercise. Because the company does not want to falsely accuse an employee who tests positive but is not a drug user, the company retests the group of those who test positive on the first try. The same test is used as before. Extend your model to determine the percentage of false positives to expect in this group.

6. There are many useful means that are defined in mathematics, and several of these have geometric interpretations associated with them (see Eves (1990), pp. 200–201). Create an *Excel* model to enter values a and b and compute the following means of a and b:

a. Arithmetic: $A = (a + b)/2$

b. Geometric: $G = \sqrt{ab}$

c. Harmonic: $H = 2ab/(a + b)$

d. Heronian: $H_e = (a + \sqrt{ab} + b)/3$

e. Contraharmonic: $C = (a^2 + b^2)/(a + b)$

f. Root-mean-square: $R = \sqrt{(a^2 + b^2)/2}$

g. Centroidal: $C_e = 2(a^2 + ab + b^2)/(3(a + b))$

7. Create a spreadsheet model to display graphically via a trapezoid with bases a and b the following various means (see Eves, (1990) p. 201). Sample output illustrating the first three means is provided in Figure 6.64. Provide animation in your graph via a scroll bar.

a. Arithmetic: bisects the sides of the trapezoid

b. Geometric: divides the trapezoid into two similar figures

c. Harmonic: passes through the intersection of the diagonals

d. Heronian: lies 1/3 of the distance from the arithmetic to the geometric mean

e. Contraharmonic: as far below the arithmetic mean as the harmonic mean is above the arithmetic mean

f. Root-mean-square: bisects the trapezoid's area

g. Centroidal: passes through the centroid of the area

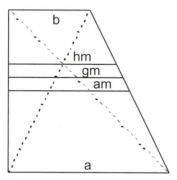

Figure 6.64

8. Design a spreadsheet model to determine those times when the minute and hour hands of a clock are perpendicular, point in the same direction, or point in opposite directions (see Figure 6.65). You might want to use the solver in your model. Try to create an effective graph to illustrate the ideas behind your solutions. In addition, design a clock face linked to a scroll bar. Make sure that the hands move in the correct direction.

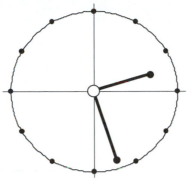

Figure 6.65

9. [*] Sports often focus on *winning streaks*, which are also called *runs* in statistics. If a team has probability p of winning any game and the outcomes of games are independent, use a spreadsheet to find the probability that in a sequence of n games the team has a winning streak of at least r games. Hint (see Uspensky, (1937) pp. 77–78): Denote the probability of a winning streak of r games in n games by $P(n,r)$. To find the probability $P(n+1,r)$ for Game $n+1$, either one of two mutually exclusive events must occur: (a) there already is a winning streak of r games by Game n, with probability $P(n,r)$, or (b) the winning streak of r games is only achieved in Game $n+1$. In the latter case, there is no winning streak of r games in the first $n-r$ games (with probability $1 - P(n-r,r)$), the team must lose game $n-r+1$ (with probability $1-p$), and the team must win the next r games (with probability p^r).

10. [*] Amplify the model of the previous exercise to find the probability that the longest winning streak in a sequence of n games is r.

11. Create a model to solve the following pursuit problem: Four flies start at the corners of a square table. Each moves continuously toward the fly on its left. Determine the paths that result. Include an animated graph showing the progress of the flies. A sample illustration is provided in Figure 6.66.

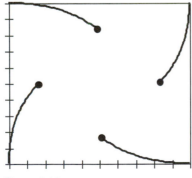

Figure 6.66

Combinatorics and Probability

CHAPTER OUTLINE

7.1   Birthday Problem

Suppose that a group of 8 people are gathered in a room. Do you think that it is very probable that among the eight at least two will have the same birthday? While that seems to be rather unlikely, what would your answer be if there were 20, or 50, or 100 people in the room? Certainly the probability of duplicate birthdays occurring grows as the number of people in the room increases. What would you estimate to be the minimum number of people that must be in the room before the probability of there being at least one duplicate birthday reaches 0.5? Until it reaches 0.8?

If you do not know the answer to any of these questions, then try to make some estimates before we arrive at the answer mathematically. Also try to remember your estimates and then compare them to what we will calculate in this section.

Unfortunately, the problem of what to do with people who were born on February 29 in a leap year causes some difficulties. As a result, for the sake of simplicity, we initially assume that there are $N = 365$ days in a year. In analyzing the problem, it is easier to first determine the probability, $P(n)$, that in a group of n people all have different birthdays. Then the probability that at least two people in a group of n people share the same birthday is then $1 - P(n)$.

In our spreadsheet model we use N as a parameter by placing its value in a separate cell. This will allow us to investigate a number of related situations with the same model.

To solve this problem we can think of identifying each of the n people in the room by an integer $1, 2, 3, \ldots, n$. Then we can consider sequences of n birthdays and seek to find the proportion of all such sequences that contain no duplicate dates.

First, if there is only one person in the group, then there are 365 possible dates for his or her birthday, and each of these generates no duplicates within the group. Thus, the probability, $P(1)$, of having no duplicates in the group is

$$\frac{365}{365} = 1 \text{ or } \frac{N}{N} = 1.$$

Consequently, we begin to build our model by using the first column to count the number of people, n, by our standard counting process (Figure 7.1). We then enter 1 as the value for $P(1)$ and the formula for at least one duplicate as $1 - P(1)$.

Figure 7.1

Next, we observe that if there are two people in the room, then there are $N \cdot N = 365 \cdot 365$ possible sequences for their birthdays, since there are $N = 365$ possible birthdays for the first person, and for each of these the same number of possible birthdays for the second person.

To determine the number of sequences in which both birthdays differ, we observe that there would be $N = 365$ possibilities for the first birthday but only $N - 1 = 364$ possibilities for the second birthday because the second must be different from the first. Therefore the number of sequences of two dates in which both dates are different is $N \cdot (N - 1) = 365 \cdot 364$. From this it follows that the probability that both are different is

$$\frac{365 \cdot 364}{365 \cdot 365} = \frac{365}{365} \cdot \frac{364}{365} \text{ or } \frac{N \cdot (N - 1)}{N \cdot N} = \frac{N}{N} \cdot \frac{N - 1}{N}.$$

Thus, we can obtain the probability, $P(2)$, for two people as

$$\frac{364}{365} \text{ or } \frac{N - 1}{N}.$$

We now enter that as our next calculation and copy the formula in the right column (Figure 7.2).

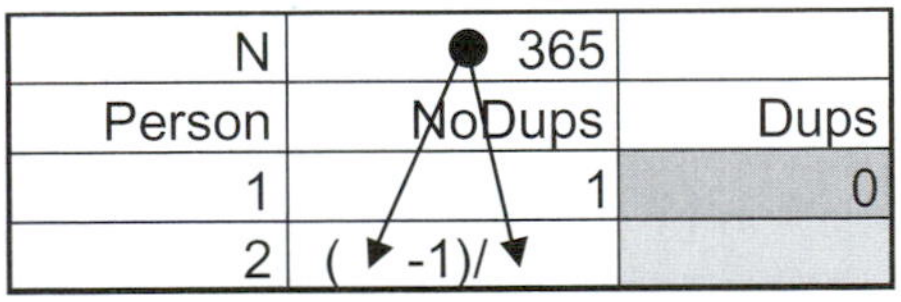

Figure 7.2

In the same way, if there are 3 people in the room, then there are a total of

$$N \cdot N \cdot N = 365 \cdot 365 \cdot 365$$

possible sequences for their birthdays. If all of the days are to be different, then we have $N - 365$ possibilities for the first, $N - 1 = 364$ possibilities for the second since it must be different from the first, and $N - 2 = 363$ possibilities for the third since it must be different from the first two. Therefore the number of sequences in which all three dates are different is

$$N \cdot (N - 1) \cdot (N - 2) = 365 \cdot 364 \cdot 363.$$

Consequently, the probability that all three birthdays are different is

$$\frac{365 \cdot 364 \cdot 363}{365 \cdot 365 \cdot 365} = \frac{365 \cdot 364}{365 \cdot 365} \cdot \frac{363}{365} \ \text{or} \ \frac{N \cdot (N - 1) \cdot (N - 2)}{N \cdot N \cdot N} = \frac{N \cdot (N - 1)}{N \cdot N} \cdot \frac{N - 2}{N}.$$

Here we can see that we can get the probability that all three have different birthdays, $P(3)$, by multiplying the probability, $P(2)$, that the first two are different by

$$\frac{363}{365} \ \text{or} \ \frac{N - 2}{N}.$$

This expression can be entered into the spreadsheet straightforwardly (see Figure 7.3). Since the 2 that we subtract refers to the group size in the previous computation, we will use a cell to provide a reference rather than entering a constant. In creating our formulas we use an absolute reference to N, since it is a parameter of the model.

N		365	
Person		NoDups	Dups
1		1	0
2		0.997	0.00274
3		*(▼ - ▲)/	

Figure 7.3

By now perhaps you see the pattern for determining the next probability. First, the initial n sequences must contain no duplicates. This has the probability $P(n)$ that is computed in the previous row. Second, the next birthday in the sequence must fall on a day that is distinct from the n already used. Thus, there are only $N - n = 365 - n$ days that are still available for the new entry if there are to be no duplicates. The probability that person number $n + 1$ has a birthday on one of these days is $(N - n)/N$ or $(365 - n)/365$. The probability that both are true is their product,

$$P(n + 1) = P(n) \frac{N - n}{N} = P(n) \frac{365 - n}{365}.$$

Consequently, we just have to copy our previous formulas in order to complete the model (Figure 7.4). After we do this, we also might notice that we could have computed the probability $P(2)$ in the same way.

N	365	
Person	NoDups	Dups
1	1	0
2	0.997	0.003
3	0.992	0.008
4	0.984	0.016
5	0.973	0.027

Figure 7.4

Partial output of our model is presented in Figure 7.5. Notice what our computations reveal to us. When there are as few as 23 people in the room, the probability that at least two of them have the same birthday already exceeds 0.5! Most people assume that the number required is much higher, perhaps as many as 50 or 100. However, our model shows us that if there are 50 people in the room, then it is overwhelmingly likely that there will be duplicate birthdays, since by then the probability of having duplicate birthdays has risen to 0.97. It is interesting to carry out such an investigation by looking up the dates of various groups of events, say the birthdays or death dates of groups of presidents or scientists, in an almanac or on the World Wide Web.

N	365	
Person	NoDups	Dups
1	1.0000	0.0000
2	0.9973	0.0027
3	0.9918	0.0082
4	0.9836	0.0164
22	0.5243	0.4757
23	0.4927	0.5073
50	0.0296	0.9704
80	0.0001	0.9999

Figure 7.5

Once we have created our model, it is then possible to construct a graph that shows nicely how the desired probability is related to n. The graph in Figure 7.6 uses the first column for the x-axis series and the third column for the y-series. From the graph we can easily see how many people are needed for the probability to reach various levels.

By using N as a parameter of the model, it is easy to investigate a number of related questions. For example, if n people each write down an integer between 1 and 100, what is the probability that there will be at least one duplicate? How many people do

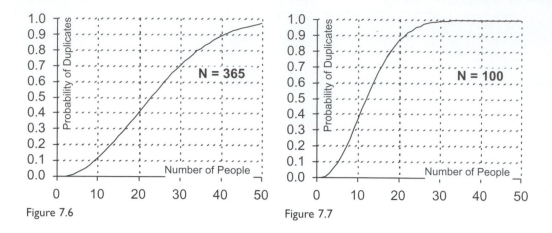

Figure 7.6

Figure 7.7

we need for the probability to reach 0.5? Here we set $N = 100$ and use the same model. In this way it is possible to use the problem effectively in demonstrations with smaller groups of people. The outcome is shown for $N = 100$ in Figure 7.7. For example, with 20 people there is nearly a 90 percent chance that at least two people will have written the same number. This is an excellent opportunity to create a scroll bar with which to vary the value of N.

Since we can apply this model to dates of deaths as well as to births, it might be of interest to note that of the first three presidents of the United States, two died on the same day. In fact, both John Adams and Thomas Jefferson died on exactly the same day, July 4, 1826, precisely 50 years after each had signed the U.S. Declaration of Independence.

In a later section we will return to this problem when we use the spreadsheet to do some Monte Carlo simulations.

Construction Summary: Birthday Problem

1. Enter the number of days, N.
2. Generate a counter column.
3. Enter 1 as the probability of no duplicates when $N = 1$.
4. Generate the next probability by multiplying the previous value by $(N - n)/N$.
5. Find the probability of at least one duplicate by subtracting the probability at the left from 1.
6. Copy as indicated.

	N	1	365	
	Person		NoDups	Dups
2	1	3	1	5	0
	2	4	0.997	0.003
	3		0.992	0.008
	4		0.984	0.016
	5		0.973	0.027
	6		0.960	0.040
	7		0.944	0.056
	8		0.926	0.074

Exercises

1. A teacher asks her class to write down an integer between 101 and 200, inclusive. If there are 20 students in the class, what is the probability that least two students will have written the same number? If the teacher wants to be 95 percent certain that a duplicate will occur, from how many numbers should she have her students choose?

2. Consult an almanac, an encyclopedia, or a Web site to obtain interesting lists to check for duplicate dates or similar data.

3. Modify the basic birthday spreadsheet model to include a data table to generate the number of people that are required for the probability of duplicate numbers occurring to reach a given probability, say $p = 0.5$, $p = 0.75$, or $p = 0.9$ if the numbers come from a set of N numbers ($N = 365$ is the classical birthday problem).

7.2 Birth Control: The Geometric Distribution

Suppose that a married couple uses a particular means of contraception. Because there is no one means of contraception that is 100 percent effective, they want to know the likelihood that the method that they use might fail within a given number of months. Abstracting from this concrete problem we can formulate that we perform an experiment with a given probability and want to know the probabilities that the event does not happen once when we perform this experiment a given number of times.

In this section we examine this topic and see that it gives rise to a probability distribution that is quite useful and yet differs markedly from the well-known normal and similar distributions with their "bell-shaped" curves. As in the previous section, it can also lead to some counterintuitive results. This example seeks to answer the question, "What is the probability that if a couple uses a particular means of contraception, then it will not fail for a given number of months?" We will determine the probability that conception will occur for the first time in Month n as well as the probability that the first conception will occur within the first n months.

In this model we first assume that if no method of contraception is used then the natural probability of conceiving in any month is a constant. This value varies considerably among different individuals, but studies indicate that probability values between 0.10 and 0.25 are not unusual. We will illustrate our model using a natural probability of 0.15.

Next, we assume that a particular contraception method has a monthly efficiency rate of e, where $0 \leq e \leq 1$. For example, having an efficiency rate of $e = 0.9$ means that conception is prevented in 90 percent of the months in which it would have occurred naturally. Again, studies give reasonable estimates for values of efficiency rates for different methods of contraception. Thus, the effective probability of conception occurring in any given month, p, is found by multiplying the natural probability by $(1 - e)$. We will denote the probability that a conception does not occur in a given month by $q = 1 - p$.

Note that in the study of probability, p generally represents the probability of a "success" in an event. Thus, in such a usage "success" would mean the occurrence of conception, an event that undoubtedly would be considered a failure in the actual application.

Our model in Figure 7.8 uses 0.15 as the natural probability of conceiving in a given month, together with a birth control method of effectiveness 0.9. We enter these values as shown in the figure and compute the effective probability by multiplying the natural probability by $(1 - e)$. In this case $p = (1 - 0.9)(0.15)$, or $p = 0.015$.

In computing the monthly probabilities, the first column is used to count months in the usual way. In the first cell we enter 1, and in the cell below we enter a formula that adds 1 to the value of the cell above it.

Natural prob		●0.15	
Efficiency rate	●	0.9	
Effective prob	(1-▼) * ▼		
	Month	Conceive	Cumulative
●	1		
1+ ▼			

Figure 7.8

The second column computes the probabilities that conception first occurs in a given month (see Figure 7.9). For the first month this is simply the effective probability, p.

Natural prob		0.15	
Efficiency rate		0.9	
Effective prob	●	0.015	
	Month	Conceive	Cumulative
	1	▼	
	2		

Figure 7.9

For conception to first occur in the second month, two things must happen: (1) there is no conception in the first month (probability q), and (2) there is conception in the second month (probability p). Thus, the probability sought is the product, pq. This can be obtained by multiplying the previous month's probability by q in the form of $1 - p$ (see Figure 7.10). We have indicated using the effective rate as an absolute reference for reasons that will soon become clear. In the third column we find

Natural prob	0.15	
Efficiency rate	0.9	
Effective prob	0.015	
Month	Conceive	Cumulative
1	0.015	
2	(1- ▼) * ▼	

Figure 7.10

the cumulative probability that the first conception occurs within the first n months. For the first month this is simply the probability that it occurs in the first month.

Next, we observe that for the first conception to occur in the third month, again two things must happen: (1) there is no conception in the first two months (probability q^2), and (2) there is conception in the third month (probability p). Thus, the probability sought is the product, pq^2. As before, this can be obtained by multiplying the previous month's probability by q in the form of $1 - p$. In fact, perhaps we can begin to see the general pattern by now. For the first conception to occur in any given month after the first, there must be one more month of failure to conceive than is the case for the previous month. Therefore, we can obtain the probability for a given month by multiplying the previous month's probability by $(1 - p)$. Because we have used an absolute reference for p, this can be implemented in our model simply by copying down the previous expression in the second column.

To compute the cumulative probabilities beyond the first month, we add the probability of conception in the current month to the previous cumulative probability. All that remains to complete the main portion of our model is to use a copy or fill command (Figure 7.11).

Natural prob	0.15	
Efficiency rate	0.9	
Effective prob	0.015	
Month	Conceive	Cumulative
1	0.015	0.015
2	0.014775	▼ + ▲

Figure 7.11

After we have extended our model to include, say, 100 months, we might want to compute the mean time until the first pregnancy occurs. This can be a challenge to do since the number of months to be considered extends indefinitely. The value that we need to find is

$$\mu = 1p_1 + 2p_2 + 3p_3 + \ldots + np_n + \ldots .$$

If we were to decide to terminate this sum after 100 months, or even 200 months or a longer period, then we can approximate its value by using the SUMPRODUCT function as indicated in Figure 7.12. However, if we were to do this and then experiment in extending the number of months, we will find that this value keeps changing noticeably until a very large number of months have been included.

Natural prob	0.15	Mean	SUMPRODUCT(↗ ↗)
Efficiency rate	0.9	Median	
Effective prob	0.015		
Month	Conceive	Cumulative	
1	0.0150	0.0150	
2	0.0148	0.0298	
1	0.0146	0.0443	
2	0.0143	0.0587	

Figure 7.12

Later we will provide a mathematical derivation to show that the mean, or average, time until the first conception is $\mu = 1/p$ months. In Figure 7.13 we have used this formula rather than the SUMPRODUCT. The output of our model shows that with our chosen values for the parameters the mean time until the first conception is 66.7 months, or over 5½ years. This can be interpreted as the average time it would take a very large population of couples satisfying the same hypotheses to have their first conception.

Natural prob	0.15	Mean	66.67
Efficiency rate	0.9	Median	
Efficiency prob	0.015		
Month	Conceive	Cumulative	
1	0.0150	0.0150	
2	0.0148	0.0298	
3	0.0146	0.0443	
4	0.0143	0.0587	
12	0.0127	0.1659	
45	0.0077	0.4934	
46	0.0076	0.5010	
66	0.0056	0.6312	
67	0.0055	0.6367	

Figure 7.13

However, as you look at the resulting table, you may be surprised to see that far more than half of the couples (almost 2/3) will have conceived by the 67th month. Thus, this is one setting in which the concept of the mean can be somewhat misleading.

A "better" measurement in this case is the median. Thus median can be thought of as the number of months before which half of the conceptions will occur and after which the other half will occur. The cumulative probability can be used to find the median number of months until the first conception. This is the time by which the probability of conception within the first n months reaches 0.5. This corresponds to a cumulative probability of 0.5, which from the table in Figure 7.13 occurs by Month 46.

This example provides us with a good illustration of a situation in which the concepts of mean and median differ significantly. Incidentally, although spreadsheets have built-in average and median functions, these are designed to be used with specific sets of data rather than with probability distributions. Consequently, that cannot be used here.

We can compute the median by using the spreadsheet's LOOKUP(value,array1,array2) function (see Figure 7.14). This function finds the largest value in array1 that does not exceed the sought-for value (here 0.5), and returns the corresponding value in array2. Here this would be 45, so we add 1 onto this position to find the first month in which the cumulative probability exceeds 0.5.

Natural prob	0.15	Mean			66.67
Efficiency rate	0.9	Median	LOOKUP(0.5, ▲ , ▲)+1		
Effective prob	0.015				
Month	Conceive	Cumulative			
1	0.0150	0.0150			
2	0.0148	0.0298			
3	0.0146	0.0443			
45	0.0077	0.4934			
46	0.0076	0.5010			

Figure 7.14

The resulting output is provided in Figure 7.15, where we see that 46 is the median number of months until the first conception occurs.

Natural prob	0.15	Mean	66.67
Efficiency rate	0.9	Median	46
Effective prob	0.015		

Figure 7.15

We can create an xy-graph (Figure 7.16) of the distribution by using the first column of the model (month) as the x-axis series and the second column (probability of first conception in that month) as the y-series. The y-coordinates of the graph of the distribution give the probabilities that the first conception occurs in the corresponding month on the x-axis. The graph, which of course extends infinitely to the right, shows the distribution to be quite non-normal, and not at all bell-shaped. The median of the distribution is indicated by the dotted vertical line at the right that will divide the region under the curve into two portions having equal areas. The existence of the possibilities for some extremely long waits until the first conception occurs has the effect of moving the mean farther to the right. A scroll bar can be incorporated to vary the value of p.

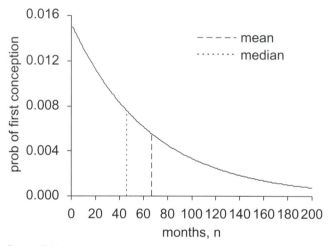

Figure 7.16

The graph of the cumulative probabilities provides us with a way to illustrate the concept of the median. The median is the value of n at which the cumulative probability reaches 0.5. Thus, we can determine the median from the graph by moving a finger on a horizontal line from 0.50 on the y-axis and finding the corresponding value for n on the x-axis. This graph, as well as the model's numerical output, also highlights some other nonintuitive facts that arise in the *geometric distribution*, as this distribution is called. For example, using the values with which we started, approximately 25 percent of the couples will conceive within the first 20 months, even though the mean time is over 5 years (see Figure 7.17).

At this time we note that we can also derive the results of this section using a slightly different approach. The probability that the first conception occurs in the first month is $p_1 = p$. In general, for the first conception to occur in Month $n + 1$, there must be no conception in the first n initial months (probability q^n) followed by a conception that occurs in the $(n + 1)$st month (probability p). Thus, the probability that the first conception occurs in the month n is $P_{n+1} = pq^n$. From this it also follows that the recurrence relation that we have been using, $P_{n+1} = qP_n$, holds.

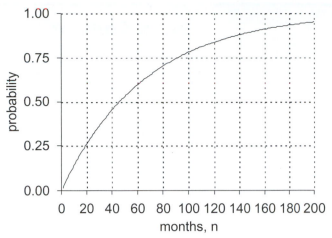

Figure 7.17

This model also allows us to carry out investigations for the geometric distribution as it occurs in other settings. For example, we can flip a coin until the first head occurs ($p = 1/2$), roll a die until the first 6 appears ($p = 1/6$), roll a pair of dice until the first 7 is attained ($p = 1/6$), or shoot free throws in basketball until the first miss takes place (assuming a constant success rate). Further, many of the ideas of this model can be readily adapted to use similarly with other discrete probability distributions.

To finish our examination of this topic, it can be helpful to algebraically compute the average, or mean, time, μ, until the first conception. Intuitively, if the probability for any given month is, say, $p = 1/40$, then we might expect the mean time to be $\mu = 1/p = 40$ months. This turns out to be the case, as can be developed through the definition of *expected value*, or mean, from statistics:

$$\mu = 1p_1 + 2p_2 + 3p_3 + \ldots + np_n + \ldots$$
$$= 1p + 2pq + 3pq^2 + \ldots + npq^{n-1} + \ldots .$$

If we do not worry about the convergence of the infinite series, then we can use a little algebra to show that

$$\mu = (p + pq + pq^2 + \ldots) + (pq + 2pq^2 + 3pq^3 + \ldots)$$
$$= p(1 + q + q^2 + \ldots) + q(1p + 2pq + 3pq^2 + \ldots),$$

or

$$\mu = \frac{p}{1 - q} + \mu q = \frac{p}{p} + \mu q = 1 + \mu q,$$

where we have summed the geometric series $1 + q + q^2 + \ldots = 1/(1 - q)$. Thus, by solving the equation $\mu = 1 + \mu q$ for μ we obtain $(1 - q)\mu = 1$, so that $\mu = 1/(1 - q) = 1/p$.

1. Enter parameter values.
2. Compute effective probability as (1–efficiency rate) × natural probability.
3. Generate a counter column.
4. Reproduce effective probability.
5. Reproduce initial probability to start the cumulative column.
6. Multiply the previous probability by (1-effective probability), copy.
7. Add the current probability to the cumulative probability, copy.
8. Use *Microsoft Excel* functions to find mean, median.

Nat prob	1	0.15	Mean	8	66.7
Eff rate		0.9	Median		46
Eff prob	2	0.015			

	Month		Conceive		Cumulative	
3	1	4	0.0150	5	0.0150	
	2	6	0.0148	7	0.0298	
	3		0.0146		0.0443	
	4		0.0143		0.0587	
	5		0.0141		0.0728	
	6		0.0139		0.0867	

Exercises

1. A basketball player makes free throws with probability $p = 0.8$. What is the probability that the first miss will occur on the first, second, third, . . . free throw shot that is attempted?

2. The *mean* or expected number of successes of a discrete probability distribution is computed by $\mu = 0P(0) + 1P(1) + 2P(2) + 3P(3) + \ldots$. Compute this sum through a number of terms in your spreadsheet model for the geometric division. To what value does it appear to converge?

    _____ Suppose that an experiment has two possible outcomes that we shall denote as *success* (S) and *failure* (F), with the probability of success of p and the probability of failure of $q = 1 - p$. Such an experiment is called a *Bernoulli experiment*. Further, suppose that the outcomes of repetitions of the experiment are independent, so that the probabilities do not change. Then we can represent n repetitions of the experiment via a tree diagram, with the probabilities of an end branch determined by multiplying branch probabilities, as illustrated in the tree diagram in Figure 7.18 with $n = 4$. The solid circles represent successes, and the open circles failures. Many discrete probability distributions can be explained and examined in terms of a series of Bernoulli experiments. Some of these are studied in the following exercises.

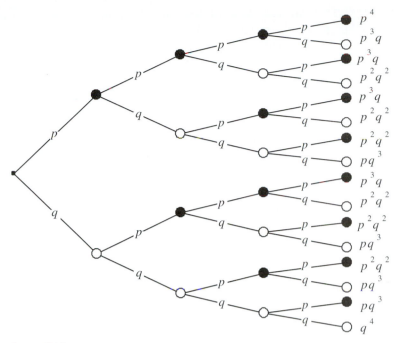

Figure 7.18

3. Suppose that a Bernoulli experiment is repeated n times. The probabilities of obtaining k successes, where $0 \le k \le n$, gives a *binomial distribution*. For small values of n we can determine these directly from the diagram in Figure 7.18. Binomial probabilities can either be displayed in two columns for a given value of n, or in a two-dimensional table format for a range of values of n. These two approaches are shown in Figure 7.19.

n/k	0	1	2	3	4
0	1				
1	q	p			
2	q^2	$2pq$	p^2		
3	q^3	$3pq^2$	$3p^2q$	p^3	
4	q^4	$4pq^3$	$6p^2q^2$	$4p^3q$	p^4

k	P(k)
0	q^4
1	$4pq^3$
2	$6p^2q^2$
3	$4p^3q$
4	p^4

Figure 7.19

There are many ways to carry out these computations in a spreadsheet. In standard texts the formulas and their derivations usually are presented using binomial coefficients. Although we can use that approach, it is usually more convenient to develop a recurrence relation to implement them in a spreadsheet. For the two-dimensional case, notice that in general we obtain $k + 1$ successes in $n + 1$ repetitions either by (*) having $k + 1$ successes in n repetitions (cell above) followed by a failure (q) or (*) k successes in n repetitions (cell above and to the left) followed by a success (p). This can be implemented in a spreadsheet as shown through our standard diagram in Figure 7.20. The initial cells in each row are special cases with probabilities given by q^n, but they can be created using the same construction if we leave a blank column at the left, since the value of a blank cell is treated as 0. Observe that the probability of obtaining 0 successes in 0 repetitions is 1.

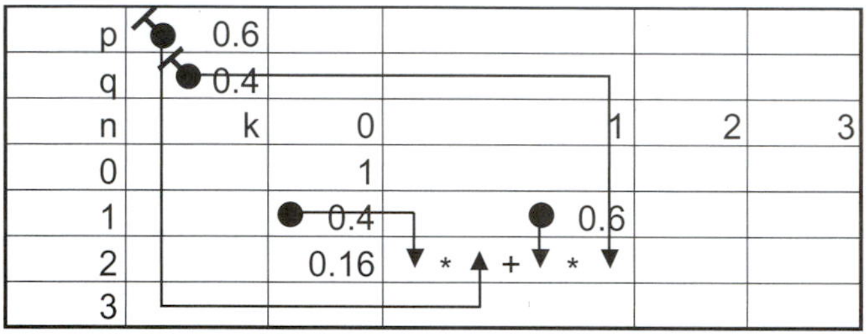

Figure 7.20

    a. Design a two-dimensional spreadsheet model for the binomial distribution with p a parameter. Write the recurrence relation in mathematical notation.

    b. [*] Design a one-dimensional spreadsheet model for the binomial distribution with n and p as parameters. The write the recurrence relation in mathematic notation. Note: This is a very challenging exercise.

    c. Create the tables using the spreadsheet's *Excel* built-in binomial probability function.

    d. Use the approach of Exercise 2 to compute the mean number of successes for given values of n and k in a binomial distribution.

4. Use the model of Exercise 3 to solve the following problems.

    a. A basketball player makes free throws with probability $p = 0.8$. Find the probability of making k shots out n of attempts for $k = 0,1,2,3, \ldots , n$.

    b. Assuming that the probability of giving birth to boys and girls is the same, what is the probability of having $k = 0,1,2,3,4$ girls in a family of four children?

    c. Suppose that the proportion of people supporting a certain proposition is $p = 0.53$. If nine people are chosen at random, what is the probability that most of them support the proposition?

5. [*] Using a series of Bernoulli experiments, let k be the number of failures that occur before the nth success. The probability distribution of k is called the *negative binomial distribution*. Again we can analyze this from diagrams. One format is illustrated in Figure 7.21, with the event of $n = 2$ indicated by boxes.

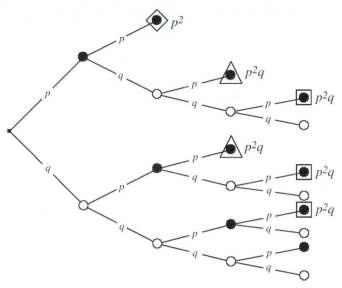

Figure 7.21

If we extend the tree diagram to include a fifth repetition, we will obtain the probabilities shown in Figure 7.22. Notice that the blank cells ultimately need to be completed. Use the table and/or the diagram to discover a recurrence relation and create a spreadsheet model for this distribution.

n/k	0	1	2	3	4
1	p	pq	pq^2	pq^3	pq^4
2	p^2	$2p^2q$	$3p^2q^2$	$4p^2q^3$	
3	p^3	$3p^3q$	$6p^3q^2$		
4	p^4	$4p^2q$			
5	p^5				

Figure 7.22

6. Modify your binomial probability model of Exercise 3 to generate Pascal's triangle and its binomial coefficients by replacing each of p and q by 1. Note that this is equivalent to adding "the cell above and to the left" to "the cell above."

7. Modify your binomial probability model to generate the coefficients that arise in the expansion of $(ax + by)^n$ by replacing p and q by a and b.

8. [*] In one version of the game of Lotto, players select k numbers from the range 1, 2, 3, . . . , n. Then k numbered balls are drawn from a set consisting of the same n numbers. Create a two-dimensional spreadsheet model similar to the binomial example to find the probability that the player has selected i of the chosen balls, for $i = 0,1,2, . . . ,k$. This is called the *hypergeometric distribution*. Note: This is a very challenging problem that may be the hardest exercise in the book.

9. Two teams play a best 4-out-of-7-game playoff series. If the probability of Team I winning any given game is p, determine the probability that Team I wins the series. Compute values for p in steps of size 0.01 over the interval $0 \leqslant p \leqslant 1$, and then create a graph showing the probabilities. Also compute these probabilities for a best 2-of-3, 3-of-5, and 5-of-9 series. What observation can you make about how the length of such a series affects the probability that the best team wins the series?

7.3 Genetics: The Hardy-Weinberg Principle

A diverse assortment of biological subjects that involve probability can be modeled effectively on a spreadsheet through the use of difference equations. One of these topics is the manner in which certain genetic traits are maintained within a large population. An English mathematician, G. H. Hardy, and a German physician, Wilhelm Weinberg, developed much of the overlying theory during the early twentieth century.

A genetic characteristic in an individual diploid organism typically is determined by a pair of genes. We assume that a gene for a certain characteristic (such as eye color) can be of one of two forms, or alleles, A (say brown) or B (say blue). In this case each individual will have a pair of genes of one of three genotypes, AA, AB, or BB.

> A fundamental principle of elementary population genetics concerning proportions of alleles and genotypes is the *Hardy-Weinberg principle*: "Suppose that the alleles of a certain genetic characteristic are of two types, A and B, and that, of the total number of genes for this characteristic in a large population, the proportion of Allele A is $P(A) = p$ and the proportion of Allele B is $P(B) = q$, where $p + q = 1$. If mating is random within the population, both sexes have the same proportions of the alleles, and all individuals have identical reproductive rates, then the allele and the genotype proportions within the population remain constant throughout future generations."

In this section we examine some of the ideas involved in this principle. Our more general approach allows for variable reproduction rates among the different genotypes, and we start with the assumption that we know the proportions of genotypes in a population rather than the proportions of alleles. We denote the proportions of AA, AB, and BB individuals by $P(AA)$, $P(AB)$, and $P(BB)$, respectively.

Instead of all genotypes reproducing with the same fertility rate, we assume that the proportions of each genotype that reproduce are modified by fitness coefficients, e, where each satisfies $0 \leq e \leq 1$. If $e = 1$, then that genotype reproduces at full capacity, while if $e = 0$, then individuals in that genotype cannot reproduce. When $e < 1$, we say that there is selection against that genotype: the smaller the value of e, the greater the degree of selection against the genotype. In our model the fitness coefficients for each genotype are parameters. By setting $e = 1$ for all genotypes we obtain the classical conditions for the Hardy-Weinberg principle. Hence, our model is a generalization of the standard derivation.

We begin by entering the fitness coefficients and the initial proportions of the various genotypes, here using $P(AA) = 0.5$, $P(AB) = 0.3$, and $P(BB) = 0.2$, respectively. We use the first column to count generations, using our standard construction technique of adding 1 to the entry in the previous row. Since each genotype may not reproduce with equal efficiency, before computing the proportions the alleles of the next generation, we first must compute the proportion of each genotype within the reproductive total of individuals. The relative reproductive proportions of each genotype will be $P(AA)e_{AA}$, $P(AB)e_{AB}$, and $P(BB)e_{BB}$, respectively, so that the total proportion of those who reproduce will be their sum. Thus, the proportion of AA individuals in the reproductive group will be

$$\frac{P(AA)e_{AA}}{P(AA)e_{AA} + P(AB)e_{AB} + P(BB)e_{BB}}.$$

We first compute the denominator of this expression by using the SUMPRODUCT function to compute the sum of the pairwise products of the current genotype proportions and the fitness coefficients (see Figure 7.23). Since we will be repeating this in future generations, the array of fitness coefficients is an absolute reference.

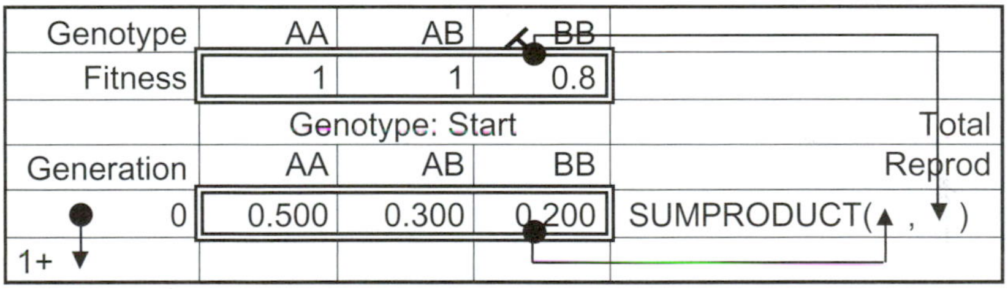

Figure 7.23

We next compute the proportion of reproducing AA individuals by dividing the relative number of reproducing AA individuals (the product of the proportion of AA individuals and the corresponding fitness coefficient) individuals by the total reproducing proportion. By making the denominator an absolute column reference, we can copy the expression to the right to obtain the corresponding proportions for the other genotypes (Figure 7.24). By making the fitness coefficients represent absolute row references, we will be able later to copy these expressions downward for future generations.

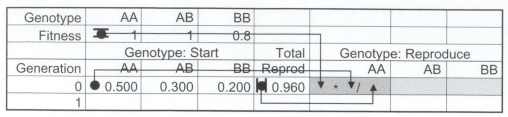

Genotype	AA	AB	BB					
Fitness	1	1	0.8					
	Genotype: Start			Total	Genotype: Reproduce			
Generation	AA	AB	BB	Reprod		AA	AB	BB
0	0.500	0.300	0.200	0.960		*	/	
1								

Figure 7.24

Now we can compute the proportion of alleles in the next generation. We note that all of the genes from *AA* individuals and half of the genes of the *AB* individuals will be of Allele *A*, while all of the genes from *BB* individuals and half of the genes from *AB* individuals will be of Allele *B*. Thus, the proportion of *A* genes is found by finding the sum of the proportion of the *AA* genotype and half of the proportion of the *AB* genotype. Similarly, the proportion of *B* genes will be the sum of the proportion of the *BB* genotype and half of the proportion of the *AB* genotype. These observations provide the formulas shown in Figure 7.25.

Genotype	AA	AB	BB						
Fitness	1	1	0.8						
	Genotype: Start			Total	Genotype: Reproduce			Alleles: Next Gen	
Gen	AA	AB	BB	Reprod	AA	AB	BB	A	B
0	0.500	0.300	0.200	0.960	0.521	0.313	0.167	▲ + ▲/2	▼ + ▲ /2
1									

Figure 7.25

Next, we find the proportions of the three genotypes in the succeeding generation (see Figure 7.26). Here we observe that an *AA* individual must receive two *A* genes, so that the probability of producing such an individual is

$$P(AA) = P(A)P(A) = pp = p^2.$$

Similarly, the probability of obtaining a *BB* individual is

$$P(BB) = P(B)P(B) = qq = q^2.$$

Finally, an *AB* individual obtains an *A* gene from one parent and a *B* gene from the other. There are two ways that this can happen, so that

$$P(AB) = P(A)P(B) + P(B)P(A) = pq + qp = 2pq.$$

Genotype	AA	AB	BB						
Fitness	1	1	0.8						
	Genotype: Start			Total	Genotype: Reproduce			Alleles: Next Gen	
Gen	AA	AB	BB	Reprod	AA	AB	BB	A	B
0	0.500	0.300	0.200	0.960	0.521	0.313	0.167	● 0.677	●0.323
1	▲ ^2	2* ▲ *	▼ ^2						

Figure 7.26

To complete the table we simply use a fill down or copy command (Figure 7.27).

Genotype	AA	AB	BB						
Fitness	1	1	0.8						
	Genotype: Start			Total	Genotype: Reproduce			Alleles: Next Gen	
Generation	AA	AB	BB	Reprod	AA	AB	BB	A	B
0	0.500	0.300	0.200	0.960	0.521	0.313	0.167	0.677	0.323
1	0.458	0.437	0.104	0.979	0.468	0.447	0.085	0.692	0.308
2	0.478	0.427	0.095	0.981	0.487	0.435	0.078	0.705	0.295
3	0.497	0.416	0.087	0.983	0.506	0.423	0.071	0.717	0.283

Figure 7.27

Our model produces the output shown in Figure 7.28. Here we notice that in successive generations the proportion of *AA* individuals increases, while that for *BB* decreases. The reason for this effect is the selection against *BB* individuals. Perhaps surprisingly, the proportion of *AB* individuals also decreases after the first generation. If continued for a large number of generations, eventually few other than the *AA* individuals will exist.

Genotype	AA	AB	BB						
Fitness	1	1	0.8						
	Genotype: Start			Total	Genotype: Reproduce			Alleles: Next Gen	
Generation	AA	AB	BB	Reprod	AA	AB	BB	A	B
0	0.500	0.300	0.200	0.960	0.521	0.313	0.167	0.677	0.323
1	0.458	0.437	0.104	0.979	0.468	0.447	0.085	0.692	0.308
2	0.478	0.427	0.095	0.981	0.487	0.435	0.078	0.705	0.295
3	0.497	0.416	0.087	0.983	0.506	0.423	0.071	0.717	0.283
4	0.515	0.405	0.080	0.984	0.523	0.412	0.065	0.729	0.271
5	0.532	0.395	0.073	0.985	0.539	0.401	0.060	0.740	0.260
6	0.547	0.385	0.068	0.986	0.555	0.390	0.055	0.750	0.250
7	0.563	0.375	0.062	0.988	0.570	0.380	0.051	0.760	0.240
8	0.577	0.365	0.058	0.988	0.584	0.370	0.047	0.768	0.232
9	0.591	0.356	0.054	0.989	0.597	0.360	0.043	0.777	0.223
10	0.603	0.347	0.050	0.990	0.609	0.350	0.040	0.785	0.215

Figure 7.28

We can also show long-term trends by creating graphs of the genotype and the allele proportions as shown in Figure 7.29 and Figure 7.30.

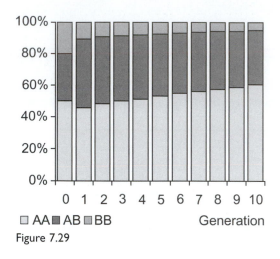

AA ■ AB BB   Generation

Figure 7.29

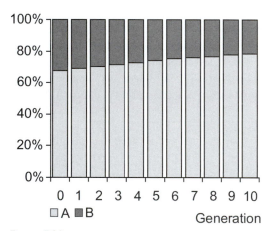

A ■ B   Generation

Figure 7.30

We can employ our model to observe that if all of the fitness coefficients are identical, then equilibrium is achieved within the one generation, as shown in the output in Figure 7.31. This phenomenon is called the *Hardy-Weinberg equilibrium*. We also can use our model to note that more than one initial genotype distribution may give rise to the same proportion of alleles (for example, 0.4, 0.5, 0.1) but that in subsequent generations once p and q are determined, then the genotype proportions are determined as well.

Genotype	AA	AB	BB						
Fitness	1	1	1						
	Genotype: Start			Total	Genotype: Reproduce			Alleles: Next Gen	
Generation	AA	AB	BB	Reprod	AA	AB	BB	A	B
0	0.500	0.300	0.200	1.000	0.500	0.300	0.200	0.650	0.350
1	0.423	0.455	0.123	1.000	0.423	0.455	0.123	0.650	0.350
2	0.423	0.455	0.123	1.000	0.423	0.455	0.123	0.650	0.350
3	0.423	0.455	0.123	1.000	0.423	0.455	0.123	0.650	0.350

Figure 7.31

Also, notice that even if the fitness coefficient for the genotype *BB* is set to 0 (see Figure 7.32), many *BB* individuals will still be in existence after several generations, even though there are none in the reproducing column. Can you see how this happens?

In practice, all of the assumptions needed for the Hardy-Weinberg equilibrium are seldom satisfied. Consequently, it can be illuminating to create models that allow for additional modifications of this problem. For instance, you can consider the case of three alleles rather than two, allow for the effects of mutation, examine the result of migration between two or more different populations, or consider what happens when the different genders have different distributions of genotypes.

Genotype	AA	AB	BB						
Fitness	1	1	0						
	Genotype: Start			Total	Genotype: Reproduce			Alleles: Next Gen	
Generation	AA	AB	BB	Reprod	AA	AB	BB	A	B
0	0.500	0.300	0.200	0.800	0.625	0.375	0.000	0.813	0.188
1	0.660	0.305	0.035	0.965	0.684	0.316	0.000	0.842	0.158
2	0.709	0.266	0.025	0.975	0.727	0.273	0.000	0.864	0.136
3	0.746	0.236	0.019	0.981	0.760	0.240	0.000	0.880	0.120

Figure 7.32

Further Mathematics: The Algebra of Hardy-Weinberg

For those who are interested, here is a brief algebraic derivation of the Hardy-Weinberg result. Suppose that we know that in a given generation that $P(A) = p$ and $P(B) = q$, where $p + q = 1$. Then the proportion of *AA* individuals is $pp = p^2$, the proportion of *AB* individuals is $2pq$, and the proportion of *BB* individuals is $qq = q^2$. If all genotypes have the same fitness coefficients, then the number of *A* genes in the next generation can be found from all of the genes of *AA* individuals and half of the genes of *AB* individuals. Therefore the next value of $P(A)$ will be

$$p^2 + pq = p(p + q) = p \cdot 1 = p.$$

The value of $P(B)$ can be determined similarly from *BB* and *AB* individuals as

$$q^2 + pq = q(p + q) = q \cdot 1 = q.$$

Construction Summary: Genetics (Hardy-Weinberg Principle)

	Genotype		AA	AB	BB			
	Fitness Coeff:		1	1	0.8	1		
	Init Proportion:		0.4	0.4	0.2			
	Alleles		Genotype: Start			Genotype: Reproduce		
n	A	B	AA	AB	BB	AA	AB	BB
2 0			3 0.40	3 0.40	3 0.20	4 0.42	0.42	0.17
1	5 0.63	6 0.38	7 0.39	8 0.47	9 0.14	0.40	0.48	0.12
2	0.64	0.36	0.41	0.46	0.13	0.42	0.47	0.10
3	0.66	0.34	0.44	0.45	0.12	0.45	0.46	0.09

1. Enter parameter values.
2. Generate a counter column.
3. Reproduce the initial genotype proportions.
4. Generate the genotype reproduce proportion for *AA* and copy to the right.
5. Compute the new *A* proportion: all of *AA* and half of *AB*.
6. Compute the new *B* proportion: all of *BB* and half of *AB*.
7. Compute the new *AA* proportion: the product of *A* and *A*.
8. Compute the new *AB* proportion: twice the product of *A* and *B*.
9. Compute the new *BB* proportion: the product of *B* and *B*.
10. Copy as indicated.

Exercises

1. Create a spreadsheet model for the Hardy-Weinberg principle with selection for a genetic trait that has three alleles, *A, B, C*.

2. Examine the fitness coefficients in the model of Exercise 1 and determine which have the most profound effect on the ultimate composition of the population.

3. Create a spreadsheet model for the Hardy-Weinberg principle that incorporates the possibility of mutation. Assume that in every generation a certain percentage of the *B* alleles mutate into *A* alleles.

4. Create a spreadsheet model for the Hardy-Weinberg principle that incorporates the possibility of migration. Maintain the distributions for two separate populations, but in each generation assume that a certain percentage of the second pop-

ulation migrates into the first population. Make an assumption about the comparative sizes of the two populations. Observe the long-run distributions of the alleles and genotypes.

5. Modify Exercise 4 so that there is migration each way between the populations. Experiment with equal and different migration rates. Observe the long-run distributions of the alleles and genotypes.

7.4 Simulation

Usually when we analyze a problem or describe a process we try to produce a mathematical model or even a formula that provides a direct solution. However, at times either the problem or the underlying mathematics proves to be too difficult or complex to do so. When this occurs we frequently can gain insights into the problem and its mathematics by generating a series of simulations with some random variations and discover patterns. In this section we illustrate the process by creating *Excel* simulations of several elementary events that, in fact, we can also deal with in a more traditional fashion. This way we can illustrate the simulation approach and compare the insights that we obtain from the simulation with the exact solution and see how close we come, or what we might miss in using the approach.

The key tool in these examples is the spreadsheet's library random function, RAND(). This function generates a real number at random so that $0 \leq$ RAND() <1. This is done in such a way that any number in this interval is equally likely of being generated as a random number. The numbers generated in this fashion are said to come from the uniform probability distribution on the interval from 0 to 1.

Numbers generated by a computer normally are not real random numbers in the sense of being totally unpredictable. Therefore, the correct technical term for this kind of number is pseudo-random numbers.

Studies by mathematicians who are concerned about the quality of random numbers have shown that *Excel's* random number generator has some problems. Therefore, it should not be used for studies needing high scientific accuracy and fulfilling higher requirements for the quality of random numbers. The quality of *Excel's* random number generator, however, is good enough for the kind of small projects we will do in this section.

A simple look at the RAND() function is given in Figure 7.33 through Figure 7.36. The five formulas in Figure 7.33 generate random numbers. We simply enter the first formula and use a fill down or copy command to copy it into the other four cells. Sample output from the formulas is shown in Figure 7.34. Each time the spreadsheet is recalculated, either by changing an entry in the spreadsheet or by pressing the recalculation key, a new set of random numbers is formed. The results of two additional recalculations are shown in Figure 7.35 and Figure 7.36 Notice that each time we modify any cell in the spreadsheet, the spreadsheet will be recalculated and a new set of random numbers will be generated.

RAND()	0.990463	0.517948	0.182333
RAND()	0.103861	0.451818	0.583237
RAND()	0.231759	0.272196	0.598755
RAND()	0.317502	0.116695	0.223235
RAND()	0.394744	0.877294	0.737299
Figure 7.33	Figure 7.34	Figure 7.35	Figure 7.36

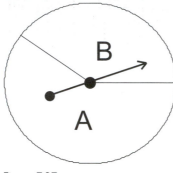

Figure 7.37

We use the spreadsheet's random function to simulate a variety of probabilistic events. The use of random numbers in this way is called *Monte Carlo simulation.*

We start with a simple example. Suppose that we have a process that generates random output, such as with the spinner shown in Figure 7.37. Here the circle is divided so that the area of the section for A forms 60 percent of the circle, with the section for B forming the remaining 40 percent. Thus, to simulate the spinner we want to create a spreadsheet procedure that produces the output A with probability 0.6 and, consequently, produces B with probability 0.4.

We can simulate spinning the arrow with the statement

$$\text{IF}(\text{RAND}() < 0.6, "A","B").$$

If the random number generated is less than 0.6, then an A is produced; otherwise a B results. Since each number between 0 and 1 is equally likely of being generated, and 60 percent of them lie to the left of 0.6 on the real number line, this produces the desired simulation. The picture in Figure 7.38 shows one way of visualizing the generation of random numbers. Suppose that the rectangle is divided vertically along the segment $0 \le x < 1$ at $x = 0.6$. Then a vertical line is drawn from the location of the random number. In the case illustrated, the random number is 0.7231, so that B is the resulting output.

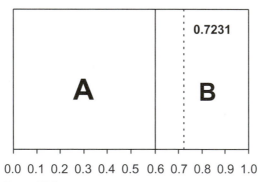

0.0 0.1 0.2 0.3 0.4 0.5 0.6 0.7 0.8 0.9 1.0

Figure 7.38

Figure 7.39 and Figure 7.40 show one format that can be used in a model for this simple event. In fact, we create a more flexible model by using a parameter p with $0 \le p \le 1$ rather than the constant 0.6. If the random number generated is less than the value of p, then A is produced, otherwise B is.

p	0.6 ●
Output	IF(RAND()< ▼,"A", "B")

Figure 7.39

p	0.6
Output	B

Figure 7.40

Notice that in the previous design we do not see the random number that is generated. If this feature is desired, then we can reformulate our model as shown in Figure 7.41 and Figure 7.42.

p	●	0.6
rand	●	Rand()
Output	IF(▼ < ▼,"A", "B")	

Figure 7.41

p	0.6
rand	0.1052
Output	A

Figure 7.42

1. Simulating Bernoulli Trials and the Binomial Distribution

We now extend the previous example. Again, we assume that we have an experiment that generates exactly two outcomes. Perhaps we shoot at a target (hit or miss), flip a coin (heads or tails), or select a member of a class at random (female or male). This time one outcome, S, is often called a success and has a probability of p. The other outcome, F, is then called a failure and has a probability of $q = 1 - p$. We might think of shooting at a target so that hitting a target is a success and missing it is a failure. Or we might consider getting a head in flipping a coin as a success and getting a tail as a failure. The resulting probability distribution is called a *Bernoulli distribution*.

If the experiment is repeated, with the probabilities p and q remaining constant in each repetition and the outcomes independent of each other, then the resulting process is called a series of Bernoulli trials. If we repeat the experiment a fixed number of times, then the number of successes is said to have a binomial distribution.

Our model in Figure 7.43 and Figure 7.44 generates a set of 10 repetitions of a simple Bernoulli experiment whose probability of success is a parameter of the model. To use the model, we enter a number between 0 and 1 as the probability of success in any given repetition. Here we use 0.6 as this probability.

The first column of our model counts the repetitions of the simple event. We want to simulate 10 random and independent repetitions so that each will represent a success or a failure. The formulas in the second column generate either a 1 (for a success) if a random number RAND() is less than the given probability or a 0 (for a failure). We can interpret the corresponding IF statement as

IF the random number is less than the selected probability

THEN return a 1 to represent a success

ELSE return a 0 to represent a failure.

We enter the indicated formula at the top of the column and then copy the formulas as indicated. The number of successes that are produced is generated at the top of the model by summing the second column. In doing this we have employed a standard mathematical technique: Summing a list of 1s and 0s produces the same result as counting the number of 1s—but is easier to do. This sum gives us the number of successes.

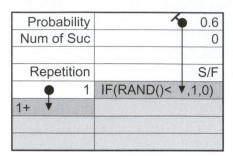

Figure 7.43   Figure 7.44

A partial set of typical output is shown in Figure 7.45. We will obtain a different set of simulated values each time that we press the recalculation key (see Appendix).

Probability	0.6
Num of Suc	5
Repetition	S/F
1	1
2	0
3	1
4	1

Figure 7.45

The next two columns of the model are created via the data table procedure that we described in an earlier section and in the Appendix (see Figure 7.46 and Figure 7.47). The data table is used to list the number of successes in 100 repetitions of this experiment. We first create a column to count the iterations of the trials 1, 2, . . . , 100 in the usual manner. In the cell at the top of the second column of the data table, we enter a formula to reproduce the number of successes obtained in the set of 10 repetitions at the left. Issuing the data table command, using any empty cell as the column input location, then will carry out 100 iterations of the process and generate a summary list in the second column.

We can also use built-in spreadsheet functions to compute various statistics from the results of the simulations. In Figure 7.46 and Figure 7.47 we have found the mean and the standard deviation for the number of successes.

Probability	0.6	Mean:	
Num of Suc	5	SD:	
			Num
Repetition	S/F	Trial	Suc
1	1		
2	1	1	
3	0	2	
4	0	3	

Figure 7.46

Mean:	AVERAGE(▲)	
SD:	STDEV(▲)	
		NUM
Trial		Suc
		5
1		7
2		6
3		6

Figure 7.47

We can obtain any numbers of sets of these simulations simply by pressing the recalculation key repeatedly. We see partial results from two repetitions of our model in Figure 7.48 and Figure 7.49.

Mean:	6.05
SD:	1.527
	Num
Trial	Suc
	8
1	5
2	7
3	5
4	7

Figure 7.48

Mean:	5.77
SD:	1.613
	Num
Trial	Suc
	5
1	4
2	4
3	7
4	9

Figure 7.49

One of the ways in which we can examine the nature of the output of this process is through a graph showing the frequencies of different numbers of successes in a set of 100 (or more, if you prefer) trials. To do this we create a column to list the number of possible outcomes, ranging from 0 through 10 successes (see Figure 7.50). We then use the spreadsheet's COUNTIF (*array,number*) function. This counts the number of cells in the array of the numbers of successes that match a given number of successes. Thus, the first of these counts the number of trials with 0 successes. As usual, we enter the formula into the top cell with the reference to the array absolute. We then copy the formula down the column to complete the calculations. Typical output for another set of trials is shown in Figure 7.51.

Mean:	6.12	Num				Num	
SD:	1.546	Suc		Freq		Suc	Freq
	Num					0	0
Trial	Suc	0	COUNTIF(,)			1	0
	7	1				2	0
1	4	2				3	5
2	5	3				4	11
3	6	4				5	20
4	8	5				6	25
5	4	6				7	22
6	6	7				8	12
7	5	8				9	4
8	5	9				10	1
9	6	10					

Figure 7.50   Figure 7.51

We can now use the last two columns to create a column graph to show the frequencies of the number of successes in the 100 trials, with the values used as labels on the bars (Figure 7.52). By repeatedly pressing the recalculation key, we can get a better picture of what our Monte Carlo simulations produce (Figure 7.53).

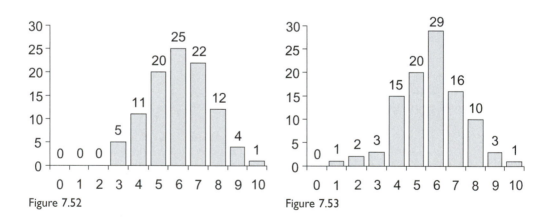

Figure 7.52   Figure 7.53

Finally, we note that in this example we have generated random real numbers in the range $0 \leq x < 1$ by RAND(). To form random real numbers in the range $a \leq x < b$, we use $a + (b - a)^* $ RAND().

Construction Summary: Binomial Simulation

1. Enter probability of success, p.
2. Generate counters in columns.
3. Use =IF to generate 1 (success) if RAND() is less than p, or 0 (failure) otherwise. Copy down.
4. Use =SUM to count successes.
5. Generate a data table to count successes in repeated trials. In the top cell, reproduce the number of successes (Box 4). Use any blank cell as the column input cell.
6. Compute the average of the column.
7. Compute the standard deviation of the column.
8. Use =COUNTIF function to find the number of appearances of the value of the cell to left within the success column. Copy down.

ProbSuc	1	0.6	Mean:	6	6.00	Num	
NumSuc	4	4	SD:	7	1.57	Suc	Freq
					Num		
	Trial	S/F	Trial		Suc	2 0 8 0	
2	1 3	0		5	4	1	0
	2	0	2 1		6	2	1
	3	0	2		8	3	3
	4	1	3		4	4	17
	5	1	4		7	5	16
	6	1	5		8	6	24
	7	0	6		6	7	18
	8	0	7		7	8	18
	9	0	8		3	9	3
	10	1	9		7	10	0

2. Rolling a Pair of Dice

In the previous examples of this section we generated random real numbers. However, frequently we will want to generate random integers instead. Here we illustrate the process for doing this by simulating rolling two dice, each of which generate an integer, 1, 2, 3, 4, 5, or 6. To see how we might do this, we consider the following steps. First, since the number generated by RAND() lies in the interval $0 \leq$ RAND $() < 1$, it follows that the expression 6*RAND() generates a real number in the range $0 \leq 6*$ RAND $() < 6$. If we now convert this value to an integer by INT(6*RAND()), then INT() truncates the decimal part of the number, to generate 0, 1, 2, 3, 4, or 5. Finally, to obtain values in the proper range, we add 1, to obtain INT(6*RAND()) + 1. This formula is implemented in the model shown in Figure 7.54 and Figure 7.55.

Again we have chosen to expand our model by generating pairs of random integers that are within the range from 1 through a positive integer n that is a parameter of the model. Here we have set this value to simulate the dice setting. The model uses the word "faces" to indicate the number of sides on the dice that are being used. In

Figure 7.54, the first column provides values for an iteration counter. We also enter the formula to simulate throwing one of the dice and then use fill or copy commands to produce the outcomes of the remaining tosses of both dice. We copy far enough to produce 100 iterations.

In Figure 7.55, the fourth column is used to obtain the sum of each pair of dice. We can readily expand our model to examine the frequency of occurrence of a number of phenomena. Here, in the fifth column we generate a 1 if a "double" is obtained (that is, the numbers on both dice are the same) or a 0 otherwise. This allows us to examine the frequency of obtaining doubles.

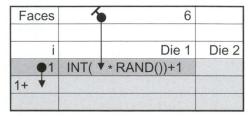
Figure 7.54

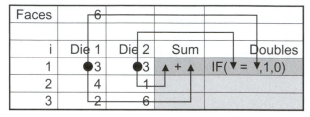
Figure 7.55

We then obtain a count of the number of doubles by summing the last column. Again we have used the technique that was discussed in a previous example of this section (see Figure 7.56). A portion of our output for one set of simulations is shown in Figure 7.57. Recall that there are 100 tosses of the pair of dice.

Faces		6	Num. Doubles:	SUM(▲)
i	Die 1	Die 2	Sum	Doubles
1	3	3	6	1
2	4	1	5	0
3	2	6	8	0

Figure 7.56

Faces		6	Num. Doubles:	15
i	Die 1	Die 2	Sum	Doubles
1	3	3	6	1
2	4	1	5	0
3	2	6	8	0

Figure 7.57

It would clearly be helpful for us to generate a summary of the various outcomes that are obtained in the 100 iterations. This can be done by employing the COUNTIF function. We illustrate its use in Figure 7.58. The expression COUNTIF(*array, value*) counts the number of occurrences of the number *value* in the block *array*. In the model, we first generate a list of possible outcomes, 2, 3, . . . , 12.

Sum:	2	3	4	5	6	7	8	9	10	11	12
Count:											

Figure 7.58

We next enter the indicated function under 2 to count the number of 2s produced in the simulation, and then copy the formula to the right (Figure 7.59).

Faces	6	Num. Doubles:		11					
i	Die 1	Die 2	Sum	Doubles	Sum:	2		3	4
1	4	2	6	0	Count:	COUNTIF(,)			
2	2	6	8	0					
3	5	2	7	0					

Figure 7.59

The resulting output is displayed in Figure 7.60.

Sum:	2	3	4	5	6	7	8	9	10	11	12
Count:	3	4	4	7	17	21	15	14	8	6	1

Figure 7.60

We can also create graphs for the summary output using the two rows displayed in Figure 7.60. Performing this experiment will typically produce something like the general "bell-shaped" curves you might expect, but frequently this will be without complete symmetry because we are dealing with small random samples from a distribution and they can vary noticeably. We can see this by repeatedly pressing the recalculation button to observe the results of many samples. Two of these are shown in Figure 7.61 and Figure 7.62.

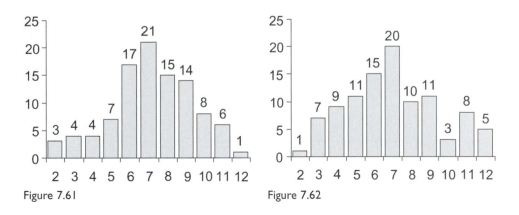

Figure 7.61

Figure 7.62

Construction Summary: Dice Simulation

1. Enter the number, N, of die faces.
2. Generate a counter column.
3. Simulate a die roll using =INT(number*RAND())+1, copy down and right.
4. Add the two cells to the left, copy.
5. Use =IF to produce 1 if doubles appear, 0 otherwise.
6. Compute the column sum to find the number of doubles.

	N	1	6	Num. Doubles:	6	19
	i		Die 1	Die 2	Sum	Doub
2	1	3	1	3	4	4	5	0
	2		4	2	6	0
	3		4	4	8	1
	4		3	4	7	0
	5		5	5	10	1
	6		4	2	6	0
	7		5	5	10	1

3. The Birthday Problem Revisited

We will now pursue one more Monte Carlo simulation. In an earlier section we determined the probability that in a room of n people there will be at least two people with the same birthday. Now we can test our results by simulating the birthdays of a group of n people. In the model that follows we use $n = 30$, but it is easy to vary this to whatever number you wish.

To build the model we use the first column as a counter for the number of people in the room (Figure 7.63). Next, we identify every possible birthday with the integers from 1 to 365 (or 366 if you prefer—simply change the parameter for the number of days). We then generate the first random integer in the correct range at the top of the second column and copy it down the column. Of course, the reference to the number of possible days must be absolute.

Days		365
Person		Birthday
1	1+INT(	* RAND())
2		
3		

Figure 7.63

Next, in another column we create a counter for all of the days in the year (see Figure 7.64). To the right of the first day we use the COUNTIF function to determine the number of birthdays in our group that fall on that day. If the value is 2 or larger, then we have located a duplicate. Consequently, we enter another COUNTIF function at the top to count how many of the 365 days contain a number greater than 1 (see Figure 7.65). This gives the number of duplicates in the list of 30. In the same way, it is just as easy to find the number of triplicates (that is, groups with three or more persons who share the same birthday). The latter probability is difficult to compute directly.

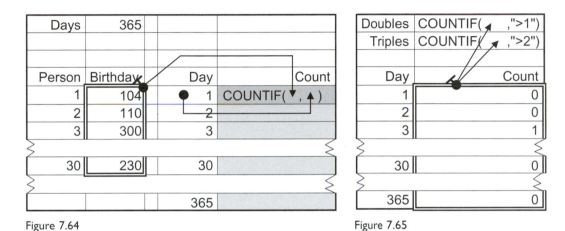

Figure 7.64

Figure 7.65

As usual, we will generate a new set of birthdays for the group each time that we press the recalculation button. Partial outputs of two repetitions of our simulation are shown in Figure 7.66 and Figure 7.67.

Days	365	Doubles	3
		Triples	0
Person	Birthday	Day	Day
1	65	1	0
2	162	2	2
3	82	3	0
4	282	4	1

Figure 7.66

Days	365	Doubles	0
		Triples	0
Person	Birthday	Day	Day
1	164	1	0
2	212	2	0
3	346	3	1
4	347	4	0

Figure 7.67

Finally, we use two more columns at the right to create a data table that will allow us to generate a summary of any number of trials of this experiment (see Figure 7.68). Here we have simulated a set of 100 trials, although you will probably want to do more. The formula at the top of the data table produces a 1 if the corresponding simulation contains any duplicates. Thus the data table consists of a list of 0s (no duplicates) and 1s (at least one duplicate).

Doubles	2	Num	
Triples	0		
		Trial	Doubles
Day	Count		IF(>0,1,0)
1	1	1	
2	0	2	
3	1	3	
4	0	4	

Figure 7.68

The SUM() function is used at the top to find the number of 1s by adding the column (Figure 7.69). This is the same technique that we have used twice before in this section.

One set of partial output is shown in Figure 7.70. We can now check the numbers produced here with our theoretical results in an earlier model to see if they seem to agree with each other. Notice that in 100 sets of 30 people, 72 of them contained people with duplicate birthdays.

Num	SUM( )	
Trail		Doubles
		1
1		1
2		0
3		1
4		1

Figure 7.69

Num		72
Trial	Doubles	
	1	
1	1	
2	0	
3	1	
4	1	

Figure 7.70

A number of modifications are possible in this last model. We can generate the Boolean expressions True and False instead of 0 or 1, we can vary the number of people, and we can vary the number of "dates" in order to simulate a variety of related problems.

Construction Summary: Birthday Problem Simulation

1. Enter the number of days in a year.
2. Generate counter columns.
3. Generate a random birthday as =1+ INT(days*RAND()), copy.
4. Use the =COUNT function to find the frequency of days, copy.
5. Count values greater than 1 in a column to give the number of doubles.
6. Count values greater than 2 in a column to give the number of triples.
7. Create a data table in a double outlined block. In the top cell, reproduce the number of doubles. Choose any blank cell as the column input cell.
8. Compute the sum of the doubles column.

Days	1	365	Doub	5	3	Num	8	72
			Triple	6	1			
						Trial	Doub	
Person		B'day	Day	Count			7	1
2	1	3	80	2	1	4	0	2	1	0
	2	290	2	0			2	1
	3	239	3	0			3	1
	4	35	4	0			4	1
	5	362	5	0			5	1
	6	129	6	0			6	1
	7	340	7	0			7	1
	8	234	8	0			8	0

Exercises

1. Create a spreadsheet model to simulate rolling three dice for a total of 100 times, finding the sums of each of the three dice. Then use a data table to record the results of 1,000 repetitions of this event both numerically and graphically.

2. Repeat the experiment of Exercise 1, except this time determine the number of the 100 rolls that result in a triple, a double, and three distinct singles that are generated in each of the 100 rolls of the three dice. Finally, use a data table to generate a summary of the results both numerically and graphically.

3. Simulate drawing a five-card poker hand from a standard deck of cards. Use the spreadsheet's text functions to create nice output.

4. Use a spreadsheet to simulate a lottery in which six numbers are drawn at random out of a set of 50 numbers, 1, 2, 3, . . . , 50. There can be no duplicates.

5. Simulate the birthday problem with a set of n people to determine the probability that at least two people have the same birthday. Use the integers 1, 2, 3, . . . , 365 to represent the dates. Then use a data table to find the results of 1,000 repetitions of the experiment. See if your simulation result agrees with the probability from our earlier birthday problem model.

6. Repeat Exercise 5, but in addition find the proportions of the sets that have three or more people with birthdays on the same day, or have more than one day with two or more people with birthdays on that day.

7. Simulation can be used to approximate areas. Use the spreadsheet's RANDOM function to generate 1,000 pairs of real numbers *(x,y)* in the region $0 \leqslant x \leqslant 1$, $0 \leqslant y \leqslant 1$. Design your model to count the number of pairs that lie in the first quadrant of the unit circle, $x^2 + y^2 = 1$ (see Figure 7.71). The ratio of those points that lie in the quarter-circle to all points gives us an approximation to $\pi/4$. Use this observation to obtain an approximation for π. Construct a spreadsheet model using this approach, together with a data table to obtain a large number of these approximations of π, whose average gives an improved approximation for π.

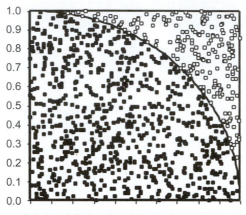

Figure 7.71

8. [*] In the game of *craps* a participant rolls a pair of dice. If the result of the first roll is either 7 or 11 then the participant wins, while if it is 2, 3, or 12 then he or she loses. If another number is generated, then it becomes the participant's "point" and he or she rolls again. On the second roll he or she wins if the roll produces his or her point and loses if the roll yields a 7. Otherwise the second step is repeated until either a win or a loss results.
   a. Design a spreadsheet model to compute the probability of winning or losing the game.
   b. Design a spreadsheet model to simulate one game of craps. Then incorporate a data table to generate the results of a large number of simulations. Determine if the result approximates the answer determined in Part a.
   c. Expand your model of Part b to determine the average number of rolls until the game ends.
   d. Mathematically derive the probability of winning, as well as various subprobabilities, such as winning if your point is 4.
   e. Casino wagering allows bets on various subevents in craps. Simulate a day of wagering on one of these events, incorporating together the odds that casinos pay.

9. Suppose that the probability of a team winning a sports event remains constant at the value *p*, say *p* = 0.550, for an entire season. Simulate a season of 162 games and find the longest winning streak during the season. Then use a data table to provide a summary of the lengths of the longest winning streaks over 1,000 seasons.

8

Finding Optimal Solutions

CHAPTER OUTLINE

8.1   Optimal Bridge Location

In business, entrepreneurs often want to minimize costs by finding the best, or optimal, location for a warehouse, store, or similar facility. This may require finding the location that will minimize the cost of delivering goods from a warehouse to stores, or to minimize the distance traveled in getting to different facilities. The fields of both calculus and operations research develop mathematical techniques to apply to these situations, and many of these techniques lend themselves naturally to spreadsheet implementations. However, frequently we can get good estimates of a solution simply by examining large numbers of possible locations in an efficient manner. In this section we use this approach to determine the optimal location for a bridge spanning a river on a road that connects two towns.

One of the advantages of using a spreadsheet to investigate a mathematical topic lies in the possibilities that it provides us for developing more than one approach for the analysis, presentation, and visualization of the topic. Here we examine an elementary optimization problem and see how it can be analyzed in various ways on a spreadsheet. Techniques like those used here can be adapted for use with a variety of more complex problems. This example is also discussed in Arganbright (1978).

Consider the situation of two towns whose centers are located at coordinates (a,b) and (c,d), with distances given in kilometers, on opposite sides of a wide, vertical river of width w kilometers. The area government wants to construct a road, including a bridge over the river, to connect the towns so that the road between them has the minimum length possible. If the bridge must be built so that it is perpendicular to the riverbanks, then where should it be located? Of course, the land segments of the road will be straight-line segments. Two possible locations are presented in Figure 8.1 and Figure 8.2.

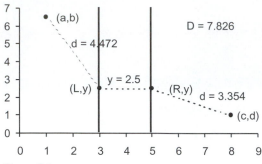

Figure 8.1

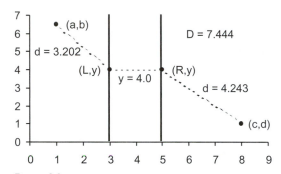

Figure 8.2

These figures show the towns, *A* and *B*, with their respective centers located at the points and $(a,b) = (1.0, 6.5)$ and $(c,d) = (8.0, 1.0)$. The riverbanks are shown as the vertical lines given by the lines $x = 3$ and $x = 5$. In general, the x-values of the left and right endpoints of the bridge are parameters, denoted by L and R, and we let (L,y) and (R,y) denote the coordinates of these points. The value y gives the location of the bridge. We enter values for the model's parameters in the top rows and then enter a trial value for the optimal value of y. The resulting length of the road is computed as

$$\sqrt{(a - L)^2 + (b - y)^2} + w + \sqrt{(c - R)^2 + (d - y)^2}.$$

Because the length of the bridge, w, is the same for all locations, we will eliminate it from our calculations and work only with the total road distance, excluding the bridge. This distance is given by

$$D = \sqrt{(a - L)^2 + (b - y)^2} + \sqrt{(c - R)^2 + (d - y)^2}.$$

Our model is described in Figure 8.3.

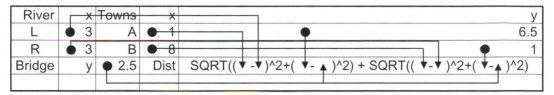

Figure 8.3

Using this model with an estimate of $y = 2.5$ for the bridge's location produces a road length of $D = 7.826$, as shown in Figure 8.4. By systematically adjusting the value of y we can obtain better and better answers. For example, in Figure 8.5 we find that $y = 4.3$ gives a smaller value of $D = 7.433$. Continuing in this manner we can eventually determine the optimal location to the degree of precision that is desired. We also can incorporate a solver effectively at this point to vary the bridge location.

River		x	Towns	x	y
L		3	A	1	6.5
R		5	B	8	1
Bridge		y		2.5	Dist	7.8262

Figure 8.4

River		x	Towns	x	y
L		3	A	1	6.5
R		5	B	8	1
Bridge		y		4.3	Dist	7.433

Figure 8.5

We can augment our model effectively by incorporating the spreadsheet's graphic capabilities to help us visualize various aspects of our solution technique. The display in Figure 8.6 shows the format that was used to supplement our model in order to draw our diagrams using an xy-graph. Each of the values is either a constant or is generated by a simple formula that reproduces the value of a cell in the model. As our estimates of y change, the bridge and the road are redrawn. We can attach interior

x	y path	y river
1	6.5	
3	2.5	
5	2.5	
8	1	
3		0
3		7
5		0
5		7

Figure 8.6

labels to the graph to show lengths and coordinates that are updated with the changes. The model can be enhanced through animation by incorporating a scroll bar to vary the bridge location.

Using the graphic display we can observe what is happening as we approach the optimal value for y. It seems that the two line segments of the road at the optimal point are parallel. This actually is the case.

We can get a better idea of this by computing the slopes of the road line segments (see Figure 8.7). Recall that the slope of a line is given by the change in the y-coordinates (rise) divided by the change in the x-coordinates (run).

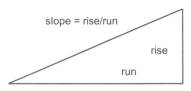

Figure 8.7

The next step is to include the computation of these slopes (Figure 8.8 and Figure 8.9).

River	x	Towns	x	y	slope
L	3	A	1	6.5	-2.00
R	5	B	8	1	-0.50
Bridge	y	2.5	Dist	7.826	

Figure 8.8

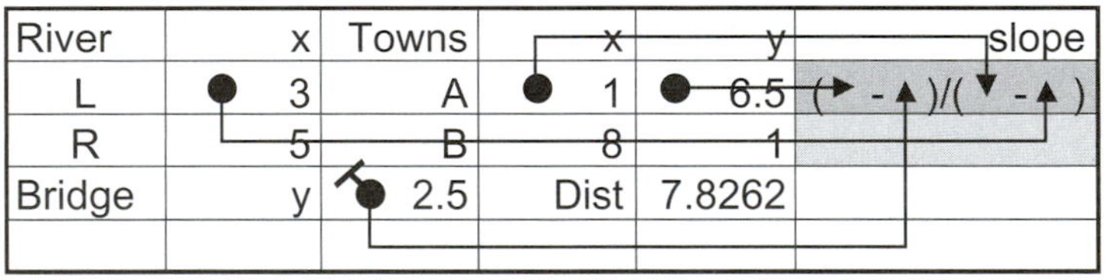

River	x	Towns	x	y	slope
L	● 3	A	● 1	● 6.5	(➤ - ▲)/(▼ - ▲)
R	5	B	8	1	
Bridge	y ↖ ● 2.5	Dist	7.8262		

Figure 8.9

It is possible to see geometrically that these line segments must have the same slope. That is, if the values of the two angles do not match at your candidate for the optimal point, then you have not found the exact minimum. To see this, we can consider drawing our bridge layout on a sheet of paper. Since the bridge width is the same for any location, we eliminate it by folding the paper to eliminate the river. The shortest distance between the two towns is then simply a straight line. We can draw this and then unfold the paper to see the solution! (Refer to Figure 8.10 through Figure 8.13.)

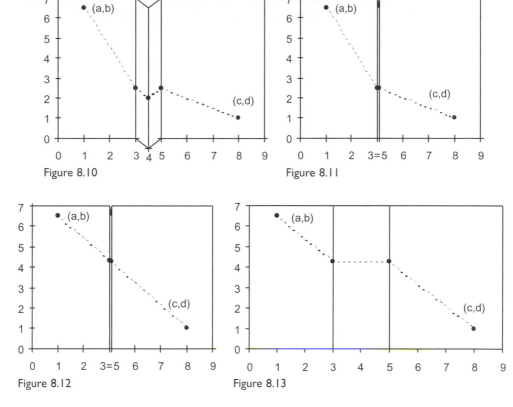

Figure 8.10

Figure 8.11

Figure 8.12

Figure 8.13

While the spreadsheet approach that was used in the preceding model is quite valid, its need for a manual trial-and-error iteration is somewhat inefficient. In addition, more complex problems will not necessarily allow us to devise such a clever geometric solution as the paper-folding scheme. Thus, we next consider some other mathematical methods.

Our first alternate approach is to replace the manual computations used in the foregoing example by employing the spreadsheet's solver command. Although the details of these commands vary among the different spreadsheets, in *Microsoft Excel* we select a cell that contains a formula (here the cell for D) whose value we want to minimize and a cell that contains an independent variable that is changed (here the cell for y) in order to optimize (here to minimize) the value of the selected formula. Doing this with our model produces the output in Figure 8.14, with the optimal length (to three decimal places) being $D = 7.433$ and the bridge at $y = 4.300$. In doing this we also observe that the slopes of the two segments are equal. We could also use either the solver or the goal seek command to set the slopes of the two line segments equal.

River	x	Towns	x	y	slope
L	3	A	1	6.5	-1.10
R	5	B	8	1	-1.10
Bridge	y	4.300	Dist	7.433	

Figure 8.14

Another way to approach the solution of the problem is by generating a large number of values of bridge locations, y, at regular increments over the interval $b \leq y \leq d$ and then computing the corresponding road lengths, D, for each of these. In the next model we first enter the indicated labels and values for the parameters (see Figure 8.15). We will generate y values in equal steps of size 0.1, starting with the y-coordinate of the first town, y_0, and ending with the y-coordinate of the second town, y_1. We first reproduce the initial value of y in the top cell as y_0 and then in the cell beneath it add 0.1 onto the value of the cell above. Next, as the initial distance formula, we enter the expression used earlier,

$$D = \sqrt{(a - L)^2 + (b - y)^2} + \sqrt{(c - R)^2 + (d - y)^2}.$$

Because of its complexity we do not reproduce the formula for "dist" in the display in Figure 8.15. However, it is essentially the same construction that we used in the first model of this section, except that all of the references other than for y must be absolute to allow us to copy the formula. Notice that we have counted the iterations, n. This can help us in our planning when we later design the functions that will locate the minimum distance.

River		x	Towns	x	y
L		3	A	1	6.5
R		5	B	8	1
Optimal Bridge:			Distance	List Rank	y value
n		y	Dist		
1			(see text)		
1+	0.1+				

Figure 8.15

We then copy these values down their columns as indicated. After this we can find the minimum length of D that occurs in this column by using the spreadsheet's MINIMUM function (see Figure 8.16).

Although we have found a good approximation of the minimum distance, we still need to determine the corresponding value of the bridge location, y. Of course, one way in which we can do this is to scan down the output list manually to locate the minimum distance that was found. However, we can also arrange for the spreadsheet to do this for us. First we use an *Excel* spreadsheet function to find where in the list the sought-for value first appears. We can interpret the function, MATCH(value,list,0), as returning the location within the list where the value first appears. This corresponds to finding the appropriate value of n. Here the function returns 34, meaning that 7.433 is the 34th term in the column (Figure 8.17).

River	x	Towns	x	y
L	3	A	1	6.5
R	5	B	8	1
Optimal Bridge:		Distance	List Rank	y value
		MIN(▲)		
n	y	Dist		
1	1.0	8.852		
2	1.1	8.760		
3	1.2	8.671		
4	1.3	8.586		

Figure 8.16

Optimal Bridge:		Distance	List Rank	y value
		● 7.433	MATCH(➤ , ➤ ,0)	
n	y	Dist		
1	1.0	8.852		
2	1.1	8.760		
3	1.2	8.671		
4	1.3	8.586		

Figure 8.17

We can now locate the *y*-value corresponding to the minimum value by using another spreadsheet function. We can interpret INDEX(*array,num_rows*) as giving the value of the cell that is located *num_rows* rows down from the top in the column array *array*. Here, the sought-for *y*-value is down 34 rows from the top of the *y*-column (see Figure 8.18).

Optimal Bridge:		Distance	List Rank	y value
		7.433	● 34	INDEX(➤ , ▼)
n	y	Dist		
1	1.0	8.852		
2	1.1	8.760		
3	1.2	8.671		
4	1.3	8.586		

Figure 8.18

The resulting output is shown in Figure 8.19.

River		x	Towns	x	y
	L	3	A	1	6.5
	R	5	B	8	1
Optimal Bridge:			Distance	List Rank	y value
			7.43303	34	4.3
	n	y	Dist		
	1	1.0	8.852		
	2	1.1	8.760		
	3	1.2	8.671		
	4	1.3	8.586		

Figure 8.19

Using the columns shaded in Figure 8.19, we can create a graph of the road length, D, as a function of bridge location, y, as shown in Figure 8.20. From the graph we can see where the optimal location occurs, as well as get a feeling for the fact that the points y that are not far from the optimum location will produce a distance that is not very much longer than the optimum.

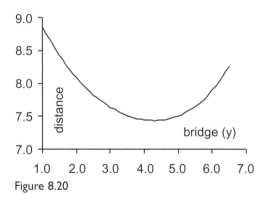

Figure 8.20

Some observations should be made at this time. First, in the tabular approach of the last model we have used only some of the values for y that are possible. We well may have missed the absolute minimum. Thus, we could reduce the step size to obtain better approximations. Nonetheless, our graph shows that we already have come close to the optimal solution.

As an alternative approach, the data table concept discussed in an earlier section can be used effectively here also. It saves us the hassle of needing to enter the distance formula for each location. We will present this approach as our last solution technique.

We first create two columns, the first systematically listing the various values for y and the second left empty for subsequent computations of the corresponding distances (see Figure 8.21). At the top we include a row that is initially empty. Then into the distance column of that row we enter a formula to reproduce the value of D for the current location, y. We then issue the data table command, using the cell containing the value for y as the column input variable. The results are shown in Figure 8.21 at the right. From these we can use techniques developed earlier to locate the minimum.

River	x	Towns	x	y		y	Dist		y	Dist
L	3	A	1	6.5						7.8262
R	5	B	8	1		0.0			0	9.963
Bridge	y	2.5	Dist	7.8262		0.1			0.1	9.8373
						0.2			0.2	9.7147
						0.3			0.3	9.5952

Figure 8.21

Finally, we observe that once they have been created, we can use our models to experiment with the town locations as well as with the location and width of the river. As an interesting project, you might try to redesign the models of this section to allow for a river of a different shape. In addition, a number of other possible solution techniques can be discovered and implemented on your spreadsheet. One of these is outlined in Part 2 of the following Construction Summary. In the fourth column we use an IF statement to reproduce the y-value of the minimum distance and a blank otherwise.

Construction Summary: Optimal Bridge Location, Part I

1. Enter the river, town locations.
2. Enter a bridge location estimate.
3. Compute the length of the route.
4. Compute the slope of the left line, copy.
5. Minimize *distance* via solver.

River		x	Towns		x	y		slope
L	1	3	A	1	1	6.5	4	-2.00
R		5	B		8	1		-0.50
Bridge	y	2 2.5	Dist	3		7.826		

Construction Summary: Optimal Bridge Location, Part 2

1. Enter the river, town locations.
2. Create a counter column.
3. Compute the bridge location, y, in incremental steps.
4. Compute the length of the route, copy.
5. Compute the minimum distance.
6. Enter a formula to reproduce y when the distance equals the minimum, and 0 otherwise. Copy.
7. Compute the maximum value in the fourth column.

River		x	Towns		x	y
L	1	3	A	1	1	6.5
R		5	B		8	1
Bridge		y	7	4.3	D	5	7.433

	n		y		dist	min pt	
2	0	3	1.0	4	8.852	6	
	1		1.1		8.760		
	2		1.2		8.671		
	3		1.3		8.586		

Exercises

1. An explorer wants to walk from a point on one side of a swamp that consists of a rectangular region that is 15 miles across to another point 50 miles farther along on the other side of the swamp. The explorer can walk at the rate of 5 kilometers per hour on the dry land and 3 kilometers per hour through the swamp. How should she plan the trip to minimize her time of travel? The diagram in Figure 8.22 shows the results at an intermediate time from two possible routes. The underlying model can use a slider connected to time to compare visually the progress of the two solutions.

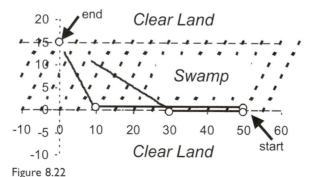

Figure 8.22

---

2. Modify Exercise 1 so that the starting and finishing points are on the clear land and each is a given distance from the swamp. When you have solved the problem, try to detect a pattern.

3. A thin rigid pipe is carried horizontally down two corridors that intersect at right angles. One corridor is 5 feet wide and the other one is 10 feet wide. What is the longest rod that can be carried through the corridors if it must be carried horizontally? See Figure 8.23 for an illustration.

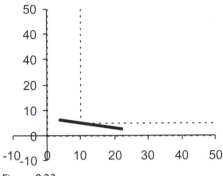

Figure 8.23

4. Repeat the same question but this time carrying a long rectangular box that is 3 feet wide.

5. What is the largest rectangle that can be inscribed in a right triangle with base 5 feet and height 4 feet? Create a spreadsheet model that contains a drawing and a scroll bar that is used to vary the rectangle (see Figure 8.24 and Figure 8.25).

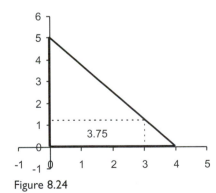

Figure 8.24

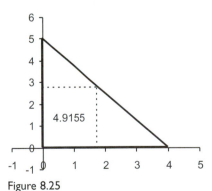

Figure 8.25

6. What is the volume of the largest right circular cylinder that can be inscribed in a right circular cone with a base radius of 3 and a height of 5? In addition to solving the problem, incorporate an interactive graph illustrating the problem (see Figure 8.26).

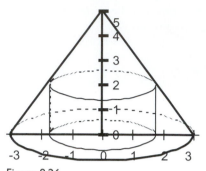

Figure 8.26

7. At noon Ship I is 40 miles due north of Ship II. Ship I is sailing south at the rate of 20 miles per hour. Ship II is moving west at the rate of 15 miles per hour. When will the distance between the ships be a minimum? What is the minimum distance? In addition to determining a solution, create an animated graph to illustrate the motion of the ships as depicted in Figure 8.27 through Figure 8.29.

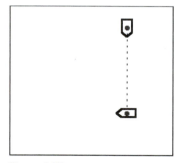

Figure 8.27

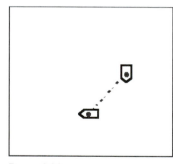

Figure 8.28

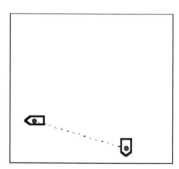

Figure 8.29

8. Determine the dimensions of the rectangle with perimeter 100 whose area is a maximum.

9. Determine the dimensions of a rectangle with area 250 whose perimeter is a minimum.

10. A political subdivision wishes to create a waste-to-energy plant to serve its four towns. The centers of the towns are located at coordinates (10,20), (−8,18), (−23,2), (4,−10), where distances are in kilometers (see Figure 8.30). Determine the location of the point (a,b) for which the sum of the distances to the towns is a minimum.

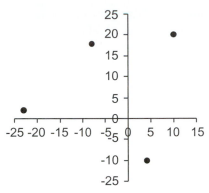

Figure 8.30

11. Solve Exercise 10 if we weight the distances in proportion to the towns' populations, which are, respectively, 4,000, 20,000, 2,000, and 7,000.

12. If one apple tree is grown on a plot of land it produces 1,000 apples. For each additional apple tree planted in the plot, the production per tree drops by 25 apples. What number of trees will result in a maximal production of apples? How many apples will be produced?

8.2 Apportionment Algorithms

Sometimes interesting and profound mathematical applications arise in surprising places. Deciding how to apportion the seats in a legislative body among states or political parties is such a case. Because of a mathematically imprecise statement in the U. S. Constitution regarding how to apportion its House of Representatives, over the years different mathematical models that were used to do so gave rise to some exasperating consequences. Both the fascinating history and the mathematical descriptions of this subject are found in Balinsky and Young (2001). Since similar issues still arise in the apportionment of the parliaments of nations, it remains a topic of practical interest. In this section we create some spreadsheet implementations of models that have been used or proposed for solving apportionment problems.

In many cases related to elections there is a very typical problem: people can cast their votes for a fixed number of alternatives (for example, political parties), and after counting these votes a number of representatives has to be assigned for each of the alternatives. The same problem also arises when a legislative body (for example, a parliament) has to distribute its seats according to the number of inhabitants for certain parts of a country. The problem is finding a "fair" distribution, meaning that the percentage distribution of the assigned seats should be as close as possible to the percentage distribution of the votes. Under normal circumstances there are many more votes than seats.

In our examples we will suppose that a national legislature has a number of seats that are to be apportioned among the country's primary political subdivisions, called states. Each state is to be allocated a number of seats in the legislature that is in proportion to its population. However, these models also can be applied to those countries in which the seats in the legislature are apportioned among political parties based upon how may votes each received in the most recent election.

Perhaps surprisingly, such a process is not as straightforward as it at first might seem. In fact, several distinct apportionment methods have been used in different countries at various times. Here we examine two primary approaches that have been used. The names of these methods vary from country to country. The ones used here are those that arose historically in the United States. For a further discussion of these topics see Balinsky and Young (2001); Nielsen and de Villiers (1997); Garfunkel (2000).

1. Hamilton's Method

Sometimes the process of apportioning seats for representatives in the legislature seems to be quite obvious. In fact, in our initial illustrations we first pursue this topic without using a spreadsheet, although we will display our work in the normal table format of a spreadsheet.

For example, suppose that there are three states, A, B, and C, whose respective populations are 5000, 3000, and 2000, and that the legislature is to contain 10 seats. How many seats should each state receive? One method to determine this number is to divide 10 into the total population of 10,000 to obtain 1,000 as the "group size," or the ideal number of people a population group to be represented by a single representative. Often this size is called a "standard divisor," the term that we will use in this section. From this we can then obtain each state's "fair" allotment, or quota, by dividing its population by the standard divisor 1,000, as shown in Figure 8.31, to obtain the integer values of 5, 3, and 2.

Number Seats		10
Population		10,000
Standard Div		1,000
State	Pop	Seats
A	5000	5
B	3000	3
C	2000	2

Figure 8.31

However, this process practically never gives integer results, because division of the large population numbers by the number of seats in a legislative body almost always produces a remainder. For example, using the populations 5300, 3300, and 1400 in Figure 8.32, similar computations produce quotas of 5.3, 3.3, and 1.4, respectively.

Since only whole numbers of seats can be allocated to states, we can begin the process by allocating 5, 3, and 1 seats to the states. However, now who should get the seat that remains unassigned so far?

One standard procedure, called the method of greatest remainders, or Hamilton's method, is to award any additional seats beyond the initial integer assignments to those states with the largest remainders until all of the seats are allotted. In this example state C gets the extra seat, since the remainder for C (0.4) is larger than the remainders for each of the other two states (0.3 and 0.3).

However, it turns out that this method produces many anomalies, or paradoxes. For example, if the size of the legislature is increased from 10 to 11 seats, then even with one more seat available one of the states, C, actually loses a seat in the process! We see this in Figure 8.33.

Number Seats	10	
Population	10,000	
Standard Div	1,000	
State	Pop.	Quota	Seats
A	5300	5.3	5
B	3300	3.3	3
C	1400	1.4	2

Figure 8.32

Number Seats	11	
Population	10,000	
Standard Div	909	
State	Pop.	Quota	Seats
A	5300	5.83	6
B	3300	3.63	4
C	1400	1.54	1

Figure 8.33

To study Hamilton's method we now create a spreadsheet model for it (see Figure 8.34). We begin with a three-state country and 10 seats in the legislature. We enter the state populations in the second column and calculate the total population as the sum of the state populations. In addition, our model computes the standard divisor, by dividing the total population by the number of seats (Figure 8.35).

Seats		10
State		Pop
A		524
B		329
C		147
TOTALS		SUM(▼)

Figure 8.34

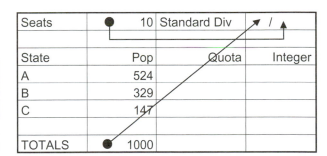

Figure 8.35

Next, in the third column we generate the quota for State *A* by dividing its population by the standard divisor, and copy the expression for the other states (Figure 8.36).

Then, in the fourth column the preliminary seat assignment for a state is calculated from the quota by using the function INT that "rounds down" the quota to the largest integer that does not exceed the quota (Figure 8.37). Thus, this expression "strips off" the decimal part of the quota. We then copy these expressions down their columns.

Seats	10	Standard Div		100
State	Pop		Quota	Integer
A	524		/	
B	329			
C	147			
TOTALS	1000			

Figure 8.36

Seats	10	Standard Div		100
State	Pop		Quota	Integer
A	524		5.24	INT( )
B	329		3.29	
C	147		1.47	
TOTALS	1000			

Figure 8.37

We now introduce a new column to compute the fractional remainder of each state's quota. This represents the portion of the state's entitlement that so far is "wasted," or unallocated. We first enter a formula to calculate this for State *A* as the difference between the quota and the integer seat assignment. Again we then fill this expression down the column (Figure 8.38).

Next, we copy the SUM function across the last row (Figure 8.39). The sum of the remainders will give the total number of additional seats that need to be assigned.

Seats	10	Standard Div	100	
State	Pop	Quota	Integer	Remain
A	524	5.24	5	-
B	329	3.29	3	
C	147	1.47	1	
Totals	1000			

Figure 8.38

Seats	10	Stand Div	100	
State	Pop	Quota	Integer	Remain
A	524	5.24	5	0.24
B	329	3.29	3	0.29
C	147	1.47	1	0.47
Totals	1000	10	9	1

Figure 8.39

We pause now in our development of a model and examine the display in Figure 8.40. With this we can manually examine the remainders and see which state gets the extra seat. It is interesting to do this and then see what happens if we increase the number of seats to 11. Also, with the number of seats kept at 10, you might try to predict what would happen if each state lost 20 residents, or try to compare the difference it would make if 12 people were to move from *C* to *A* with the situation if 11 people were to move from *C* to *A* and one person from *C* to *B*.

Seats	10	Standard Div	100	
State	Pop	Quota	Integer	Remain
A	524	5.24	5	0.24
B	329	3.29	3	0.29
C	147	1.47	1	0.47
Totals	1000	10	9	1

Figure 8.40

To complete the model, we create a new column in which we use the RANK function (see Figure 8.41). The expression RANK(*number,array*,0) finds the rank of *number* within the block *array*. The 0 listed as the third argument of the function indicates that the rank is with respect to descending order. Thus, the largest number in array has Rank 1, the second largest has Rank 2, and so on. In the new column we first enter the expression that determines the rank of State *A*'s decimal remainder among the array of all state remainders. The reference to the array must be absolute. We then can copy the formula down its column to determine the remainders for the other states.

Seats	10	Standard Div	100		
State	Pop	Quota	Integer	Remain	Rank
A	524	5.24	5	0.24	RANK(▼ , ▲ ,0)
B	329	3.29	3	0.29	
C	147	1.47	1	0.47	
Totals	1000	10.00	9	1	

Figure 8.41

Finally, we create two additional columns to determine which states receive the extra seats. In the penultimate column we use the rank of that state's remainder in an IF statement to determine if a state receives an extra seat by the following scheme:

IF rank ≤ number seats not yet assigned THEN allocate 1 extra seat
ELSE allocate 0 extra seats.

In the final column we then add the value for the extra seat (either 0 or 1) to the integer number of seats that was determined earlier as the preliminary allocation. We finish our model's construction by using a fill down or copy command for the formulas in the last column, and by copying the sum formula into the last two columns of the bottom row (not illustrated).

The output of our model for 10 seats is shown in Figure 8.42. Notice that in this case State C is allocated two seats in the legislature.

Seats	10	Standard Div	100				
						Extra	
State	Pop	Quota	Integer	Remain	Rank	Seat	Seats
A	524	5.24	5	0.24	3	0	5
B	329	3.29	3	0.29	2	0	3
C	147	1.47	1	0.47	1	1	2
Totals	1000	10.00	9	1		1	10

Figure 8.42

Now we can use our model to observe that if the size of the legislature is increased to 11 seats, then the allocation of State C actually is decreased to one seat (Figure 8.43).

Seats	11	Standard Div	90.91				
						Extra	
State	Pop	Quota	Integer	Remain	Rank	Seat	Seats
A	524	5.764	5	0.764	1	1	6
B	329	3.619	3	0.619	2	1	4
C	147	1.617	1	0.617	3	0	1
Totals	1000	11.00	9	2		2	11

Figure 8.43

We can use the model just described to find the seat assignments for the states for any given number of seats. However, by using the spreadsheet's data table feature we can construct a table of all of the allocations for a whole series of values for the total number of seats, 8, 9, 10, ... We create the data table in columns directly to the right of our existing model.

To create our model, we enter values for the number of seats across one row and label the adjacent rows with the state names, initially leaving a blank column at the start (Figure 8.44). Into this first, or dummy, column of the data table we then enter simple formulas to reproduce the state allocations. We then select the block indicated by the double lines and issue the data table command, with the cell containing the number of seats designated as the row input cell.

By scanning across the resulting data table in Figure 8.45, we can find two places where a state's allotment of seats drops even as the number of seats increases. There are others farther along in the list. This anomaly has been called the *Alabama paradox* after an instance involving the state of Alabama during one early reapportion-

Seats	11					8	9	10	11
State	Pop	Seats	Seats						
A	524	6	A						
B	329	4	B						
C	147	1	C						
Totals	1000		11						

Figure 8.44

Seats		8	9	10	11	12	13	14	15	16	17	18	19	20	21	22	23
A	6	4	5	5	5	6	6	7	7	8	9	9	9	10	10	11	12	12
B	4	3	3	3	4	4	4	4	5	5	6	6	6	7	6	7	8
C	1	1	1	2	1	2	2	2	2	2	2	3	3	3	3	3	3

Figure 8.45

ment of the U.S. House of Representatives. Historically, such investigations were carried out by hand when deciding upon the number of representatives in that assembly for reapportionment following a census during part of the nineteenth century.

Notice that neither of these models takes into account the possibility of ties. In actual practice the occurrence of a tie is not very likely. It has been shown mathematically that there is no "fair" solution for ties! One refinement that you should pursue, however, is to modify the model to ensure that every state will be allocated at least one seat, regardless of how small the state's population is.

There are many other topics that arise in using Hamilton's method that can be pursued. For example, by entering the data shown in Figure 8.46 into a model, we observe that with Hamilton's method, between one census and the next, one state may lose population and still gain a seat, while another state may gain population and lose a seat! This situation is shown in our model using 435 seats. For this and other similar reasons, Hamilton's method was abandoned as a means for apportioning the U.S. House of Representatives many years ago. You are encouraged to look up actual historical data and experiment with the models presented (see Balinsky and Young (2001)).

State	Old Census		New Census		Change	
	Pop,	Seats	Pop.	Seats	Pop.	Seats
A	1001	77	1036	78	+35	+1
B	111	8	110	9	-1	+1
C	4000	307	4060	306	+60	-1
D	555	43	560	42	+5	-1

Figure 8.46

1. Enter the number of seats.
2. Enter state populations.
3. Compute the total population as sum.
4. Compute the population per seat.
5. Find the state quota.
6. Sum the column and copy right.

Seats	1	10	Pop/S	100	4					
									Ex	
State		Pop	Quota	Int	Rem	Rank		Seat	Seats	
A		524	5	5.2	7	5	8	0.2	9	3	10	0	11	5
B	2	329	3.3	3	0.3	2	0	3
C		147	1.5	1	0.5	1	1	2
Tot	3	1000	6	10	q	1		1	10

7. Compute integer number of seats.
8. Subtract to obtain the remainder.
9. Use =RANK to determine rank of remainders in the column.
10. Use =IF to add one seat if the rank does not exceed the total number of remainders.
11. Add extra seats to get allocation.
12. Copy as indicated.

2. Jefferson's Method

There are several alternate apportionment schemes that avoid the problems that are caused by the greatest remainders procedure, or Hamilton's method. Most of the apportionment systems in actual use are members of a category called *divisor methods*. We now work with the most straightforward of these, which is called Jefferson's method in the United States and dHondt's method in Europe.

Perhaps the most direct way to see how this method works and why it might be adopted is by examining a small example. Suppose that we have a number of seats to allocate, either to three states whose populations are shown here, or in another political system to three political parties whose votes in an election are shown. The numbers on which our allocations will be based are

A: 5400, *B:* 2100, *C:* 1500.

We will award seats systematically and "fairly" to the states one at a time until all of the available seats have been allocated. Who should get the first seat to be allocated? If it is allocated to *A*, then *A* will have one seat for each 5400 people. If it is awarded to *B*, then *B* will have one seat for each 2100 people; while if it is given to *C*, then *C* will have one seat for each 1500 people. Thus, to be "fair," it would seem that we should award the first seat to represent the most people, so we award the seat to *A*.

Now we look at which state should get the second seat. If it goes to A, then A will have two seats for 5400 people, or one for each $5400/2 = 2700$ people. If it goes to B, then B will have one seat for each 2100 people. If it goes to C, then C will have one seat for each 1500 people. Consequently, in order for the seat to represent the most people, we award the second seat to A as well.

Next we consider the third seat. If it is assigned to A, then A will have three seats for 5400 people, or one for each $5400/3 = 1800$ people. If it goes to B, then B will have one seat for 2100 people; while if it is allocated to C, then C will have one seat for 1500 people. Therefore, using the same principle as before, we award the third seat to B.

We continue in this way until all of the available seats are awarded. We summarize the computations for the first five seats in Figure 8.47.

State			A	B	C	
Pop			5400	2100	1500	
Seat				Calculations		Result
1	prev assn		0	0	0	
	if next		1 for 5400	1 for 2100	1 for 1500	Seat to A
2	prev assn		1	0	0	
	if next		1 for 2700	1 for 2100	1 for 1500	Seat to A
3	prev assn		2	0	0	
	if next		1 for 1800	1 for 2100	1 for 1500	Seat to B
4	prev assn		2	1	0	
	if next		1 for 1800	1 for 1050	1 for 1500	Seat to A
5	prev assn		3	1	0	
	if next		1 for 1350	1 for 1050	1 for 1500	Seat to C

Figure 8.47

To discover a mathematical algorithm for this method, we observe that at each step we examine fractions that are obtained by dividing each state population by the number of seats it would have if it were allocated the next seat. From this list the largest one remaining determines who is allocated the next seat. Thus, we first need to list the fractions of the form population/seats from each state that are used in the process:

$$A: 5400/1, 5400/2, 5400/3, \ldots B: 2100/1, 2100/2, 2100/3, \ldots$$
$$C: 1500/1, 1500/2, 1500/3, \ldots$$

We next need to find a way to order these fractions and make seat assignments accordingly. There are several ways to do this on a spreadsheet. We will illustrate one of these using three states and six seats (see Figure 8.48). However, our method can be expanded immediately to any number of states or seats.

First we enter the populations of the states and the number of seats. We generate a counter for the seats to be awarded down the first column and find the total population of the country at the top. We ensure that the column counter extends to 6, although it is unlikely that we will need to use all of the rows in the allocation process (for then one state would get all of the seats).

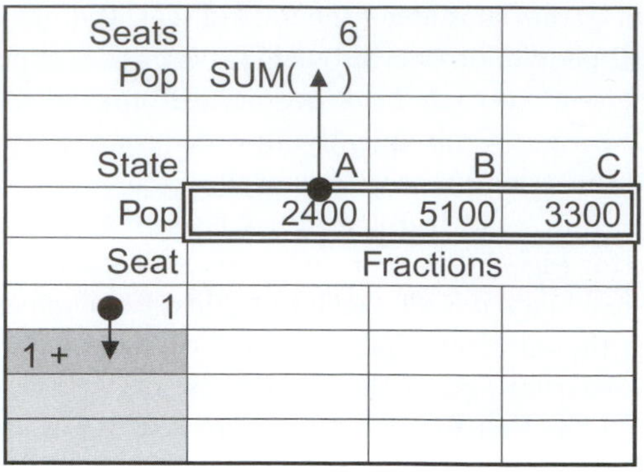

Figure 8.48

Next, in the top row we divide the population of the first state by the number of seats in that row (which of course is 1). After fixing the reference to the number of seats to come from the first column, we then copy the expression into the remainder of the block (Figure 8.49).

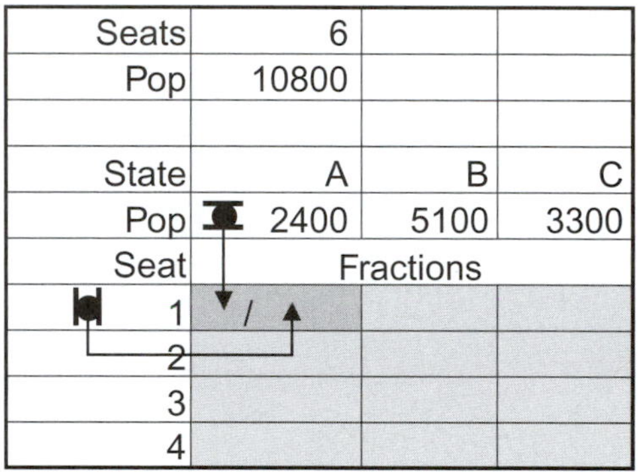

Figure 8.49

We now use three columns farther to the right to compute the relative size of each of the quotients. To do this we use the RANK function (Figure 8.50). The expression RANK(*value*, *array*) finds the position of *value* in an ordered list of the numbers in *array*, with the largest item having Rank 1, the second largest having Rank 2, and so on.

Next we use the COUNTIF function (see Figure 8.51) to count the number of fractions in a column that are less than or equal to the total number of seats (here, 6). The expression COUNTIF(array, condition) counts the number of values in the array that satisfies the condition. The component for condition must be a string. With six seats to be awarded, in this case the condition would be "<=6". However, to make this term completely general we convert the number of seats into a string with no decimals. We do this by using the function TEXT(value, 0) and the technique of concatenation (that is, appending text together) to build the expression: "<=6"&TEXT(number_seats, 0).

Once we have entered this expression we copy it across to get the count for each state, observing that the number of seats is an absolute reference. This will provide the states' allocations of seats.

Seats	6						
Pop	10800						
State	A	B	C		A	B	C
Pop	2400	5100	3300				
Seat	Fractions				Fraction Ranks		
1	2400	5100	3300		RANK(➤ , ▼)		
2	1200	2550	1650				
3	800	1700	1100				
4	600	1275	825				
5	480	1020	660				
6	400	850	550				

Figure 8.50

Seats	6						
Pop	10800						
State	A	B	C		A	B	C
Pop	2400	5100	3300	COUNTIF(▲,"<"&TEXT(▼ ,0))			
Seat	Fractions						
1	2400	5100	3300		4	1	2
2	1200	2550	1650		8	3	6
3	800	1700	1100		13	5	9
4	600	1275	825		15	7	12
5	480	1020	660		17	10	14
6	400	850	550		18	11	16

Figure 8.51

We see the output of our model showing each state's allocation in Figure 8.52.

Seats	6						
Pop	10800						
State	A	B	C		A	B	C
Pop	2400	5100	3300		1	3	2
Seat	Fractions						
1	2400	5100	3300		4	1	2
2	1200	2550	1650		8	3	6
3	800	1700	1100		13	5	9
4	600	1275	825		15	7	12
5	480	1020	660		17	10	14
6	400	850	550		18	11	16

Figure 8.52

We have now completed a model that finds the apportionment for us. However, we can also use a little insight into doing the computations in another, more efficient way. First we observe in Figure 8.53 that all of the seats allocated went to fractions that were at least as large as the one for the last seat (here the sixth seat) that was allocated. Suppose that we call this nth largest fraction the standard divisor. Here this is 1650. If we divide this number into each state's population, then the resulting quotient will be an integer for the last state that was allocated a seat (Figure 8.54).

In addition, the resulting quotients for the other states will be larger than the number of seats allocated to those states, since their last allocation produced fractions that were larger than the standard divisor. However, it will be smaller than the next largest integer; otherwise, dividing the state's population by such a larger integer would give a fraction that would have been used to allocate more seats to the state.

Seats	● 6	St Div	LARGE(▲ , ▲)	
Pop	10800			
Quotient				
Seats				
State	A	B		C
Pop	2400	5100		3300
Seat		Fractions	●	
1	2400	5100		3300
2	1200	2550		1650
3	800	1700		1100
4	600	1275		825
5	480	1020		660
6	400	850		550

Figure 8.53

Seats	6	St Div	● 1650
Pop	10800		
Quotient	▲ / ← ●		
Seats	INT(▼)		
State	A	B	C
Pop ●	2400	5100	3300
Seat		Fractions	
1	2400	5100	3300
2	1200	2550	1650
3	800	1700	1100
4	600	1275	825
5	480	1020	660
6	400	850	550

Figure 8.54

Thus, the number of seats that will be awarded to each state can be found by dividing the population by the standard divisor as the nth largest fraction, where n is the total number of seats. So the seat allocation row takes the greatest integer of the quotient row just above it.

Thus, we have been able to reduce the size of our model to just the columns of the states, as shown in the left four columns of Figure 8.55. From this we can use the data table approach to find allocations for other sizes of the legislature. In the right columns we generate the number of seats in the left column. In the dummy row at the top we copy the allocations computed for each of the states.

We then select the indicated block and issue the data table command, with the number of seats as the column input variable. Figure 8.56 lists the output for legislatures of 1 through 11 seats. To be sure that we have enough fractions, eventually we will have to expand the model in Figure 8.55 to include more seats.

Seats	6	St Div	1650	Seats			
Pop	10800						
Quotient	1.4545	3.0909	2.0000	1			
Seats	● 1	3	2	2			
State	A	B	C	3			
Pop	2400	5100	3300	4			
Seat		Fractions		5			
1	2400	5100	3300	6			
2	1200	2550	1650	7			
3	800	1700	1100	8			
4	600	1275	825	9			
5	480	1020	660	10			
6	400	850	550	11			

Figure 8.55

Seats	A	B	C
	1	3	2
1	0	1	0
2	0	1	1
3	0	2	1
4	1	2	1
5	1	3	1
6	1	3	2
7	1	4	2
8	2	4	2
9	2	4	3
10	2	5	3
11	2	6	3

Figure 8.56

A number of other similar divisor methods have been proposed and put into operation at different times and in different places. It has been shown mathematically that each of the divisor methods avoids the troublesome paradoxes that arise in Hamilton's method. However, these new schemes in turn introduce other new drawbacks. In fact, it turns out that there is no ideal apportionment method that can fulfill every property that people generally want an apportionment plan to incorporate. You are invited to read about the various methods and to design spreadsheet models for them.

Construction Summary: Apportionment Algorithms (Jefferson)

1. Enter the number of seats.
2. Enter the state populations.
3. Compute the sum of the state populations.
4. Generate a counter column.
5. Divide the state population by the seats, copy down and right.
6. Use =RANK to compute the rank of numbers in the fractions block.
7. Count the number of values in the column below that do not exceed the number of seats, and copy to right.

Seats	6	1						
Pop	10800	3						
			2					
State		A	B	C		A	B	C
Pop		2400	5100	3300	7	1	3	2
Seat			Fractions					
4	1	5	2400	5100	3300	6	4	1	2
	2		1200	2550	1650		8	3	6
	3		800	1700	1100		13	5	9
	4		600	1275	825		15	7	12
	5		480	1020	660		17	10	14
	6		400	850	550		18	11	16

Exercises

1. There is an alternate way of describing Jefferson's method that is presented in many discussions of the topic. This approach can use the layout presented in Figure 8.57 and Figure 8.58 with 30 seats to be allocated among four states. We first estimate an ideal number of people to be represented by a seat and then divide this number into the population of each state. We then round the resulting number of seats down to obtain an integer allocation. If we obtain too few or too many seats using the initial estimate, as illustrated in Figure 8.57, then we adjust the size of the ideal district appropriately until the desired number of seats allocated is reached (Figure 8.58). Create a model for this representation of Jefferson's method.

Pop/Seat		300	
State	Pop	Seats	Seats
A	780	2.600	2
B	1100	3.667	3
C	2113	7.043	7
D	5249	17.497	17
		Total	29

Figure 8.57

Pop/Seat		290	
State	Pop	Seats	Seats
A	780	2.690	2
B	1100	3.793	3
C	2113	7.286	7
D	5249	18.100	18
		Total	30

Figure 8.58

2. Adams's apportionment method can be implemented similarly to the one in Exercise 1, except that in this case each estimate is rounded *up* to the next highest integer. Create a model for Adams's method.

3. Webster's method can also be implemented in a similar manner but by using another method for rounding. Here we first drop any fractional component to get a state's preliminary allocation, n. We then look at the remaining fraction, adding an additional seat to each state whose remainder is 0.5 or larger. Create a spreadsheet model for this version of Webster's method. We note that this method for allocating an additional seat is equivalent to adding an additional seat if the decimal number of seats exceeds the arithmetic mean of n and $n + 1$, or $\dfrac{n + (n + 1)}{2}$, where n is the preliminary allocation obtained from dropping the decimal. This method is illustrated in the output in Figure 8.59.

Pop/Seat		314			
State	Pop	Seats	Int	Rem	Int
A	780	2.484	2	0.484	2
B	1100	3.503	3	0.503	4
C	2113	6.729	6	0.729	7
D	5249	16.717	16	0.717	17
				Total	30

Figure 8.59

4. Two other methods can be implemented similarly by incorporating different rounding criteria. In Hill's method we allocate an additional seat if the decimal number of seats exceeds the geometric mean of n and $n + 1$, $\sqrt{n(n + 1)}$. In Dean's method we allocate an additional seat if the decimal number of seats exceeds the harmonic mean of n and $n + 1$, $\dfrac{2}{\dfrac{1}{n} + \dfrac{1}{n + 1}} = \dfrac{2n(n + 1)}{2n + 1}$. Implement each of these methods on a spreadsheet.

5. Design spreadsheet implementations of the methods of Adams, Webster, Hill, and Dean using the approach used in this section for Jefferson's method.

6. Modify your models to ensure that each state receives at least one seat regardless of how small the state's population is.

7. Use your models for the legislatures or parliaments of various states and nations known to you.

8.3 Medicine Dosage

In treating some illnesses it is important for medical personnel to maintain the concentration of a particular medical drug in a patient's bloodstream within a certain range. In this section we examine how mathematics can play a part in determining an optimal dosage strategy to accomplish this. Further discussion of this problem is found in Garfunkel (2000), Giordano et al (2003) and Horelick and Koont. We want to determine a strategy for the dosage amount and frequency of administration that will keep the concentration of the drug in a patient's bloodstream within the prescribed levels while subjecting the patient to no more injections than are necessary.

Like many other natural processes that we have studied earlier in this book, after a dosage of a drug has been administered to a patient, the concentration of the drug present in the patient's bloodstream tends to decrease over time at a rate proportional to the concentration.

For example, suppose that each hour after a drug has been administered the level of concentration falls to 90 percent of its level of 1 hour earlier. We can examine this by using the spreadsheet model outlined in Figure 8.60 and Figure 8.61 where the initial concentration of the drug in the bloodstream is 5 units (measured in appropriate units), and where for each hour thereafter the concentration falls to 90 percent of its previous level. We will call this percentage the retention rate for the given drug and patient. Throughout this section we assume that this rate is known. At the end of the section we describe a method whereby it can be estimated experimentally. The first column of our model counts hours, while the second column repeatedly multiplies each hour's previous concentration by the rate of retention.

Retn Rate		0.9
	Time	Conc
	0	5
1+		*

Figure 8.60

Retn Rate	0.9
Time	Conc
0	5
1	4.5
2	4.05
3	3.645
4	3.2805
5	2.95245

Figure 8.61

Plotting a graph of the values gives us the graph shown in Figure 8.62, where we have extended the x-axis to 30 hours. Frequently the function is referred to as a *decaying exponential function*. Notice that although the drug level decreases continuously and eventually becomes very small, the level never reaches 0.

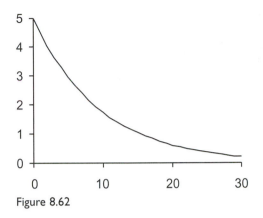

Figure 8.62

Usually it is important to give constant dosages of a medicine at regular time intervals so as eventually to keep the level of concentration of the drug in the bloodstream within certain prescribed levels, say between 12 and 16. We will create a first draft of a model (Figure 8.63) using the assumptions from the earlier example. As one of the parameters we have the dosage level of the medicine (5 units). This is the amount by which the concentration in the patient's bloodstream is increased. The other two parameters are the hourly drug retention percentage (90 percent) and the administration frequency, or the time between successive administrations in hours (3 hours).

In this first model the time between administrations of the drug is an integer. In a later example we will modify this somewhat artificial constraint.

Dose	5	Ret Rate	0.9	Lo Lim	10
Freq	3			Hi Lim	16
Count	Time	Dose	Conc	Lo Lim	Hi Lim
0	0.000	5	5.000	10	16
1	1.000	0	4.500	10	16
2	2.000	0	4.050	10	16
3	3.000	0	3.645	10	16
0	3.001	5	8.654	10	16
1	4.001	0	7.781	10	16
2	5.001	0	7.002	10	16
3	6.001	0	6.302	10	16
0	6.002	5	11.302	10	16

Figure 8.63

The first column counts the hours in a dosage cycle, with 0 denoting the time of administration of the drug and the last nonzero entry designating the end of the period before the next administration. The second column maintains the overall time. In order for the end of a cycle and the administration of the medicine to occur at different times, we introduce a brief gap of 0.001 hours between the end of a cycle and a new injection that increases the concentration of the medicine in the bloodstream.

The next column gives the increase of concentration that results when the drug is administered. The value of this column is 0 except when the drug is administered. The next column keeps track of the amount of the drug in the bloodstream. Notice that there is a discontinuous "jump" in the concentration at the start of each cycle. The final two columns are used to draw vertical lines demarking the lower and upper limits of the acceptable levels of concentration of the drug in the bloodstream.

In the first column (Figure 8.64), we enter 0 to indicate the start of the first cycle, a time when the drug is administered. We then enter a formula that counts in 1-hour increments in the administration cycle until the count reaches the length of the period. After this another 0 is generated to denote the administration of the drug and the start of a new cycle. For example, with a 3-hour cycle, the count is 0, 1, 2, and 3 (Figure 8.65).

In the second column time is incremented by 1 hour except when counter is 0, when it is incremented by 0.001 hour to administer a new dosage. The third column gives the increase in concentration of the drug in the bloodstream when the count is 0 (that is, when a new dose is administered) and generates the value 0 otherwise. After we enter these formulas, we copy them down their columns.

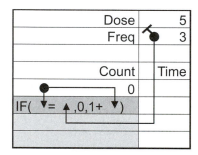

Figure 8.64

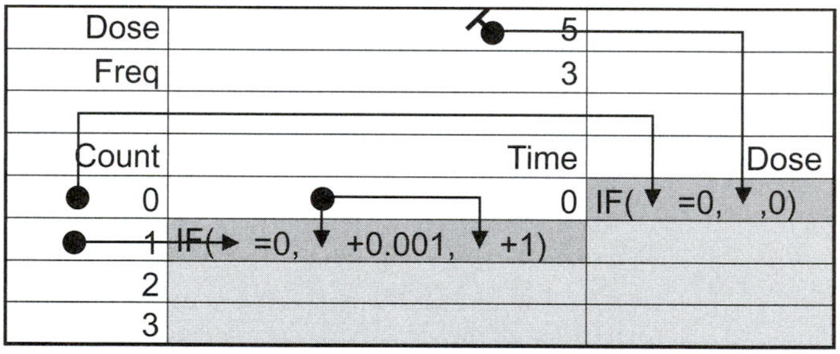

Figure 8.65

The next column calculates the level of concentration of the drug in the bloodstream. The initial existing amount is, of course, 0. From then on the IF . . . THEN . . . ELSE formula increases the concentration level by the effect of the new dosage if the count is 0, and multiplies the previous level by the retention rate otherwise. Thus, we can interpret this formula as

IF count = 0 (that is, administer medicine)

THEN add increase in drug concentration to the previous level

ELSE multiply previous drug concentration level by the retention factor.

Naturally, the retention rate is an absolute reference. We then copy the remaining formulas down the column to complete the model (Figure 8.66).

Dose	5	Ret Rate			0.9
Freq	3				
Count	Time	Dose			Conc
0	0.000	5			
1	1.000	0	IF(▼ =0, ▲ + ▼, ▼ * ▼)		
2	2.000	0			
3	3.000	0			

Figure 8.66

At this time it can be quite instructive to create an xy-graph showing what happens to the concentration level of the drug in the bloodstream. To do this, we first extend our model to include the prescribed lower and upper acceptable concentration levels, and use them to form the y-coordinates of points on horizontal lines (Figure 8.67).

Dose	5	Ret Rate	0.9	Lo Lim	10
Freq	3			Hi Lim	16
Count	Time	Dose	Conc	Lo Lim	Hi Lim
0	0.000	5	5.000		
1	1.000	0	4.500		
2	2.000	0	4.050		
3	3.000	0	3.645		

Figure 8.67

Next we create an xy-graph using the time as the x-variable and the last three columns as y-series (Figure 8.68). Here the x-series is separated from the three y-series. Perhaps the easiest way to create the graph is to select the five adjacent columns for the graph, and then in the first step of the Chart Wizard delete the unwanted series.

Dose	5	Ret Rate	0.9	Lo Lim	10
Freq	3			Hi Lim	16
Count	Time	Dose	Conc	Lo Lim	Hi Lim
0	0.000	5	5.000	10	16
1	1.000	0	4.500	10	16
2	2.000	0	4.050	10	16
3	3.000	0	3.645	10	16

Figure 8.68

The resulting graph is shown in Figure 8.69. With the values for the parameters that we have used in our model we can see that the drug concentration level in the bloodstream soon gets out of bounds. The level becomes too high at times. We can then use our model to adjust the parameters, either reducing the dosage or increasing the time between administrations. After some experimentation we might find that a dosage level of 5.5 administered every 4 hours produces the graph in Figure 8.70. From the output and the graph it seems as if we will stay within the required limits with this administration policy.

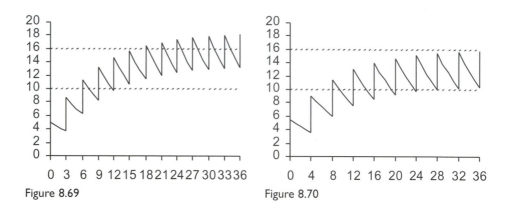

Figure 8.69   Figure 8.70

Nevertheless, there are some problems with our approach so far. Perhaps the most noticeable is that we have only considered times measured in whole-hour intervals. While this may fit with the staffing arrangements at a medical center, we would really like to be able to use any times. Also, we have found solutions only by using trial and error. It would be desirable to find a mathematical solution as well. As a result, at this time we will switch from our previous discrete modeling approach to a continuous one.

In this approach we will compute the retained concentration of the drug in the bloodstream in a different way. To do this we notice that if we start with 5 units, then after 1 hour the amount remaining is found by multiplying the previous amount by 0.9 to obtain 5×0.9^1. To find the amount remaining after another hour we multiply that quantity by 0.9 again to obtain 5×0.9^2. Continuing in this fashion, we find that after n hours the amount remaining in the bloodstream will be 5×0.9^n. This formulation has the advantage that the expression 5×0.9^t is valid for any time t as a real number. In general, if the retention rate is r, then the amount remaining after t hours will be $5 \times r^n$.

It would seem to be desirable to be able to administer dosages as infrequently as possible. This would occur if we administered the dosage when the patient's concentration was at the lowest acceptable value and then we increased the concentration to immediately bring it to the highest acceptable level. Thus, we will use a dosage concentration of hi–lo units. We will now design a second model to do this. Our new model will allow us to use noninteger time intervals between dosages. Our illustrative example in Figure 8.71 and Figure 8.72 uses 4.5 hours. The model finds only the starting and ending concentrations and does not consider intermediate points.

Thus, the first column is used as a dosage cycle counter. The second column keeps track of time by simply adding the frequency to the previous time via an absolute reference.

The third column computes the concentration of the drug at the start of each period. For the first period this is the dosage size. To obtain the ending concentration for a period in the fourth column, we multiply the starting concentration by the hourly retention rate raised to the power of the time, t, as we discussed earlier. In this example after t hours we would multiply by 0.9^t. Thus, we multiply the amount at the start of the period (in Column 3) by this factor to find the concentration of the drug that is still retained at the end of the period.

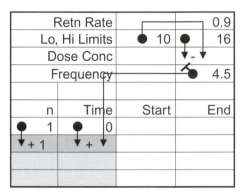

Figure 8.71

Figure 8.72

The concentration of the drug at the start of the next period is found as the sum of the new dosage (absolute reference) and the residue remaining at the end of the previous period (relative reference). When we have entered this formula we can then copy this expression to complete the model (Figure 8.73). The initial rows of output are given in Figure 8.74.

Retn Rate			0.9
Lo, Hi Limits		10	16
Dose Conc			6
Frequency			4.5
n	Time	Start	End
1	0	6	3.735
2	4.5		
3	9		
4	13.5		

Figure 8.73

Retn Rate			0.9
Lo, Hi Limits		10	16
Dose Conc			6
Frequency			4.46091
n	Time	Start	End
1	0	6	3.750
2	4.46091	9.750	6.094
3	8.92182	12.094	7.559
4	13.8327	13.559	8.474

Figure 8.74

Now if we scan down the table that we have generated we see that while our guess of a dosage of 4.5 was not too bad, it is not right either. This administration policy allows the concentration to fall below the minimum requirement (see Figure 8.75). Moreover, upon a new administration, the concentration level does not quite reach the upper limit.

	Retn Rate		0.9
	Lo, Hi Limits	10	16
	Dose Conc		6
	Frequency		4.5
n	Time	Start	End
1	0	6	3.735
2	4.5	9.735	6.059
3	9	12.059	7.506
14	58.5	15.870	9.878
22	94.5	15.891	9.891
23	99	15.891	9.891

Figure 8.75

With some more trial and error we can do better. However, at this time we can use a spreadsheet tool, Goal Seek, to assist us, although Solver also will work. As we look at the output, we see that the lower limit in steady state that has been reached in Period 22 is not quite at the target level of 10. Thus, we use Goal Seek to set this computed value to the target of 10 by changing the value of the frequency cell. The output that this produces is presented in Figure 8.76.

	Retn Rate		0.9
	Lo, Hi Limits	10	16
	Dose Conc		6
	Frequency		4.46089
n	Time	Start	End
1	0	6	3.750
2	4.46089	9.750	6.094
3	8.92178	12.094	7.559
14	57.9916	15.978	9.986
22	93.6787	16.000	10.000
23	98.1396	16.000	10.000

Figure 8.76

In addition, a little mathematics can help us at this time. If we start a period from the time of the administration of a new dosage at the *hi* level, we want to have the concentration go down until it reaches the *lo* level at the end of the period. Suppose that this takes t hours. If we let the residue factor be $r = 0.9$, then we want $r^t hi = lo$, or $r^t = lo/hi$. We next take the natural logarithm of both sides to obtain $t\ln(r) = \ln(lo/hi)$, or $t = \ln(lo/hi)/\ln(r)$. Putting this formula in the cell for time in our model gives $t = 4.46091$ and produces the output listed in Figure 8.77.

	Retn Rate		0.9
	Lo, Hi Limits	10	16
	Dose Conc		6
	Frequency		4.460909
n	Time	Start	End
1	0.000	6	3.750
2	4.461	9.750	6.094
3	8.922	12.094	7.559
14	57.992	15.978	9.986
22	93.679	15.999	10.000
23	98.140	16.000	10.000

Figure 8.77

In addition, we obtain the graph shown in Figure 8.78, indicating that we will have reached a good, steady-state drug administration policy for this patient.

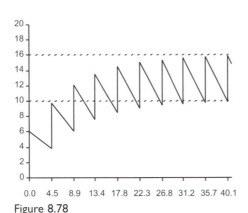

0.0   4.5   8.9   13.4 17.8 22.3 26.8 31.2 35.7 40.1

Figure 8.78

In this model we have not shown how to generate the graph in Figure 8.78. This will be left as a challenge to those interested in designing interactive graphics. The Construction Summary at the end of the section gives instructions on how to create the data needed for the graph. In addition, scroll bars provide an excellent way to investigate the effects of changing the values of the parameters. The display in Figure 8.79 illustrates a scroll bar that varies the retention rate. By moving the slider we can see how this affects the frequency of administration. A model for this example is included on the CD.

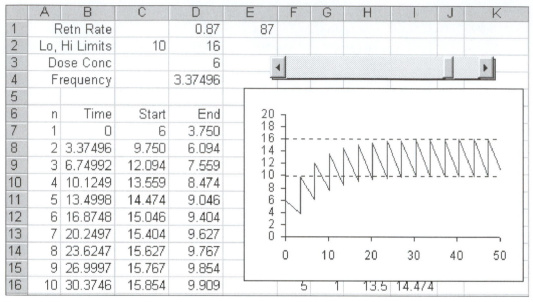

	A	B	C	D	E	F	G	H	I	J	K
1		Retn Rate		0.87	87						
2	Lo, Hi Limits		10	16							
3	Dose Conc			6							
4	Frequency			3.37496							
5											
6	n	Time	Start	End							
7	1	0	6	3.750							
8	2	3.37496	9.750	6.094							
9	3	6.74992	12.094	7.559							
10	4	10.1249	13.559	8.474							
11	5	13.4998	14.474	9.046							
12	6	16.8748	15.046	9.404							
13	7	20.2497	15.404	9.627							
14	8	23.6247	15.627	9.767							
15	9	26.9997	15.767	9.854							
16	10	30.3746	15.854	9.909							

Figure 8.79

In addition, the mathematical description of such a model is often given in terms of the exponential function. Thus, if we set $y = Cr^t$ and take the natural logarithm of both sides, then $\ln y = \ln C + t \ln r$, so that $y = e^{\ln C + t \ln r} = Ce^{t \ln r}$. For a retention factor of $r = 0.9$ this would become $y = Ce^{-0.10536t}$.

Finally, throughout this section we have assumed that we know the drug retention rate for an individual. If we do not have this information, then we can take the reading of the concentration of the drug in the bloodstream at two different times, say at Time 0 and again T hours later, and say that the respective concentrations at these times were a and b. Assuming that the drug concentration decreases exponentially over time, we have $f(t) = Cr^t$, so that $f(0) = a = C$ and $f(T) = Cr^T = ar^T$. Thus, $ar^T = b$ and $r^T = b/a$. Taking the natural logarithm of both sides, we then obtain $T\ln(r) = \ln(b/a)$, so that $\ln(r) = \ln(b/a)/T$ and $r - e^{\ln(b/a)/T}$.

For a further discussion of this topic, see Garfunkel (2000); Giordano, Weir, and Fox (2003). Also see Horelick and Koont.

Construction Summary: Medicine Dosage, Basic Model

1. Enter parameter values.
2. Subtract *lo* from *hi*.
3. Create a counter column.
4. Enter 0 as the initial time.
5. Reproduce the dose as the initial starting concentration.
6. Compute the concentration level just before next injection, copy.
7. Increment time by the frequency of administration, copy.
8. Add the standard dosage to the amount remaining at the end of the previous period, copy.

Elim Rate						0.90
Hi, Lo Limits			1	10		16
Dose Conc					2	6
Frequency						4.46

	n		Time		Start		End
3	1	4	0.00	5	6.00	6	3.75
	2	7	4.46	8	9.75		6.09
	3		8.92		12.09		7.56
	4		13.38		13.56		8.47
	5		17.84		14.47		9.05
	6		22.30		15.05		9.40
	7		26.77		15.40		9.63

Construction Summary: Medicine Dosage, Data for Graph

9. Enter 1 as the initial period counter; 0 as a toggle value; 0 as the initial drug level, *y*.
10. Look up the starting time of the current interval.
11. Maintain the value of *n* if the previous toggle value is 0, increment by 1 otherwise.
12. Add 1 modulo 2 to update toggle value.
13. Look up the start or end levels of the period depending upon the toggle value using =IF.
14. Copy as indicated.

	n		k		x		y
9	1	9	0	10	0.00	9	0.00
11	1	12	1		0.00	13	6.00
	2		0		4.46		3.75
	2		1		4.46		9.75
	3		0		8.92		6.09
	3		1		8.92		12.09
	4		0		13.38		7.56
	4		1		13.38		13.56
	5		0		17.84		8.47
	5		1		17.84		14.47
	6		0		22.30		9.05
	6		1		22.30		15.05

Exercises

1. Determine an appropriate dosage schedule for the basic model of this section if the elimination rates are 0.6, 0.7, 0.75, 0.8, 0.9, 0.95, and 0.99 while the lower and upper concentration limits remain at 10 and 16, respectively.

2. Develop the algebraic model of the drug concentration first, and then use it to design a spreadsheet implementation (see Giordano, Weir, and Fox (2003), pp. 359–366).

3. Suppose that an object whose temperature is $y_0 = 200°C$ is brought into a room whose temperature is $y = 30°$ C. Newton's law of cooling states that the rate of change of temperature of the object is proportional to the difference in temperatures of the object and the surrounding medium. Thus, over a short time span of t minutes, the difference in temperature changes by k times the temperature difference. After 10 minutes we observe that the temperature has dropped to $150°$ C. Create a spreadsheet model for Newton's law of cooling. Incorporate a cell to contain the constant of proportionality, and columns for the time (in increments of 1 minute), approximations of the temperature, and the amount of change in the temperature over the minute of time. The constant of proportionality is for an explicit unit of time, here minutes. Then enter the initial temperature and use the solver to vary k to meet the second temperature constraint.

4. Use your model in Exercise 3 to solve the following heat change problems. Notice that the same law holds when an object is brought into a warmer surrounding.
   a. A cake is taken from an oven of temperature $175°$ C and placed in the kitchen whose temperature is $24°$ C. After 5 minutes the temperature of the cake has dropped to $160°$ C. Find the temperature of the cake as a function of time in steps of 1 minute.
   b. A metal ball is taken from a deep freeze unit whose temperature is $-30°$ C and placed in a shed whose temperature is $40°$ C. After 3 minutes the temperature of the ball is $-18°$ C. What is its temperature at any given time?

5. [Calculus] Newton's law of cooling (see Exercise 3) can be expressed in terms of a differential equation as $\frac{dy}{dt} = k(y - T)$, where t is time, y is the temperature of the material, and T is the temperature of the surrounding medium. Initial conditions are typically given as temperatures at two different times, $y(t_0) = y_0$ and $y(t_1) = y_1$. Thus, in Exercise 3 we have $y(o) = 200$ and $y(10) = 150$. Solve this equation via differential equation techniques, and compare your solution to that of the model in Exercise 3.

6. A 100-gallon tank is full of saltwater that contains 1 pound of salt per gallon. A saltwater solution containing 2 pounds of salt per gallon flows into the tank at the rate of 5 gallons per minute. The tank is well stirred and its contents flow out of the tank at the rate of 5 gallons per minute. Design a spreadsheet model to approximate the amount of salt in the tank as a function of time. Notice that

since the rates of inflow and outflow are equal, the water level remains constant. However, the concentration of salt changes since the concentration of the outflow varies over time.

7. [Calculus] Use the model of Exercise 6 (which can be written as a difference equation) to develop a differential equation to model the problem, solve the resulting differential equation, and compare the solution to the approximation obtained in Exercise 6.

8. A spherical mothball loses mass by evaporation at a rate that is proportional to its surface area. Create a spreadsheet model to compute the radius of the mothball for t days. Suppose that the mothball originally has a radius of 10 millimeters and that it loses 40 percent of its mass after 100 days. For those with differential equations backgrounds, also state and solve the equivalent initial value problem, and compare that solution with the spreadsheet approximation. The model should employ a constant of proportionality that is for an explicit unit of time, here days.

8.4 Inventory Control

Suppose that a hardware store stocks a certain item for sale—perhaps 4-litre containers of white paint—that customers are known to buy at a steady rate throughout the year. How frequently should the store replenish its stock? How large should its orders be? If the annual sales are 36,500 units and the store orders only once a year, then it will need to tie up a lot of money in the stock kept on hand and in providing a facility for storing it. The store thereby incurs a sizable holding expense. However, it has to process only one order a year. On the other hand, if it orders the average daily requirement of 100 units every day, then its holding cost will go down but the firm will incur a cost of repeated orders. The sum of these two costs represents the item's inventory cost. As the order frequency is adjusted, one component goes up while the other goes down. Thus, the store needs to determine an optimal order frequency that will minimize the total inventory cost. This is a simple version of the basic inventory problems that sales and manufacturing firms encounter. In this section we create a model to solve this problem.

The model used here assumes that the unit cost for each item does not vary with the order size, that the time required for re-supply is negligible, and that you are unwilling to incur shortages of the product. There are two competing cost components that you must consider: the cost of processing an order (called the order cost), and the cost of storing items in inventory (called the holding cost). If you increase the order size and place fewer orders each year, then this reduces the cost of processing repeated orders. However, it causes the resulting holding costs to increase because of the larger inventory you will have on hand. Conversely, by ordering smaller quantities more frequently, you can reduce the holding costs but only at the expense of increasing the number of orders and the resulting total order cost over extended periods.

We will investigate the problem of determining the optimal order size and the corresponding frequency of ordering by using a spreadsheet model. It will compare the total costs for different numbers of orders each year and the corresponding order quantities. The model will find an order frequency and the corresponding order quantity (called the economic order quantity) that minimizes the total cost. We will measure time in years, although any convenient time unit will suffice.

The parameters of the model include the cost of placing an order (we assume that there is a fixed cost per order, regardless of the order size), the holding cost per item per year, and the annual demand for the item. To create the model we first enter the labels and values of the parameters as shown in Figure 8.80. For illustration, the model shows that the cost of processing an order of any number of television sets is $600, with a holding cost of $50 per item per year. Also, the firm has a demand for 1000 items per year. The first column generates the number of orders per year in the usual manner. In the next column we compute the corresponding order size by dividing the annual demand by the number of orders per year. In entering the latter formula, we note that the annual demand must be an absolute reference.

Order cost		600			
Holding cost		50			
Yearly demand		1000			
Orders/ Year	Order Size	Yr Cost Order	Average Inventory	Yr Cost Holding	Yr Cost Total
1	/				
1+					

Figure 8.80

The third column computes the yearly order cost as the product of the number of orders and the cost per order (see Figure 8.81). The order cost will be an absolute reference.

Order cost		600			
Holding cost		50			
Yearly demand		1000			
Orders/ Year	Order Size	Yr Cost Order	Average Order	Yr Cost Holding	Yr Cost Total
1	1000	*			
2					

Figure 8.81

The next column calculates the mean inventory size (see Figure 8.82). This is determined by observing that throughout each inventory period the amount of stock on hand gradually decreases linearly from the amount ordered, or order quantity, down to 0. Thus, the average stock on hand at any time will be half of the order quantity. The annual holding cost is then determined in the next column by multiplying the average inventory by the annual unit holding cost, the latter being an absolute reference.

Order cost	600			
Holding cost	50			
Yearly demand	1000			
Orders/ Year	Order Size	Yr Cost Order	Average Inventory	Yr Cost Holding	Yr Cost Total
1	1000	600	/2	*	
2					

Figure 8.82

As a last step for our first row, we find the total yearly cost in the last column as the sum of the two component costs using the two columns to the immediate left (Figure 8.83).

Order cost	600			
Holding cost	50			
Yearly demand	1000			
Orders/ Year	Order Size	Yr Cost Order	Average Inventory	Yr Cost Holding	Yr Cost Total
1	1000	600	500	25000	+
2					

Figure 8.83

Now that the basic formulas are entered, we can simply carry out either the copy or the fill down procedure as far as is appropriate. This is indicated in Figure 8.84.

Orders/ Year	Order Size	Yr Cost Order	Average Inventory	Yr Cost Holding	Yr Cost Total
1	1000.0	600.00	500.0	25,000.00	25,600.00
2	500.0	1,200.00	501.0	12,500.00	13,700.00
3	333.3	1,800.00	502.0	8,333.33	10,133.33
4	250.0	2,400.00	503.0	6,250.00	8,650.00

Figure 8.84

Several rows of the resulting output are shown in Figure 8.85. Of course we cannot order fractional television sets, so a practical value for the order quantity must be rounded. From this it is easy to locate the annual number of orders (and hence also of the order quantity) where the minimal cost occurs. In this case, the optimal plan would be to place six orders per year, with an order size of 167. It is also possible to observe from our output that the optimal value appears to be where order and holding costs are equal. This is seen even more clearly from a graph, which is shown in Figure 8.86.

Orders cost		600		
Holding cost		50		
Yearly demand		1000		
Orders/yr	Order	Yr Cost	Yr Cost	Yr Cost
n	size, q	Order	Holding	Total
1	1000.0	600.00	25,000.00	25,600.00
2	500.0	1,200.00	12,500.00	13,700.00
3	333.3	1,800.00	8,333.33	10,133.33
4	250.0	2,400.00	6,250.00	8,650.00
5	200.0	3,000.00	5,000.00	8,000.00
6	166.7	3,600.00	4,166.67	7,766.67
7	142.9	4,200.00	5,571.43	7,771.43
8	125.0	4,800.00	3,125.00	7,925.00

Figure 8.85

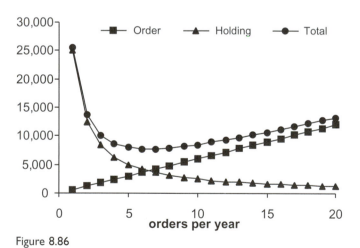

Figure 8.86

As already noted, to make our model's computations more visual, we create an xy-graph by using the first column of values of the number of orders per year as the x-axis series and designating the three cost columns as the different y-series.

We can also design a graph that uses the second column of order size as the x-axis (see Figure 8.87). The resulting graphs illustrate the changing contributions of the order and holding costs. For small order quantities (or large numbers of orders), the order cost is the dominant component of the total cost; while for large order quantities (or small numbers of orders), most of the total cost comes from the holding cost.

Of course, providing interactive tools will embellish our models. In the screen display in Figure 8.88, we have incorporated a scroll bar linked to the cost of an order. As we vary it, we will see that changes in this value affect the nature of the resulting final curve.

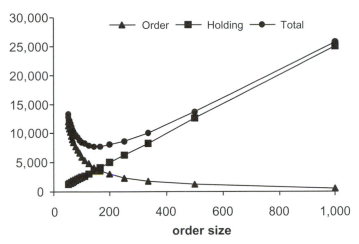

Figure 8.87

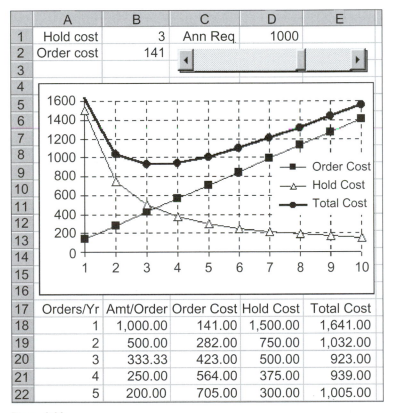

	A	B	C	D	E
1	Hold cost	3	Ann Req	1000	
2	Order cost	141			

	Orders/Yr	Amt/Order	Order Cost	Hold Cost	Total Cost
17	Orders/Yr	Amt/Order	Order Cost	Hold Cost	Total Cost
18	1	1,000.00	141.00	1,500.00	1,641.00
19	2	500.00	282.00	750.00	1,032.00
20	3	333.33	423.00	500.00	923.00
21	4	250.00	564.00	375.00	939.00
22	5	200.00	705.00	300.00	1,005.00

Figure 8.88

Another possible graph type to use is the column chart with the order and holding (but not their sum) plotted in the same columns. This is illustrated in Figure 8.89.

Figure 8.89

Another way of locating the optimal value is by using a spreadsheet's solver command (see Appendix). With this command and using the same basic layout as before (Figure 8.90), we simply select the spreadsheet's Solver option to minimize the total cost cell by modifying the orders/year size cell in the left column.

Orders cost	600			
Holding cost	50			
Year demand	1000			
Orders/	Order	Yr Cost	Avg.	Yr Cost	Yr Cost
Year	Size	Order	Invt.	Holding	Total
1	1000	600	500	25000	25600

Figure 8.90

The results of using the solver are shown in Figure 8.91. In this case we need to be able to interpret the resulting value for the number of orders per year, which is not an integer. In fact, we can place a fractional number of orders a year by not requiring order periods to fall completely within the same year. The solver technique is especially useful in more complex models in which it is hard or impossible to determine the optimal value analytically.

Order cost	600				
Hold cost	50				
Year demand	1000				
Orders/ Year	Order Size	Yr Cost Order	Avg. Invt.	Yr Cost Holding	Yr Cost Total
6.455	154.9	3873	77.5	3873	7746

Figure 8.91

Further Mathematics: Deriving Optimal Values Analytically

We can derive the optimal values of q and n analytically, too. Here we need to introduce some mathematical variable names. We let n denote the number of orders per year, q the order quantity, c the cost per order, h the annual holding cost per unit, and C the sum of the annual order and holding costs. From our analysis in creating the model, we can see that the annual order cost will be $nc = (r/q)c$. Also, the mean inventory will be $q/2$, so that the annual holding cost is $h(q/2)$. Thus, if r is the annual requirement, we can express the total annual cost as

$$C = \frac{rc}{q} + \frac{hq}{2}.$$

It follows from calculus that the minimal cost occurs where the derivative

$$\frac{dC}{dq} = -\frac{rc}{q^2} + \frac{h}{2}$$

is 0, to give an economic order quantity and its corresponding order frequency of

$$q^* = \sqrt{\frac{2rc}{h}} \text{ and } n^* = \frac{r}{q^*} = \sqrt{\frac{rh}{2c}}.$$

If we wish, we can include these in our spreadsheet models. There are a number of modifications of this model that can be pursued, too. For example, the model can be adjusted to allow for shortages of up to z units to occur during an inventory period. These shortages are filled immediately when a new supply arrives but at a cost (shortage cost) of s per unit. Although now a new cost is incurred, such a policy does decrease holding costs, since smaller amounts are held in inventory and there are times when no units are in inventory. Other inventory models can be implemented effectively on a spreadsheet as well. For example, the price of an item may vary with the quantity ordered at one time.

Construction Summary: Inventory Control

1. Enter parameter values.
2. Generate a counter column for the number of orders per year.
3. Divide yearly demand by order frequency.
4. Multiply order frequency by order cost.
5. Compute average inventory as half of the order size.
6. Multiply average inventory by annual holding cost per item.
7. Add order and holding costs.
8. Copy as indicated.

Order cost	600	
Hold cost	50	1
Year demand	1000	

Ord /Yr	Order Size	Yr Cost Order	Avg Invnt	Yr Cost Holding	Yr Cost Total
2 1 3 1000	4 600	5 500	6 25000	7 25600
2	500	1200	250	12500	13700
3	333	1800	167	8333	10133
4	250	2400	125	6250	8650

Exercises

1. A firm has a setup cost of $50,000 to produce a special type of electric motor. They have an annual demand for 200,000 units and incur a holding cost of $50 per motor per year. What is their optimal production run? How many times a year will the firm have to go through the setup process in order to minimize total cost?

2. Use your spreadsheet inventory model to determine the optimal order quantity and frequency for each of the following inventory problems. Shortages are not permitted.
   a. Order cost: $100, Holding cost: $20, Annual demand: 2000
   b. Order cost: $100, Holding cost: $40, Annual demand: 2000
   c. Order cost: $200, Holding cost: $20, Annual demand: 2000
   d. Order cost: $100, Holding cost: $20, Annual demand: 4000

3. In the main model we have used the number of orders per year as the independent variable. This in turn determines the quantity ordered. Redesign the model slightly so that the quantity ordered is the independent variable, and create a graph for reasonably spaced order quantities. This in turn determines the number of orders per year, which need not be an integer.

4. [*] It is possible to adopt an order scheme that allows for shortages to occur. In this model we will have two independent variables: Q is the order size and s is the shortage that we can allow until the next order is made. We let $S = Q - s$ denote the amount of inventory at the start of an inventory cycle. A diagram of the stock flow is provided in Figure 8.92, where a is the periodic demand. In such a plan, when a new order takes place all previous shortages are immediately

filled. In addition to order and holding costs, we assume that there is a shortage cost that is incurred. We compute the annual order cost as before. The annual holding cost is found by multiplying the holding cost per item per year by the product of the average amount of inventory at the times when there are items in inventory and the fraction of the year that items are held in inventory (S/Q). The annual shortage cost is found by multiplying the shortage cost per item per year by the product of the average amount of shortage at the times when there is a shortage and the fraction of the year that there is a shortage ($1 - S/Q$). In implementing the model, assume that there is an annual demand of 60,000 with an order cost of $8000, a holding cost of $10 per item per year, and a shortage cost of $40 per item per year. We can find the optimal order and shortage sizes by using the solver. However, because there are two independent variables, we cannot develop a graph such as the one designed in this section. However, we can do this by holding one of the two independent variables constant while varying the other one.

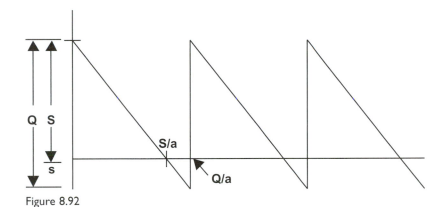

Figure 8.92

5. We have not included the cost of the items themselves in our model, since our model assumes that the price remains unchanged. However, frequently the price of purchasing an item depends upon the quantity ordered, with the sales firm offering price breaks for larger purchases. Modify our basic model to incorporate the cost of an item, and allow for quantity price breaks. Implement your model using the following parameters that were used in creating the graph of the total cost shown in Figure 8.93: annual demand: 2000; order cost: $2000; holding cost: $500 per unit per year; item cost: $50 for orders smaller than 300, $30 for orders of at least 300 but less than 500, and $20 for orders of size 500 or greater.

6. Another topic that arises in the area of operations research is linear programming. A typical example is provided by the following: A farmer has 100 hectares, 160 person-hours of labor, and $1100 in capital with which to plant two crops. Crop I costs $10 per hectare to plant, requires 1 person-hour of labor per hectare, and generates a profit of $30 per hectare. Crop II costs $20 per hectare to plant, requires 4 person-hours of labor per hour, and generates a profit of

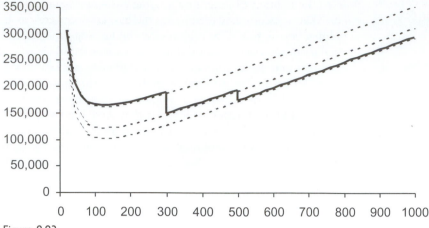

Figure 8.93

$100 per hectare. If we denote profit by P and the amounts planted in the respective crops by x_1 and x_2, then we can use the spreadsheet's solver to maximize $P = 30x_1 + 100x_1$ subject to the constraints $x_1 + x_2 \leqslant 100$ (land), $10x_1 + 20x_2 \leqslant 1100$ (capital), $x_1 + 4x_2 \leqslant 160$ (labor), $x_1 \geqslant 0$, $x_2 \geqslant 0$.

7. Design an animated graph to help in the visualization of the model in Exercise 6. Link a scroll bar to an equal-profit line as illustrated in Figure 8.94.

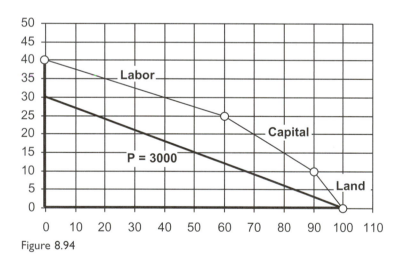

Figure 8.94

8. Design a model to solve the following minimization problem. A dietician wants to design a diet of foods A, B, C, at a minimal cost. The foods cost respectively $4, $5, and $8 per kilogram. Each kilogram of A contains 5 units of protein, 8 units of carbohydrates, and 3 units of fat. The similar amounts of B are 6, 3, and 2, while those of C are 2, 5, and 4. Each meal must contain at least 20, 30, and 10 units of these nutrients. Set up a linear program that describes this problem, and then solve it using the spreadsheet's solver utility.

9. Solve the following linear programs on a spreadsheet using the solver. This exercise is provided to show that we also can start with mathematical notation. In this case somebody else has done the modeling part. You just need to translate algebraic formulas into the spreadsheet mechanism.

a. Maximize $z = 20x_1 + 30x_2$
   subject to: $3x_1 + 2x_2 \leqslant 60$
   $2x_1 + 5x_2 \leqslant 70$
   $x_1 \geqslant 0, x_2 \geqslant 0$

b. Minimize $z = 23x_1 + 13x_2$
   subject to: $4x_1 + 3x_2 \geqslant 120$
   $2x_1 + 3x_2 \geqslant 132$
   $x_1 \geqslant 0, x_2 \geqslant 0$

c. Maximize $z = 12x_1 + 15x_2 + 8x_3$
   subject to: $2x_1 + x_2 + 3x_3 \leqslant 300$
   $3x_1 + 4x_2 + x_3 \leqslant 360$
   $x_1 + 5x_2 + 2x_3 \leqslant 200$
   $x_1 \geqslant 0, x_2 \geqslant 0, x_3 \geqslant 0$

d. Minimize $z = 7x_1 + 11x_2 + 3x_3$
   subject to: $2x_1 + 5x_2 \geqslant 100$
   $3x_1 + 4x_2 + x_3 \geqslant 120$
   $2x_2 + 3x_3 \geqslant 90$
   $x_1 \geqslant 0, x_2 \geqslant 0, x_3 \geqslant 0$

8.5   Game Theory

Most of us at some time participate in competitive games that we wish to win. In doing this we often rely on our past experience to develop strategies that we think will improve our chances of winning. Sometimes these are sports or board games in the traditional sense. However, businesses also engage in a similar activity as they vie with their competitors for customers, and military planners do so in planning for wars with opposing forces. Unfortunately, sometimes our intuition can be inadequate or lead us astray, causing disastrous results. During the twentieth century a mathematical theory of games, much of which was begun by a mathematician and an economist (see von Neuman and Morgenstern, (1953)), was developed that applies to business and military competitions as well as to ordinary games. In this section we will look at some of the underlying ideas of the subject through spreadsheet models of competition.

Suppose that two individuals, whom we will call Row and Col, play the following game. Without seeing what the other has done first, each holds up one, two, or three fingers. We refer to these respective options as *pure strategies*: Strategy 1, Strategy 2, and Strategy 3. After the fingers are selected, Row wins an amount from Col that is determined by the table in Figure 8.95. For example, if Row holds up two fingers and Col holds up one finger, then Row wins $6 from Col, and consequently Col loses $6 to Row. Negative numbers in the table represent losses to Row. Thus, if Row holds up three fingers and Col holds up one finger, then Row loses $5 to Col. Games of this type are called *two person zero-sum games*. In such a game, what one player wins the other player loses.

Row\Col	str 1	str 2	str 3
str 1	8	-4	-3
str 2	6	5	4
str 3	-5	6	2

Figure 8.95

In determining an optimal way of playing the game for each Row and Col, we will assume that they always will make rational decisions or, expressed differently, each player will try to either maximize their wins or minimize their losses. Thus, Row determines the worst that can happen if she plays each of her pure strategies. This will be the minimum value in each row, which we compute and display in the right column in Figure 8.96.

For Col, since the table gives the amounts that Row wins from Col, the worse result for each of his pure strategies will be given by the maximum of each column. We compute this in the bottom row of the table.

Row\Col	str 1	str 2	str 3	min	
str 1	8	-4	-3	-4	max of min
str 2	6	5	4	4	
str 3	-5	6	2	-5	
max	8	6	4		

min of max

Figure 8.96

From the table Row sees that if she chooses her Strategy 2, then the worst that can happen to her will be to win $4. If she chooses any other strategy, then she could do worse. Similarly, Col sees that by choosing his Strategy 3, then the best that he can do is to lose $4. If he chooses any other strategy, then he could do worse. Thus, because the resulting optimal values are the same for each player, neither Row nor Col has any incentive to choose any other strategy. When this situation occurs, the game is said to have a *saddle-point solution* (Strategy 2 for Row and Strategy 3 for Col) and we say that the *value* of the game to Row is $4.

However, not every zero-sum game has a saddle point. This is the case for the game displayed in Figure 8.97. Note that in this game the number of strategies for the two players is different.

If in this game Row were to choose the strategy with the maximum value in the last column, she would choose her Strategy 1. Then Col would notice this and would choose his best option for Row's choice, which is Strategy 3. However, Row in turn would observe this and switch to her Strategy 2. Now, to respond to this Col switches to his Strategy 1. But for this choice, Row would change back to her Strategy 1. Thus, the analysis cycles, and there is no pure strategy that is optimal to either player.

Row\Col	str 1	str 2	str 3	min
str 1	5	2	-1	-1
str 2	-2	1	2	-2
max	5	2	2	

Figure 8.97

In such a situation, it makes sense for the players to choose their strategies in a way not predictable by the opponent. Therefore, we consider *mixed strategies,* where for each game players randomly choose one of their pure strategies with given probabilities. Each mixed strategy is determined by the set of (fixed) probabilities for each pure strategy.

To see how the two players can analyze their situations in order to develop optional plans of mixed strategies, let us adjust our table layout to that shown in Figure 8.98. There we let x_1, x_2 be the probabilities that Row uses in choosing her respective strategies, while y_1, y_2, y_3 represent the similar probabilities for Col. Initially, for convenience we arbitrarily set the probability of each player's first strategy to 1 and the others to 0.

In the bottom row we will compute the expected winnings, or payoff, to Row by using x_1, x_2 with each pure strategy of Col. Similarly, in the right column we will use y_1, y_2, y_3 in computing the expected losses for Row (or the negative of Col's payoffs) from each of Row's pure strategies.

value			y_1	y_2	y_3	sum
			1	0	0	
		Row\Col	str 1	str 2	str 3	payoff
x_1	1	str 1	5	2	-1	
x_2	0	str 2	-2	1	2	
sum		payoff				

Figure 8.98

We first compute the expected winnings, or payoff, for Row for the first column strategy of Col. Since Row wins \$5 with probability x_1 and loses \$2 with probability x_2, her expected winnings will be $5x_1 - 2x_2$. This can be computed in general using the spreadsheet's SUMPRODUCT function with the probabilities and the payoffs. In doing this we must make the reference to the probabilities absolute. We then copy the formula across the bottom row (Figure 8.99).

value			y_1	y_2	y_3	sum
			1	0	0	
		Row\Col	str 1	str 2	str 3	payoff
x_1	1	str 1	5	2	-1	
x_2	0	str 2	-2	1	2	
sum		payoff	=SUMPRODUCT(▼, ▼)			

Figure 8.99

Next we compute the sum of Row's probabilities, which of course must always remain 1 (Figure 8.100).

value			y_1	y_2	y_3	sum
			1	0	0	
		Row\Col	str 1	str 2	str 3	payoff
x_1	1	str 1	5	2	-1	
x_2	0	str 2	-2	1	2	
sum	=SUM(▼)	payoff	5	2	-1	

Figure 8.100

We now repeat the corresponding process for the column player, Col (Figure 8.101 and Figure 8.102).

We obtain the output shown in Figure 8.103.

value			y_1	y_2	y_3		sum
			1	0	0		
		Row\Col	str 1	str 2	str 3	payoff	
x_1	1	str 1	5	2	-1	=SUMPRODUCT(▼ , ▲)	
x_2	0	str 2	-2	1	2		
		payoff	5	2	-1		

Figure 8.101

value			y_1	y_2	y_3	sum
			1	0	0	=SUM(▼)
		Row\Col	str 1	str 2	str 3	payoff
x_1	1	str 1	5	2	-1	5
x_2	0	str 2	-2	1	2	-2
sum	1	payoff	5	2	-1	

Figure 8.102

value			y_1	y_2	y_3	sum
			1	0	0	1
		Row\Col	str 1	str 2	str 3	payoff
x_1	1	str 1	5	2	-1	5
x_2	0	str 2	-2	1	2	-2
sum	1	payoff	5	2	-1	

Figure 8.103

To determine Row's optimal mixed strategy we want to find a mixed strategy for her and a value v so that the payoff to Row for each pure strategy of Col will be at least v. Moreover, we want v to be as large as possible. To do this we will use our spreadsheet's solver command. We illustrate the process for *Excel* through a series of screen shots.

We begin by setting aside one cell (here Cell B3) to compute v, the value of the game for Row (Figure 8.104). Initially this cell is left blank. We then want to vary the value of v and the row probabilities so that v is a maximum while all of the payoff values in the bottom row are at least as large as v.

	A	B	C	D	E	F	G
1	solve for row						
2							
3	value			y_1	y_2	y_3	sum
4				1	0	0	1
5			Row\Col	str 1	str 2	str 3	payoff
6	x_1	1	str 1	5	2	-1	5
7	x_2	0	str 2	-2	1	2	-2
8	sum	1	payoff	5	2	-1	

Figure 8.104

To do this in *Excel* we begin by issuing the command: Tools, Solver. This produces the dialog box shown in Figure 8.105. Using the techniques discussed in the Appendix, we set B3 as the target cell that we wish to maximize by changing that cell together with the probabilities for Row. We then click on the Add button to add the constraints. Here the probabilities must be nonnegative and their sum must be 1. Also, we need each of the payoffs to Row found in Row 8 to be at least as large as the value of the game, v. The resulting screen display is presented in Figure 8.105.

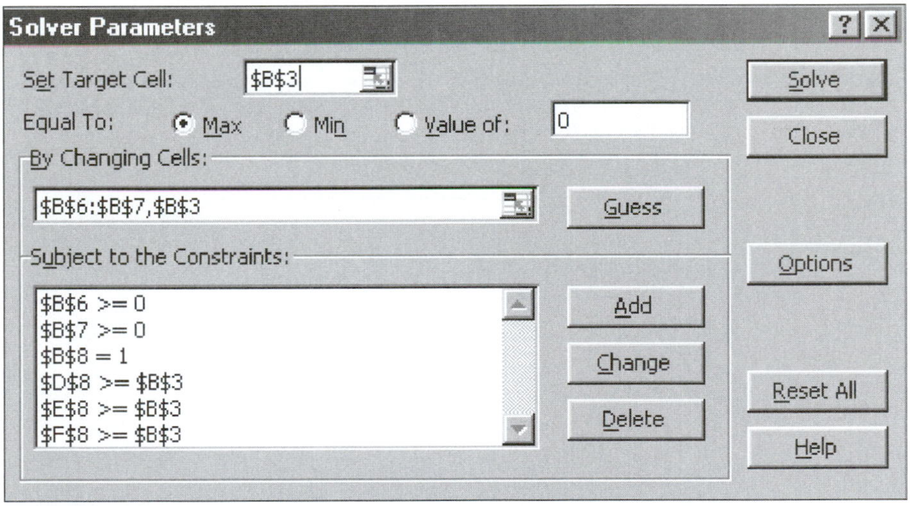

Figure 8.105

When we have completed the entries, we click on the Solve button. We will then find the output that is shown in Figure 8.106. What we have done is to locate the value of the game and the optional strategy for Row. In this case Row's optimal mixed strategy consists of randomly choosing her first pure strategy with probability 0.4 and her second pure strategy with probability 0.6. We note that in doing this the payoff to Row for each of Col's three pure strategies is at least $v = 0.8$. Moreover, our use of the solver's maximum setting ensures that no other mixed strategy of Row will result in a higher guaranteed minimum value for Row. Thus, this becomes the value of the game for Row.

	A	B	C	D	E	F	G
1	solve for column						
2							
3	value	0.8		y_1	y_2	y_3	sum
4				1	0	0	1
5			Row\Col	str 1	str 2	str 3	payoff
6	x_1	0.4	str 1	5	2	-1	5
7	x_2	0.6	str 2	-2	1	2	-2
8	sum	1	payoff	0.8	1.4	0.8	

Figure 8.106

We next need to consider the game from Col's viewpoint by carrying out a similar process for Col. However, since the table gives the payoffs to Row, in order to maximize Col's expected winnings we can think of changing the signs of the values, maximizing the value of $-v$, and ensuring that the payoffs in the right column are at least $-v$. This is equivalent to leaving the signs alone, reversing the directions of the inequalities, and minimizing the value v. To do this we use the Solver tool just as before, with the resulting settings for our implementation shown in Figure 8.107. This time we minimize the value of Cell B3. It should be noted that in the process the value of v will not change.

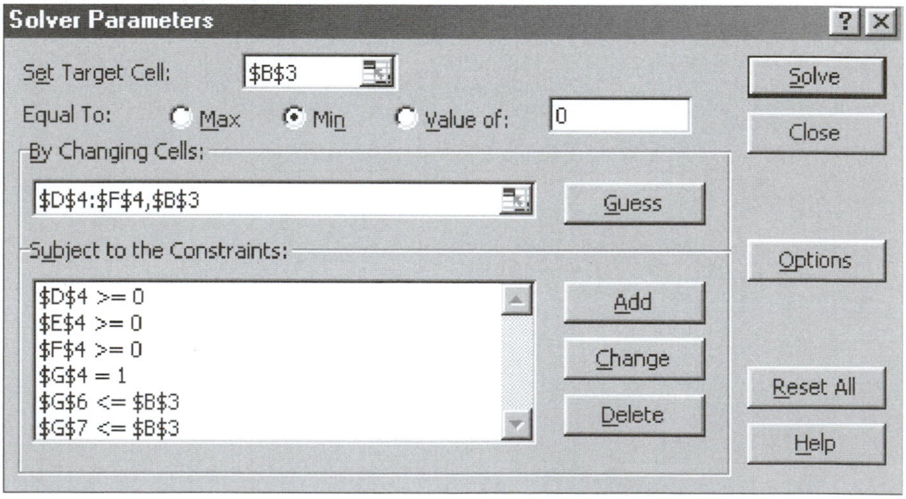

Figure 8.107

After we have issued the solver command, we find the optimal mixed strategy for Col. He should select his first strategy at random with probability 0.3 and the third one with probability 0.7. He should never select his second strategy. Under this scheme, Row's expected winnings of 0.8 remain unchanged (see Figure 8.108).

	A	B	C	D	E	F	G
1	solve for column						
2							
3	value	0.8			y_1	y_2	y_3	sum
4					1	0	0	1
5			Row\Col	str 1	str 2	str 3	payoff
6	x_1	0.4	str 1	5	2	-1	5
7	x_2	0.6	str 2	-2	1	2	-2
8	sum	1	payoff	0.8	1.4	0.8	

Figure 8.108

We can implement the same approach in games, with any number of strategies for two players. A theorem from mathematical game theory guarantees that there always is optimal strategy for both players. Our process also works just as well for games that have saddle-point solutions consisting of pure strategies. In this case, the solution our method finds will be the saddle-point solution. You can check this using the initial saddle-point example at the beginning of this section.

In the special case in which one of the players has only two options for strategies, we can also use a graphic approach to solve the problem (see Figure 8.109). When player Row has only two strategies, we draw a line for each of player Col's strategies. For a given value x with $0 \leq x \leq 1$, this line will display the payoff for player Row

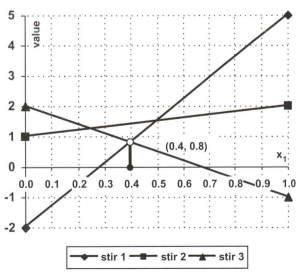

Figure 8.109

when she plays Strategy 1 with probability x (and therefore Strategy 2 with probability $1-x$). Thus, using the preceding example we draw three lines above the interval $0 \le x \le 1$ of the x-axis. For each column we plot a point at $x = 0$ with the y-value as the return from the second Row strategy and plot a second point at $x = 1$ with the y-value as the return from the first Row strategy. We then connect these points with a straight-line segment. For example, using the first column we plot the points $(0, -2)$ and $(1,5)$ and connect those points with a straight-line segment. Then for each point $x_1, 0 \le x_1 \le 1$, the y-height of the line gives the payoff for that Row strategy and the corresponding Col strategy. We draw all of these lines, and then find the highest y-value that lies on or beneath all of these lines. For the game under investigation this give us the same row strategy that we found before, $x_1 = 0.4$, while that point's y-value gives us the value of the game for Row, namely 0.8.

Construction Summary: Game Theory

1. Enter arbitrary trial probability distributions.
2. Enter a formula to compute sums.
3. Compute the expected payoffs using sumproduct, then copy.
4. Use the Solver to compute the value of the game for Row.

value	4				y_1	y_2	y_3	sum
				1	1	0	0	2	1
				Row\Col	str 1	str 2	str 3	payoff
x_1	1	{	1	str 1	5	2	-1	3	5
x_2			0	str 2	-2	1	2		-2
sum	2	1		payoff	3	5	2	-1	

Exercises

1. Modify our spreadsheet model to solve the zero-sum games that are determined by the following matrices:

   a. $\begin{bmatrix} 2 & 5 \\ 1 & 3 \end{bmatrix}$ b. $\begin{bmatrix} -2 & 2 \\ 2 & -2 \end{bmatrix}$ c. $\begin{bmatrix} 2 & -2 \\ -1 & 1 \end{bmatrix}$

   d. $\begin{bmatrix} 4 & -2 & 3 \\ -1 & 2 & 5 \end{bmatrix}$ e. $\begin{bmatrix} 5 & 2 & 3 \\ 1 & 6 & 4 \end{bmatrix}$ f. $\begin{bmatrix} 0 & -2 & 3 & -1 \\ 4 & 5 & -2 & -3 \end{bmatrix}$

   g. $\begin{bmatrix} 3 & -2 & 1 \\ 1 & 2 & -1 \\ -3 & 5 & 2 \end{bmatrix}$ h. $\begin{bmatrix} 4 & 3 & -3 & 1 \\ -2 & 1 & -2 & 5 \\ 2 & -3 & 3 & -3 \\ 0 & 1 & 4 & 2 \end{bmatrix}$

2. Design a spreadsheet model that produces the graphic solution presented at the end of this section (Figure 8.109). Use it to create graphs for Part a through Part f of the preceding problem.

3. Design an implementation of the classical "rock, scissors, paper" game. Each person displays one of these options. The outcomes "rock breaks scissors," "scissors cuts paper," and "paper covers rock" each earn 1 for the player choosing the first option and −1 for the player choosing the second option. If both players choose the same option, then neither wins (see Figure 8.110).

Row\Col	rock	scissors	paper
rock	0	1	−1
scissors	−1	0	1
paper	1	−1	0

Figure 8.110

4. Suppose that in a baseball game a pitcher can throw three pitches: fastball, curve, change-up. If a batter can guess the pitch that is coming, it increases his chance of getting a hit. These probabilities are given in the table in Figure 8.111. Find optimal strategies for the batter and the pitcher, and the resulting expected probability of getting a hit.

batter's guess\pitcher's throw	fast ball	curve	change-up
fastball	0.40	0.13	0.20
curve	0.24	0.30	0.25
change	0.18	0.23	0.40

Figure 8.111

5. The solver techniques in this section can also be used to solve linear programming problems. Design a spreadsheet model to solve the following linear programs.

a. Maximize $z = 6x_1 + 3x_2 + 4x_3$
subject to
$2x_1 + x_2 + x_3 \leq 500$
$x_1 + 2x_2 + x_3 \leq 300$
$x_1 + x_2 + 2x_3 \leq 200$
$x_1 \geq 0, x_2 \geq 0, x_3 \geq 0$

b. Minimize $z - 10x_1 + 20x_2 + 15x_3$
subject to
$4x_1 + x_2 + 2x_3 \geq 100$
$3x_1 + 2x_2 + 1x_3 \geq 80$
$x_1 + 2x_2 + 5x_3 \geq 150$
$x_1 \geq 0, x_2 \geq 0, x_3 \geq 0$

Appendix

A.1   Spreadsheet Basics and Book Conventions

This appendix describes the notational conventions that are used in the book and explains the basic spreadsheet techniques that we employ. Because different spreadsheets incorporate slightly different formats and ways of doing things, we focus on the most widely used spreadsheet, *Microsoft Excel*. Most of what we do will also work in many other spreadsheet programs like *StarCalc*, a spreadsheet that is part of the *StarOffice* program provided by *Sun Microsystems* on the Web, and its open-source sibling, *OpenCalc*, which is part of the freely available package *OpenOffice*. We will employ the methods used in *Excel* in our descriptions. The most important difference is that sliders (also called scroll bars) described later in the appendix do not exist in some other spreadsheet programs, and if they exist, using them is not as comfortable as it is in *Excel*.

The basic unit of a spreadsheet is a cell. A spreadsheet arranges these cells in a large rectangular array of rows and columns. Usually the spreadsheet screen display identifies columns by letters and the rows by positive integers. A cell's location is identified by its row and column. At any time one of the cells, sometimes called the active cell, is highlighted by a cursor. In the illustration in Figure A.1 this is Cell B3. However, except for this appendix in which we seek to unify the standard spreadsheet display with our new approaches, we will not have reason to refer to these row and column identifiers.

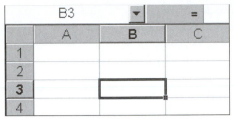

Figure A.1

We select the active cell by using the keyboard's arrow keys to move the cursor to that cell, or by pointing to the cell with the mouse and clicking on it. In *Excel* the mouse pointer is shown as a white cross.

Into the active spreadsheet cell we can enter text (often alternatively called a string), a number, or a formula that refers to other cell locations within the spreadsheet. In the latter case, the spreadsheet finds the value of the formula using the cells that are referenced and shows the result on the screen display.

Whenever an entry in the spreadsheet is changed, the entire spreadsheet is recalculated, and the screen display is updated. This feature is the basis for the spreadsheet's renowned "What if . . . ?" capabilities. We exploit this recalculation feature throughout the book's examples to study the effect of changes in data, parameters, and initial values. It is possible to turn off the automatic recalculation feature using *Excel*'s Tools, Options, Calculation command to select manual calculation. Doing this can be useful while developing models that include the generation of random numbers.

As mentioned earlier, when we enter a formula into a cell that references other cells, the spreadsheet displays the value of the formula in its cell. Over the years people have developed different ways of entering the formulas and then describing what has been entered. Unfortunately, often these styles miss the essence of what is going on. One of the primary emphases of this book is to do things in a way that supports mathematical thinking, and then to describe our spreadsheet operations and the aspects of creating a formula in a similar manner.

To present our ideas, we begin with a simple example. Suppose that we want to compute the interest earned during a year by a savings account deposit on which interest is computed annually. We might start with a display such as is shown in Figure A.2. Across the top row we enter headings for our work as text. In the next row we first enter the depositor's name. In the next two cells we enter numbers, the depositor's balance at the start of the year and the annual interest rate.

	A	B	C	D	E
1	Name	Start	Rate	Interest	
2	Tom	1000	0.05		
3					

Figure A.2

Now we need to compute the interest. Clearly we want to multiply the starting balance (1000) by the interest rate (0.05). With the cursor on the interest cell, D2, we start by typing in the = sign to indicate that a formula is being entered (see Figure A.3). Then we use the mouse to click on the location that contains the first number to multiply, 1000 (we can also select the location by using the keyboard's arrow keys). This happens to be in cell B2, but conceptually its cell address is not vital. We simply need to be able to point it out and click on it with the mouse or select it using the arrow keys. When we do this we see a dotted line appear around the cell containing the number, indicating that that indeed is the location that we are entering. Also, in the formula bar that is located just above the spreadsheet display, we will see the start of the formula, =B2.

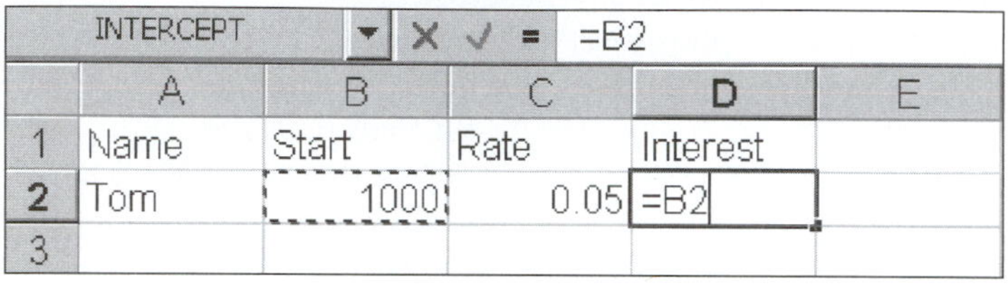

INTERCEPT		▼	✗ ✓ =	=B2	
	A	B	C	D	E
1	Name	Start	Rate	Interest	
2	Tom	1000	0.05	=B2	
3					

Figure A.3

We then type in the multiplication symbol, *, and (using either the mouse or the keyboard) point at the number that we wish to multiply by, which is in the cell that contains the rate. Again we see that cell highlighted, and in the formula bar we see that more has been appended onto the formula (Figure A.4).

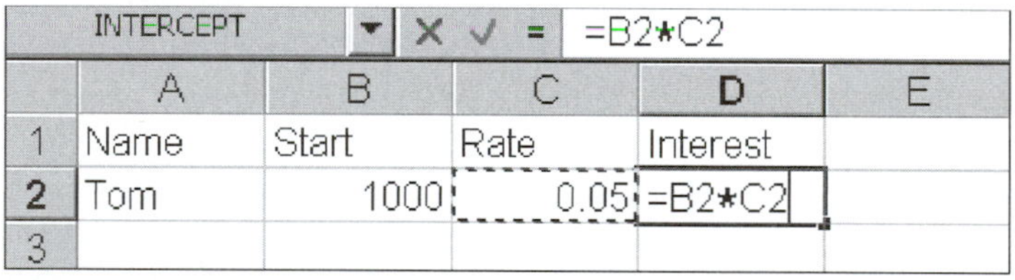

INTERCEPT		▼	✗ ✓ =	=B2*C2	
	A	B	C	D	E
1	Name	Start	Rate	Interest	
2	Tom	1000	0.05	=B2*C2	
3					

Figure A.4

We then finish the process by pressing the Enter key to generate the display shown in Figure A.5.

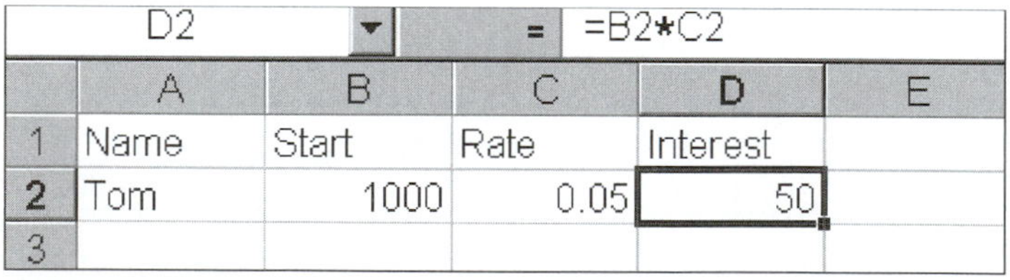

D2		▼	=	=B2*C2	
	A	B	C	D	E
1	Name	Start	Rate	Interest	
2	Tom	1000	0.05	50	
3					

Figure A.5

Notice that the spreadsheet displays the value of the cell on the screen. We can see the formula that produces the output for the active cell in the formula bar.

We might refer to our approach to entering formulas as "gesturing." From this example there are two important things to learn.

First, in entering the formulas we emphasize that we do not type in cell locations. Doing so tends to create two unfortunate problems. First, it is easy to introduce typographical errors, such as typing C1 when we mean C2. But more basically, typing in cell coordinates emphasizes a notation that the computer uses, but which is not part of our mathematical conceptualization process. When we encounter such a problem in life, we do not think about (or say) "multiply B2 by C2," but rather we would think (or say) "multiply the starting balance by the interest rate." When we use the gesturing approach, we do not worry continuously about which spreadsheet cell coordinates we have used in entering these constants in the formula. We simply point to the cells and use their contents.

Second, since we want our process to imitate the way that we create our formulas and want to concentrate on operating on the two quantities rather than worry about their cell coordinates, in this book we will adopt a notation that is in the spirit of good spreadsheet use and the nature of thinking mathematically. Thus, we will not write formulas as =B2*C2. Doing so tends to make us concentrate on following a recipe and hitting the right keys on the keyboard, perhaps without even realizing what we are doing. Instead, to describe the preceding operation, we will use the diagram shown in Figure A.6.

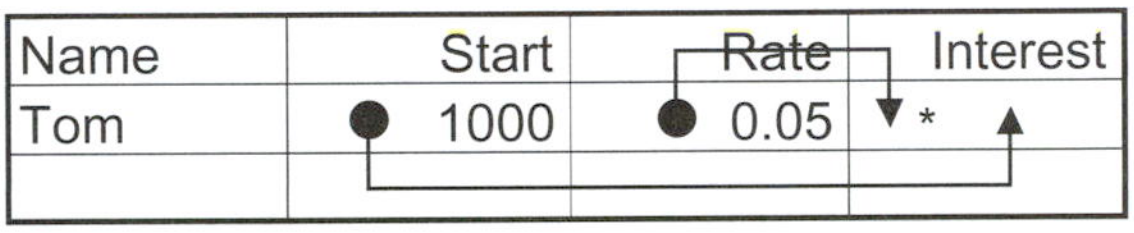

Name	Start	Rate	Interest
Tom	● 1000	● 0.05	▼ * ▲

Figure A.6

The solid circle at the start of an arrow indicates the location of the cell that we point to for an input value, and the arrow indicates where it goes in a formula. Using this notation we concentrate on the essence of what we are doing and are freed from the standard row-column notation. It should be noted that these diagrams are those of the authors. They are not part of *Excel*.

To practice this notation, we expand the preceding example and use a new column in which we compute the resulting value of the account after interest has been computed and paid. Again we do this to find the balance at the finish of the year by using the gesturing method to add the starting value to the interest.

Thus, we first select Cell E2. We next type in = and then point to the cell containing the amount at the start of the year (Figure A.7).

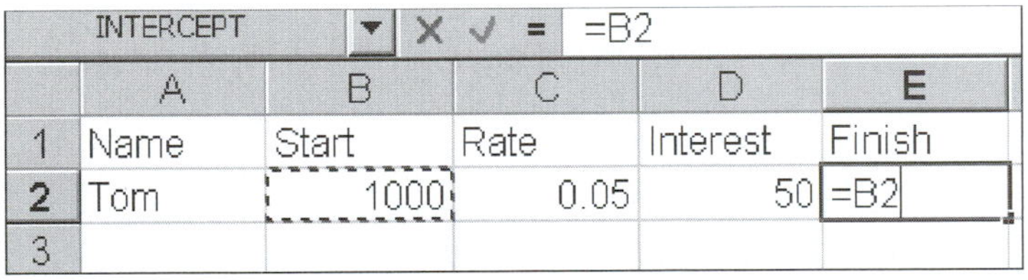

INTERCEPT	▼ X ✓ =	=B2		
	A	B	C	D	E
1	Name	Start	Rate	Interest	Finish
2	Tom	1000	0.05	50	=B2
3					

Figure A.7

We then type in + and point to the cell that contains the interest (Figure A.8).

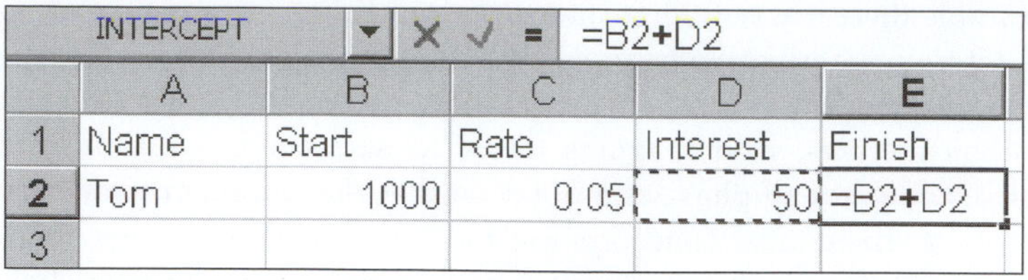

INTERCEPT	▼	X ✓ =	=B2+D2	
	A	B	C	D	E
1	Name	Start	Rate	Interest	Finish
2	Tom	1000	0.05	50	=B2+D2
3					

Figure A.8

After we have pressed the Enter key, on the screen we will see the display shown in Figure A.9.

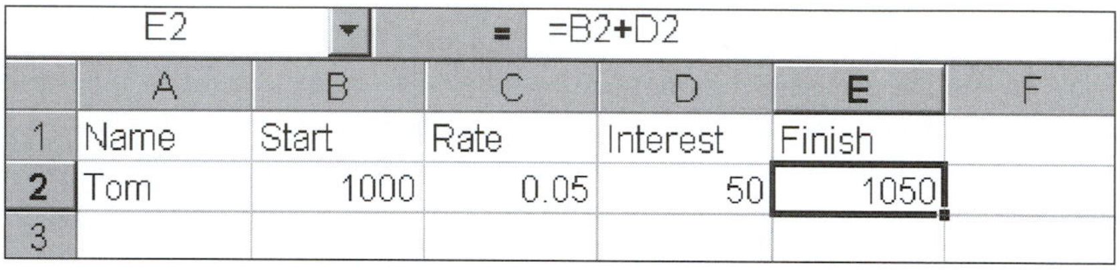

E2	▼	=	=B2+D2		
	A	B	C	D	E	F
1	Name	Start	Rate	Interest	Finish	
2	Tom	1000	0.05	50	1050	
3						

Figure A.9

This step is denoted as shown in Figure A.10.

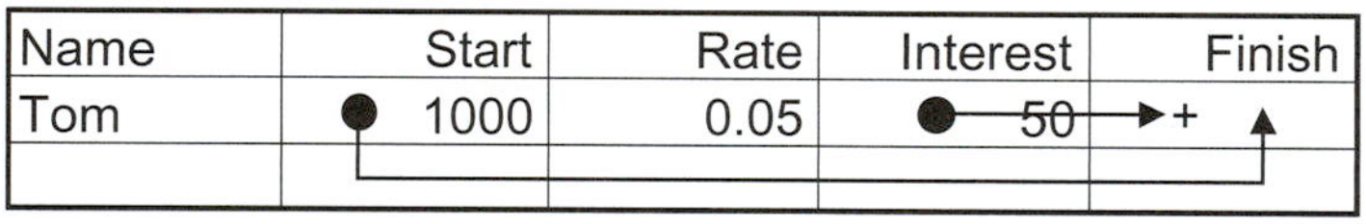

Name	Start	Rate	Interest	Finish
Tom	● 1000	0.05	● 50 → + ▲	

Figure A.10

Next, suppose that we add more people to our list of bank customers as shown in Figure A.11.

Name	Start	Rate	Interest	Finish
Tom	1000	0.05	50	1050
Mary	250.34	0.06		
Jan	3000	0.045		
Joe	2341.97	0.055		

Figure A.11

Although we could repeatedly duplicate the previous process with each person, doing this is not necessary. Instead we can copy the two formulas that we have already created. In designing models we will repeatedly use the spreadsheet's ability to do this. Usually we will be copying down a column or across a row. This is easy to do. We first click in the center of the first spreadsheet cell in the block to be copied, hold down on the left mouse button, and move the mouse to highlight, or select, the source cells containing the formula or formulas that we wish to copy. This process can also be done using the keyboard.

In our current example, suppose that we have added the new data values. We next select the two formulas giving the interest and finish values for Tom. Now we want to copy the contents of these cells. In the lower right corner of the selected block we see a marker called a fill handle (Figure A.12).

C	D	E	F
Rate	Interest	Finish	
0.05	50	1050	
0.06			
0.045	**Fill Handle**		
0.055			

Figure A.12

We can put the mouse pointer on the fill handle. When we do this the usual mouse pointer turns into a smaller, black cross. We then hold down on the left mouse button, and while holding down on the button we move the mouse to drag out a region (the destination cells) into which the formulas are to be copied. The dragging process is illustrated in Figure A.13. As we drag the mouse, the destination cells are highlighted. This particular copy method is also referred to as a *fill process*.

If we want to use the keyboard approach, we select a block by holding down on the Shift key and pressing on the arrow keys. We can use keyboard commands in place of the menu commands, as well. For example, to copy a highlighted block of cells, we type Ctrl c, while to paste an expression in a highlighted block of cells we type Ctrl v. Details of these commands can be found in your spreadsheet's help feature.

D2	▼	=	=B2*C2		
	A	B	C	D	E	F
1	Name	Start	Rate	Interest	Finish	
2	Tom	1000	0.05	50	1050	
3	Mary	250.34	0.06			
4	Jan	3000	0.045			
5	Joe	2341.97	0.055			
6						

Figure A.13

Alternatively, in *Excel* we can double-click on the fill handle and the formulas will be copied as far down as the adjacent column to the left extends.

This process can also be carried out through *Excel*'s Edit, Copy and Edit, Paste commands that are available from the main command menu displayed at the top of the screen (see Figure A.14). Thus, we first select the cells containing the formulas to be copied and select the commands Edit, Copy.

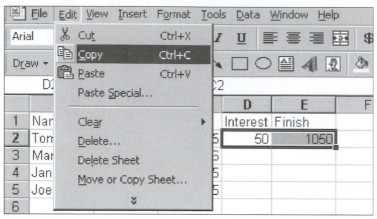

Figure A.14

Next, we use the mouse to select the cells into which we wish to copy the formulas, and select the commands Edit, Paste (Figure A.15).

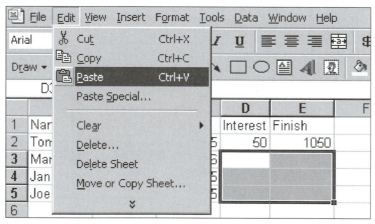

Figure A.15

In this book we denote the copy or fill operation by shading. The darker shading indicates the source cells that contain formulas that are to be copied, while the lighter shading indicates the destination cells into which they are copied (Figure A.16).

Name	Start	Rate	Interest	Finish
Tom	1000	0.05	50	1050
Mary	250.34	0.06		
Jan	3000	0.045		
Joe	2341.97	0.055		

Figure A.16

When we copied the contents of the two source cells in this example, their input cell locations moved along. If we look at the formulas that are produced in the copy process, we see that they have been adjusted (see Figure A.17). Thus, the computation for the first interest is "two cells to left times cell to left." When we look at the resulting formulas and numerical output, we see exactly the same thing. Consequently, the references to the input cell locations used in these formulas are called *relative references*.

	A	B	C	D	E
1	Name	Start	Rate	Interest	Finish
2	Tom	1000	0.05	=B2*C2	=B2+D2
3	Mary	250.34	0.06	=B3*C3	=B3+D3
4	Jan	3000	0.045	=B4*C4	=B4+D4
5	Joe	2341.97	0.055	=B5*C5	=B5+D5

Figure A.17

The output from our copying is given in Figure A.18.

	A	B	C	D	E
1	Name	Start	Rate	Interest	Finish
2	Tom	1000	0.05	50	1050
3	Mary	250.34	0.06	15.0204	265.3604
4	Jan	3000	0.045	135	3135
5	Joe	2341.97	0.055	128.8084	2470.778

Figure A.18

In its standard configuration, *Excel* displays the numerical output of a model. However, if we wish to view the underlying formulas we can issue the commands Tools, Options, View and check the Formulas box.

If we wish to examine the formula of a particular cell when we are looking at the numerical output, we can double-click on that cell. The formula is shown within the cell, with cell references given in color. The cells that are referenced are outlined in the same color (Figure A.19).

	A	B	C	D	E
1	Name	Start	Rate	Interest	Finish
2	Tom	1000	0.05	50	1050
3	Mary	250.34	0.06	15.0204	=B3+D3
4	Jan	3000	0.045	135	3135
5	Joe	2341.97	0.055	128.8084	2470.778

Figure A.19

If *Excel*'s auditing tool has been installed, we can use it to show the cell references of a formula much the same as in our descriptive diagrams. To do this we issue the commands Tools, Auditing and select the option Trace Precedents. We then will see the display shown in Figure A.20.

E3	▼	=	=B3+D3		

	A	B	C	D	E
1	Name	Start	Rate	Interest	Finish
2	Tom	1000	0.05	50	1050
3	Mary	250.34	0.06	15.0204	265.3604
4	Jan	3000	0.045	135	3135
5	Joe	2341.97	0.055	128.8084	2470.778

Figure A.20

After we have completed a model, we will probably want to format the output rather than use the rather untidy display seen in the figure. For example, we will want to align the column headings and to display fixed numbers of decimal places. However, a full description of formatting features of a spreadsheet does not fall within the purpose of this book. Usually these features can be found under the Format command option of your spreadsheet. One example of formatted output is shown in Figure A.21. In this display we have aligned and bolded the column headings, formatted dollar amounts to show two decimal places and commas to indicate thousands of units, and formatted the rates as percentages that display two decimal places.

Name	Start	Rate	Interest	Finish
Tom	1,000.00	5.00%	50.00	1,050.00
Mary	250.34	6.00%	15.02	265.36
Jan	3,000.00	4.50%	135.00	3,135.00
Joe	2,341.97	5.50%	128.81	2,470.78

Figure A.21

There are many times when we do not want some of the cell locations adjusted when we copy a formula. For example, consider the case in which all of the depositors have their money in the same bank at the same interest rate. Let us modify our previous example to apply to this new set of circumstances. We provide a separate cell for the interest rate.

To compute the interest of the first depositor we might consider creating a similar formula just as before, as is shown in Figure A.22. However, if we copy this down one row, then the cell references are adjusted to give the display in Figure A.23. This is clearly wrong, as we would be treating the word "Start" as if it were the interest rate for Mary!

Rate		0.05	
Name		Start	Interest
Tom		1000	*
Mary		250.34	
Jan		3000	
Joe		2341.97	

Figure A.22

Rate		0.05	
Name		Start	Interest
Tom		1000	
Mary		250.34	*
Jan		3000	
Joe		2341.97	

Figure A.23

Let us check to see what actually happens in the spreadsheet if we do this and copy the top formula into all of the rows below it. The original formula is shown in Figure A.24, with the formulas that are produced shown in Figure A.25. If we look at the cells that are referenced, we can clearly see that the formulas represent nonsense.

	A	B	C
1	Rate	0.05	
2	Name	Start	Interest
3	Tom	1000	=B1*B3
4	Mary	250.34	
5	Jan	3000	
6	Joe	2341.97	

Figure A.24

	A	B	C
1	Rate	0.05	
2	Name	Start	Interest
3	Tom	1000	=B1*B3
4	Mary	250.34	=B2*B4
5	Jan	3000	=B3*B5
6	Joe	2341.97	=B4*B6

Figure A.25

Figure A.26 shows what is produced in *Excel!*

	A	B	C
1	Rate	0.05	
2	Name	Start	Interest
3	Tom	1000	50.00
4	Mary	250.34	#VALUE!
5	Jan	3000	3,000,000.00
6	Joe	2341.97	586,288.77

Figure A.26

To avoid this unwanted behavior, the reference cell for the interest rate must not move. What we need to do is to "pin down" the reference to the interest rate so that it does not move along as we copy. This can be done in the following way: We begin by entering the formula as usual through our gesturing technique (Figure A.27). However, immediately after we have clicked on the cell for the interest rate, we press the F4 key at the top of the keyboard. This replaces the reference B1 by the reference $B$1 (Figure A.28). These two expressions are treated identically when the spreadsheet evaluates them. However, when an expression containing the latter expression is copied, because of the $ signs the cell location is held constant. We refer to such a reference as an *absolute reference*.

CELL	▼	✗	✓	=	=B1

	A	B	C	
1	Rate	0.05		
2	Name	Start	Interest	
3	Tom	1000	=B1	
4	Mary	250.34		
5	Jan	3000		
6	Joe	2341.97		

Figure A.27

CELL	▼	✗	✓	=	=$B$1

	A	B	C	
1	Rate	0.05		
2	Name	Start	Interest	
3	Tom	1000	=$B$1	
4	Mary	250.34		
5	Jan	3000		
6	Joe	2341.97		

Figure A.28

We then complete the formula in the usual manner, as shown in Figure A.29. After we have pressed the Enter key we copy the result using the fill handle, as shown in Figure A.30.

CELL		▼	✕ ✓ =	=$B$1*B3
	A	B	C	D
1	Rate	0.05		
2	Name	Start	Interest	
3	Tom	1000	=$B$1*B3	
4	Mary	250.34		
5	Jan	3000		
6	Joe	2341.97		

Figure A.29

C3		▼	=	=$B$1*B3
	A	B	C	D
1	Rate	0.05		
2	Name	Start	Interest	
3	Tom	1000	50.00	
4	Mary	250.34		
5	Jan	3000		
6	Joe	2341.97		

Figure A.30

Figure A.31 and Figure A.32 show the output and formulas that are produced. Notice that the reference to the interest rate has been held fixed by using the rate as an absolute reference. On the other hand, the reference to the starting deposit is a relative reference just as before and, therefore, will be moved along as we copy down the column.

	A	B	C
1	Rate	0.05	
2	Name	Start	Interest
3	Tom	1000	50.00
4	Mary	250.34	12.52
5	Jan	3000	150.00
6	Joe	2341.97	117.10

Figure A.31

	A	B	C
1	Rate	0.05	
2	Name	Start	Interest
3	Tom	1000	=$B$1*B3
4	Mary	250.34	=$B$1*B4
5	Jan	3000	=$B$1*B5
6	Joe	2341.97	=$B$1*B6

Figure A.32

In fact, rather than using the [F4] key, we could type in the $ signs manually. However, we prefer the approach of using a procedure designed specifically for adjusting references.

We denote a reference as being absolute in our diagrams by sticking a pin through the dot to indicate that it is "pinned down" in copying (see Figure A.33). When we copy a formula containing such a reference, its location will not change.

Rate	0.05	
Name	Start	Interest
Tom	1000	*
Mary	250.34	
Jan	3000	
Joe	2341.97	

Figure A.33

In addition to references that are absolute in both row and column, at times it is necessary to be able to hold the row reference fixed but allow the column reference to vary, or vice versa.

For example, suppose that we compute the interest that our four individuals would earn at each of three banks that pay different rates of interest. We enter a formula multiplying the deposit by the rate for the first person, and eventually we will copy it. However we first must ensure that the rate always comes from the same row but that the column reference for the rate can vary. To indicate this, we put a horizontal rail on the solid circle (see Figure A.34). Thus, the cell reference changes as we go horizontally to different banks but stays absolute as we compute the interest for different people (that is, different rows) from the same bank.

Similarly, the amount at the start of the year must always come from the same column. Here we put a vertical rail on the dot, so that the reference changes as we go from person to person vertically but stays absolute as we vary the banks (that is, different columns).

		Interest Rates		
		0.05	0.065	0.048
Name	Start	Bank 1	Bank 2	Bank 3
Tom	1000	*		
Mary	250.34			
Jan	3000			
Joe	2341.97			

Figure A.34

As we create our formulas in the spreadsheet we do just as before, and press the [F4] key. When we press the key once, the $ is put on both parts of the reference, as say $B$4. If we press this key repeatedly, the expression first changes to B$4, then to \$B4, then to B4, and then back to \$B\$4, and so on. Thus, we use F4 as a toggle key and simply choose the option that we want. Of course, it is still possible to manually type in the \$, too. We use Command-T on a Macintosh.

Next, we copy the formulas as indicated before. The formulas that are generated from our example are provided in Figure A.35. Here we can view the effect of using the \$ in copying our formulas. This must be done in two stages, first dragging the cell's fill handle to the right to complete the row, and then dragging the row's fill handle down to complete the columns (or vice versa).

	A	B	C	D	E
1			Interest Rates		
2			0.05	0.065	0.048
3	Name	Start	Bank 1	Bank 2	Bank 3
4	Tom	1000	=\$B4*C\$2	=\$B4*D\$2	=\$B4*E\$2
5	Mary	250.34	=\$B5*C\$2	=\$B5*D\$2	=\$B5*E\$2
6	Jan	3000	=\$B6*C\$2	=\$B6*D\$2	=\$B6*E\$2
7	Joe	2341.97	=\$B7*C\$2	=\$B7*D\$2	=\$B7*E\$2

Figure A.35

We can observe the effects in the output, too. Observe that in the display in Figure A.36 we have formatted the output to show two decimal places. However, in doing this we need to be aware that the underlying values have not been changed. If we wish to round our output so that dollar amounts will be computed to two decimal places, then we need to use the built-in rounding functions that are provided for this purpose.

	A	B	C	D	E
1			Interest Rates		
2			0.05	0.065	0.048
3	Name	Start	Bank 1	Bank 2	Bank 3
4	Tom	1000	50.00	65.00	48.00
5	Mary	250.34	12.52	16.27	12.02
6	Jan	3000	150.00	195.00	144.00
7	Joe	2341.97	117.10	152.23	112.41

Figure A.36

Although we do not make use of naming cells in our approach, it is possible to do so. Suppose that we start with the basic tax computation layout shown in Figure A.37.

	A	B
1	Income	5000
2	Tax Rate	0.1
3		
4	Tax	

Figure A.37

We first click in the cell for the income, Cell B1, and then select the menu options Insert, Name, Define. In the resulting dialog box (Figure A.38), we can either enter a name for Cell B1 or use the one that *Excel* supplies, as shown in the figure.

Figure A.38

We then do the same thing with Cell B2, naming it as tax rate. Then as we use our gesturing technique to enter the formula by clicking on the appropriate cells, *Excel* uses the names that we have defined. Thus, in Cell B4 we type the = sign and click on Cell B1 and see that the formula starts with =Income (Figure A.39). We then type in the multiplication symbol, click in Cell B2, and see the formula as =Income*Tax_Rate (Figure A.40).

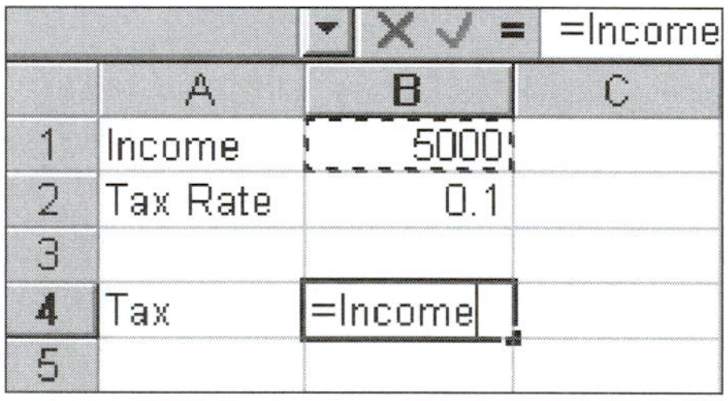

Figure A.39

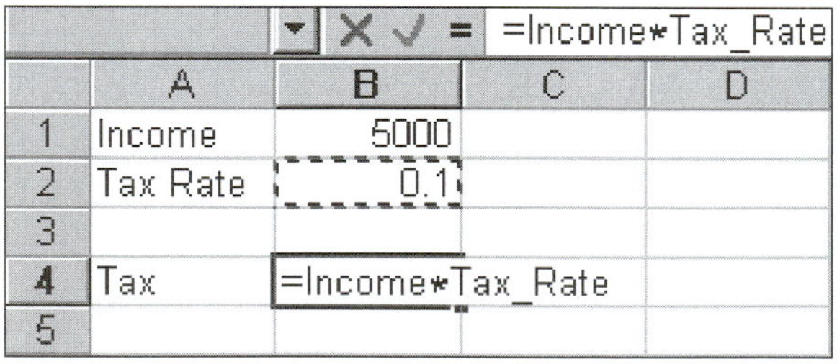

Figure A.40

When we press the Enter key, we see the computation expressed in terms of the two "variable" cells (Figure A.41).

	B4		=	=Income*Tax_Rate
	A	B	C	D
1	Income	5000		
2	Tax Rate	0.1		
3				
4	Tax	500		
5				

Figure A.41

After we have created a file we will probably want to save it. To do this for the first time in *Excel*, we issue the command File Save. Alternately, we can click on the Save icon. Then in the designated slots of the dialog box that is generated we provide the name of the file and the folder in which it is to be stored. If we provide the name Model1 for the file, then it is saved as the file Model1.xls. To save a revised version of the file under the same name, we can simply click on the Save icon or issue the command File Save. To save the modified file under a new name, we enter the command File Save As and complete the dialog box that is generated as before.

We will also want to print our work. Before doing this, we need to decide whether or not to include such features as gridlines or row and column headers in what is printed. These options are set under the command File Page Set Up (for example, gridlines and headers are under the Sheet option of the Page Set Up dialog box). To print the numerical output of our file, we then can either click on the Print icon or issue the command File Print. To print the underlying formulas, we first display the formulas by issuing the command Tool Options View and clicking on the Formulas option before issuing the File Print command.

A.2 Functions and Additional Spreadsheet Features

1. Spreadsheet Functions

In the last section we built spreadsheet formulas by using only the standard fundamental mathematical operations of addition, subtraction, multiplication, and division. However, in addition to these basic operations, spreadsheets also provide us with a wide variety of built-in functions. Some of these are so easy to remember that we generally enter them directly at the keyboard.

For instance, in the example that follows, we will use Cell B2 to compute the square root of the number that we enter in Cell A2. To do this, in Cell B2 we start by typing in

=SQRT(

We then use our gesturing technique with the mouse or the keyboard to select the cell (A2) that contains the number as shown in Figure A.42. Again, in our approach we do not type in the A2. Instead, we enter it by pointing to it, either with the mouse or an arrow key. This technique then puts the correct cell reference into the function and highlights the cell that we are referencing on the screen. As the formula is being built, it appears in both the cell and the formula bar at the top. We finish off the construction by typing the close parenthesis

$$)$$

and pressing the Enter key.

Now the spreadsheet displays the value of the expression that we have entered, with the formula (=SQRT(A2)) shown in the formula bar (Figure A.43).

INDEX	▼ X ✓ =	=sqrt(A2
	A	B	C	D
1	number	sq root		
2	5.8	=sqrt(A2		
3				

Figure A.42

B2	▼ =	=SQRT(A2)
	A	B	C	D
1	number	sq root		
2	5.8	2.408319		
3				

Figure A.43

We use our same basic notation to illustrate this process in our model descriptions, with the dot indicating an input cell, and the arrow showing where its value is used in an expression (see Figure A.44 and Figure A.45).

number	sq root
● 5.8	SQRT(↑)

Figure A.44

number	sq root
5.8	2.408319

Figure A.45

Because SQRT takes only one input, mathematically speaking it is a function of one variable. Such a function can be used with more complex expressions (including other functions) as well as with a number. For example, the hypotenuse of a right triangle is the square root of the sum of the squares of its two legs. Thus, the example displayed in Figures A.46 and Figure A.47 computes the length of the hypotenuse of a right triangle from its two legs by using the built-in function SQRT.

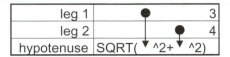

leg 1	●	3
leg 2	●	4
hypotenuse	SQRT(▼ ^2+ ▼ ^2)	

Figure A.46

leg 1	3
leg 2	4
hypotenuse	5

Figure A.47

As another example we can use functions inside of functions to compute an expression such as $f(x,y) = x^2 + \sqrt{y}$ employing the functions SUM and SQRT. The SUM function usually contains more than one input value. In *Excel* the input components of a built-in function are separate by a comma (,). The formulas and output of this example are displayed in Figure A.48 and Figure A.49.

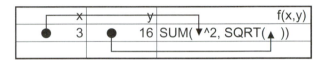

x	y	f(x,y)
● 3	● 16	SUM(▼^2, SQRT(▲))

Figure A.48

x	y	f(x,y)
3	16	13

Figure A.49

Other functions can operate on arrays of numbers that are found in blocks of spreadsheet cells. Finding the sum of an array of cells provides us with a good example. As in the previous examples, we first type in the expression

$$=\text{SUM}($$

Using either the mouse or the keyboard's arrow keys, we highlight the block of cells that we wish to add. This generates the display that is shown in Figure A.50.

SUM	▼ X ✓ =	=sum(B2:B4		
	A	B	C	D
1	Count	Numbers		
2	1	234		
3	2	356		
4	3	210		
5				
6	Sum	=sum(B2:B4		
7				

Figure A.50

Again we finish the expression by typing

)

and then pressing the ⌈Enter⌉ key.

The resulting formulas and screen display are shown in Figure A.51 and Figure A.52.

	A	B
1	Count	Numbers
2	1	234
3	2	356
4	3	210
5		
6	Sum	=SUM(B2:B4)

Figure A.51

	A	B
1	Count	Numbers
2	1	234
3	2	356
4	3	210
5		
6	Sum	800

Figure A.52

We will denote the range of cells that goes into this and the other functions that require such a range as one of the arguments by putting a double line around the block, as shown in Figure A.53.

Count	Numbers
1	234
2	356
3	210
Sum	SUM(▼)

Figure A.53

In addition to entering a function by typing it in at the keyboard, we can also use the Insert, Function command, or click on the equivalent button (called the Function Wizard or Paste Function in *Excel*.) *Excel* also provides a separate button, Σ, for entering the SUM function.

Here we will illustrate this approach to find the *n*th largest number in a set of data. Suppose that a teacher has entered a set of test scores into a spreadsheet. We use just a small set for illustration, but our method will work just as well for very large sets. If we want to find the third largest number, we enter 3 in the cell that is labeled rank (C1). Next we place the cursor on the cell in which we want the score generated (C2). We then click on the function button (Figure A.54) or choose the Insert, Function option from the main command menu.

Figure A.54

From the left side of the dialog box that is generated (Figure A.55), we select the type of function that we want (Statistical), and from the corresponding right side we select the function LARGE. We then click on the OK button.

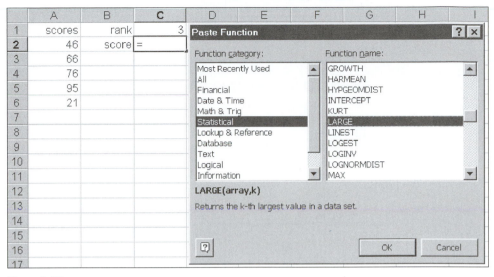

Figure A.55

This gives us the dialog box shown in Figure A.56. One at a time, we click in the various slots that are provided and fill them by clicking on the appropriate cell location. As with the SUM function, we highlight the array of scores in the first slot, and then click on the desired rank in the second slot. As we do this we can see the values of the components and the formula, as well as descriptions of the functions and their arguments. We complete this step by clicking on the OK button.

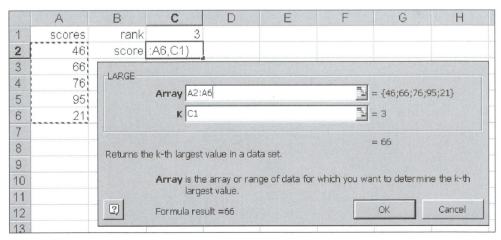

Figure A.56

The resulting function will have been entered into Cell C2 (Figure A.57). The use of the Function Wizard keeps us from needing to memorize the precise arrangements of the multitudes of functions provided by the spreadsheet.

	C
2	=LARGE(A2:A6,C1)

Figure A.57

Our notation for this simple example is given in Figure A.58.

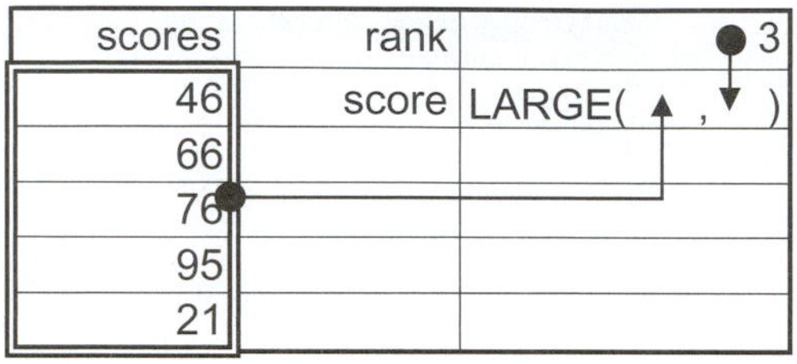

scores	rank	● 3
46	score	LARGE(▲ , ▼)
66		
76		
95		
21		

Figure A.58

We now provide one final example, illustrating the use of the powerful matrix functions that are built into *Excel*. We alert readers to the fact that this topic does require some elementary knowledge of the concepts of linear algebra. Although we use these functions only briefly in the book, some readers will want to use them in pursuing investigations in applications that employ matrices.

First we examine the matrix multiplication function, MMULT. Suppose that we want to multiply the following two matrices:

$$\begin{pmatrix} 2 & 3 \\ 4 & 6 \\ 3 & 1 \end{pmatrix} \begin{pmatrix} 3 & 7 \\ 4 & 1 \end{pmatrix}$$

After entering the components into the spreadsheet we select, or highlight, the 3×2 block for the answer and type in

$$=\text{MMULT(}$$

We then use the mouse or keyboard to highlight the first matrix (Figure A.59).

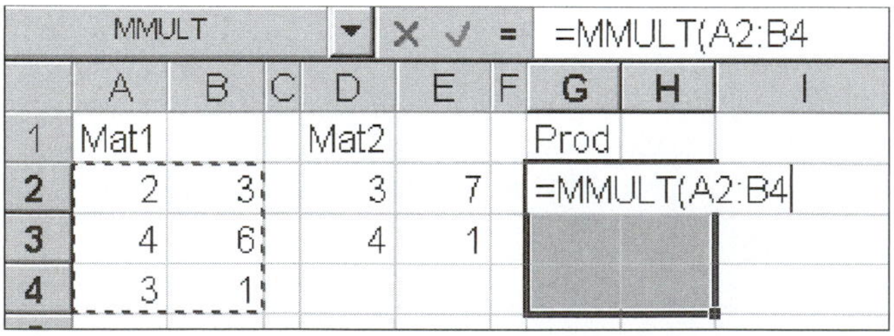

MMULT	▼	X ✓ =	=MMULT(A2:B4

	A	B	C	D	E	F	G	H	I
1	Mat1			Mat2			Prod		
2	2	3		3	7		=MMULT(A2:B4		
3	4	6		4	1				
4	3	1							

Figure A.59

We next type a comma and highlight the second matrix (Figure A.60). Finally we complete the matrix by typing the right parenthesis,

$$)$$

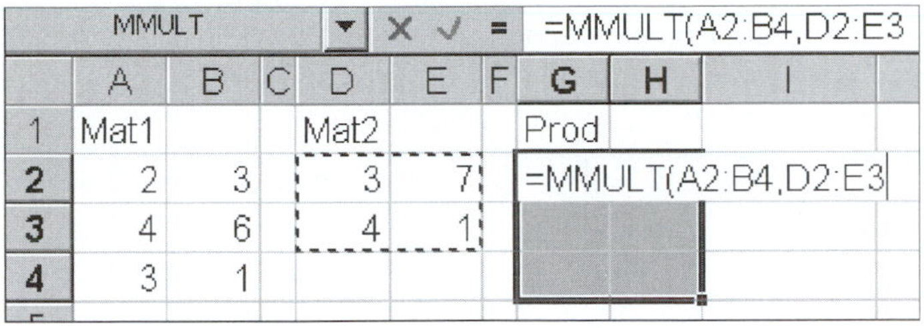

MMULT			▼	✕	✓	=	=MMULT(A2:B4,D2:E3	
	A	B	C	D	E	F	G	H	I
1	Mat1			Mat2			Prod		
2	2	3		3	7		=MMULT(A2:B4,D2:E3		
3	4	6		4	1				
4	3	1							

Figure A.60

To complete the process for this and other matrix functions, we press the combination of three keys Ctrl Shift Enter. *Excel* also allows us to use its function tools to enter the arrays into slots in dialog boxes. It is important to remember that in *Excel* we must press the combination Ctrl Shift Enter to enter the formula rather than clicking on the OK button.

The description of this process and the resulting output are illustrated in Figure A.61 and Figure A.62.

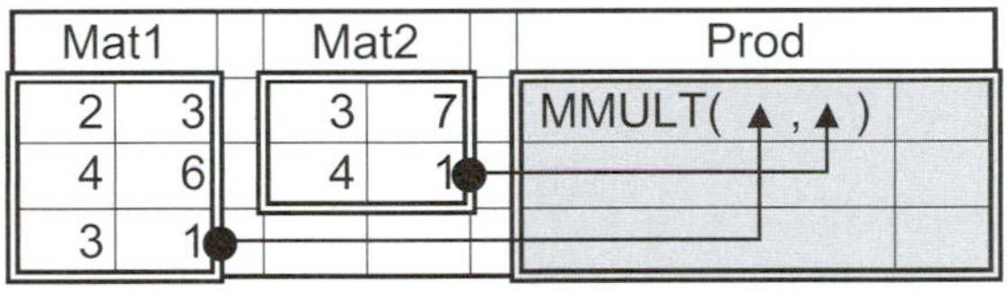

Mat1		Mat2		Prod	
2	3	3	7	MMULT(▲ , ▲)	
4	6	4	1		
3	1				

Figure A.61

Mat1		Mat2		Prod	
2	3	3	7	18	17
4	6	4	1	36	34
3	1			13	22

Figure A.62

Excel also provides us with the functions MINVERSE MDETERM to compute the inverse and determinant, respectively, of square matrices.

2. Data Table

Another powerful and useful feature that the spreadsheet provides us is the data table. This is really a hidden treasure of spreadsheets, as even many experienced spreadsheet users seem not to be aware of its usefulness. Descriptions of data table applications for mathematics are provided throughout the book. Here we will give some more specific details to illustrate its use in *Excel*.

Suppose that we wish to investigate the resulting value after 3 years of a one-time deposit of $1000 in a bank that pays annual compound interest. To do this we want to find the resulting balances for various interest rates.

For any one particular interest rate, say 0.05 (for 5%), we can use a spreadsheet model that is based on ideas discussed in the initial chapter of the book. The basic compound interest model is contained in the left side of Figure A.63. The resulting balance after 3 years is given in the bolded cell (B6).

	A	B	C	D	E	F
1	Rate	0.05			Summary	
2	Year	Bal	Intr		Rate	Bal
3	0	1000.00	50.00			
4	1	1050.00	52.50		0.05	
5	2	1102.50	55.13		0.06	
6	3	**1157.63**			0.07	

Figure A.63

In the right part of the model we use two columns that will provide us with a summary of our results. First, in the leftmost of these columns we enter the various interest rates that we want to study. For illustrative purposes we have used only three interest rates, but there is no difficulty in using more. The right column will be used to store the resulting 3-year account balances. We leave initially blank cells at the top of these rows. Then, into the cell at the top of the right column we enter a formula to provide the balance for the interest rate of the model (Figure A.64).

	F
3	=B6

Figure A.64

We describe this using our standard diagram format (Figure A.65).

Rate	0.05			Summary	
Year	Bal	Intr		Rate	Bal
0	1000.00	50.00			
1	1050.00	52.50		0.05	
2	1102.50	55.13		0.06	
3	1157.63			0.07	

Figure A.65

We next highlight the summary block and issue the command Data, Data Table from the main menu at the top of the screen. In the dialog box that is generated we click on the cell containing the interest rate (B1) to establish it as the column input cell, as illustrated in Figure A.66.

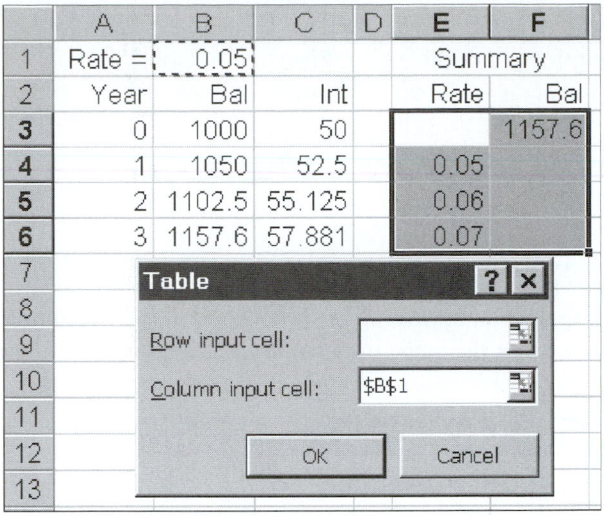

Figure A.66

We now click on the OK button. The command generates for us a summary of the resulting 3-year balances for the three interest rates that we provided (see Figure A.67). It is just as easy to do this with a larger set of rates. We need to be aware of the fact that if the data table or the underlying model is large, then the calculation process can become very slow. Once the data table is completed, we can change other parameters of the model (such as the amount of deposit), and the table will be updated.

Summary	
Rate	Bal
	1157.6
0.05	1157.6
0.06	1191
0.07	1225

Figure A.67

3. Inserting and Deleting Rows and Columns

Although it is always wise to plan ahead in laying out our spreadsheet models, occasionally we will find the need to insert some additional rows or columns to provide room for our work. As an illustration, suppose that we wish to compute the areas and the resulting costs of some sheets of metal, and that we have entered the data using the layout shown in Figure A.68. Unfortunately, we have no convenient place to enter the cost per square unit as a parameter of the model. We would like to enter this value into a cell located above the columns.

	A	B	C	D
1	Length	Width	Area	Cost
2	32	44	1408	
3	55	32	1760	
4	21	40	840	

Figure A.68

To solve our problem of creating such a cell, we can hold down on the left mouse button and drag the pointer over the row numbers at the left to highlight the first two rows. We then choose the menu options Insert, Rows as shown in Figure A.69.

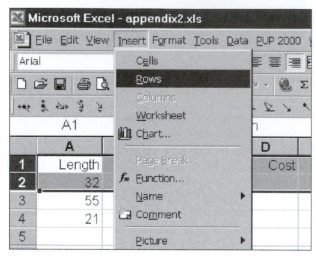

Figure A.69

This inserts two blank rows at the top of the spreadsheet page, as shown in Figure A.70. All formulas are adjusted automatically.

	A	B	C	D
1				
2				
3	Length	Width	Area	Cost
4	32	44	1408	
5	55	32	1760	
6	21	40	840	

Figure A.70

Into this area we can then enter the value of the parameter and complete the model. In Figure A.71, we show only the output but not the underlying formulas.

	A	B	C	D
1	Unit Cost	5		
2				
3	Length	Width	Area	Cost
4	32	44	1408	7040
5	55	32	1760	8800
6	21	40	840	4200

Figure A.71

Inserting columns is done in the same way. To delete entire rows we first select the rows as we have shown and then issue the command Edit, Delete, Rows. We can also do the same with columns. Again, all formulas will be adjusted automatically.

4. Conditional Formatting

Spreadsheets provide a variety of formatting features that we can use to improve the appearance of our models' output. In addition to these, there is one formatting tool that we can use to display mathematical characteristics. This is called *conditional formatting*.

Consider the following *Excel* model, in which the value of each cell in a square array of cells is given by $f(x,y) = y^2 - x^2$, where the y-values are in the left margin and the x-values are at the top. This generates a square array of numbers. To better see the pattern of these numbers, we use the mouse to select all of the interior cells, and then choose the main menu option Format, Conditional Formatting (see Figure A.72).

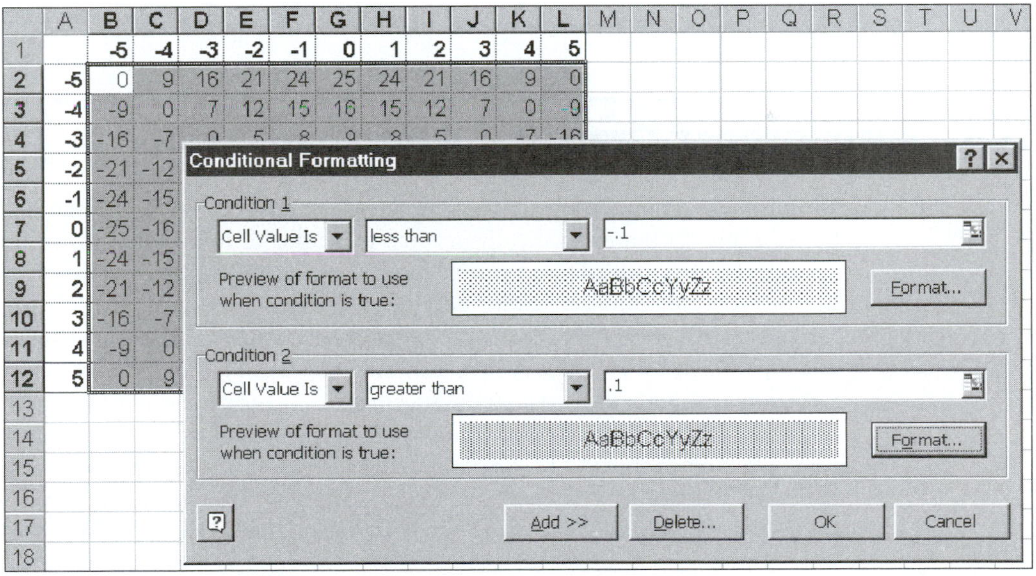

Figure A.72

This gives us a dialog box that allows us to format the cell in several ways depending upon the value of the cells. Figure A.72 shows the setting for a format for those cells for which the function is less than –0.1 and those for which the function is greater than 0.1. We select the range of numerical conditions for a certain category through the selection options provided in the dialog box, and set the display characteristics by clicking on the Format buttons. At present, *Excel* allows us to define three distinct formats in addition to the default format of the standard display.

The resulting display is presented in Figure A.73. In it we can observe the contours of a typical saddle-shaped surface. In one of the sections on graphing we show how to create a three-dimensional graph of the surface.

	-5	-4	-3	-2	-1	0	1	2	3	4	5
-5	0	9	16	21	24	25	24	21	16	9	0
-4	-9	0	7	12	15	16	15	12	7	0	-9
-3	-16	-7	0	5	8	9	8	5	0	-7	-16
-2	-21	-12	-5	0	3	4	3	0	-5	-12	-21
-1	-24	-15	-8	-3	0	1	0	-3	-8	-15	-24
0	-25	-16	-9	-4	-1	0	-1	-4	-9	-16	-25
1	-24	-15	-8	-3	0	1	0	-3	-8	-15	-24
2	-21	-12	-5	0	3	4	3	0	-5	-12	-21
3	-16	-7	0	5	8	9	8	5	0	-7	-16
4	-9	0	7	12	15	16	15	12	7	0	-9
5	0	9	16	21	24	25	24	21	16	9	0

Figure A.73

A.3 Spreadsheet Graphing Fundamentals

Graphs form an important visual component of the study of mathematics. Two sections of this book are specifically devoted to graphs of functions. This section of the appendix is designed to give some of the details of the graphing operations of *Excel*. In this program graphs are called charts.

Suppose that we wish to create the graph of a function, such as $f(x) = \cos(ax)$, where a is a parameter. To do this we use two columns (see Figure A.74). We generate the x-coordinates down the first column, starting with 0 and proceeding in steps of size 0.1 (Figure A.75). We then enter the formula for $\cos(ax)$ in the top cell of the next column, using the built-in spreadsheet function COS and treating a as an absolute reference. We then copy the formulas down their respective columns. To generate the graph, we select the columns of x- and y-values, as indicated by shading in Figure A.75.

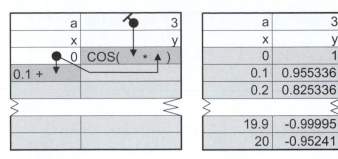

Figure A.74

a	3
x	y
0	1
0.1	0.955336
0.2	0.825336
19.9	-0.99995
20	-0.95241

Figure A.75

Part of the screen display appears in Figure A.76.

	A	B	C
1	a	3	
2	x	y	
3	0	1	
4	0.1	0.955336	
5	0.2	0.825336	
6	0.3	0.62161	
7	0.4	0.362358	
8	0.5	0.070737	

Figure A.76

Next, we either click on the Chart Wizard button (Figure A.77) or enter the command Insert, Chart. As the first step of the resulting four-part series of dialog boxes, we choose the chart type. As discussed in the graphing sections of the book, we will almost always select the xy chart type, as indicated in Figure A.78.

Figure A.77

Once we have selected the xy chart type, we then must select one of several variants as shown in the figure. The *xy*-chart plots consecutive points by their (x, y)-coordinates. If we wish to produce a graph showing only the individual points, then we choose the first subtype. This is particularly useful for creating scatter graphs in statistics.

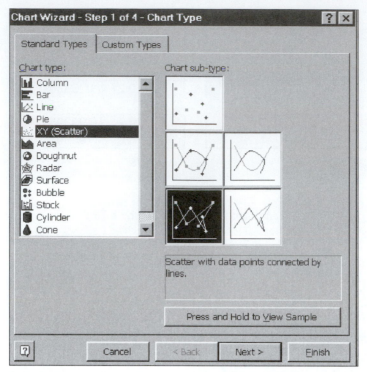

Figure A.78

The middle two subtypes use interpolation to determine curves to connect successive points, and consequently the resulting curves generally contain unwanted "bumps." The final two subtypes connect consecutively listed points with straight-line segments. When these are used to plot a large number of points that are close together, this produces a smooth-looking curve. The last subtype draws only the line segments connecting the points, while the next to the last one also provides markers at each of the points. At times we will want markers on some of the curves in a graph but not on others. Consequently, a good approach is initially to provide the markers and then later to remove those that are unwanted. We will illustrate this approach in the current example.

There are four steps in using the Chart Wizard. We will not pursue all of them here, but we do encourage you to experiment with the various options. In one of the options in the third step we chose to remove all gridlines and legends that the chart provides. The resulting output is shown in Figure A.79. In order to produce an attractive graph of appropriate dimensions, it may be necessary to use the mouse to click on one of the corners of the graph box and drag it to adjust the overall size.

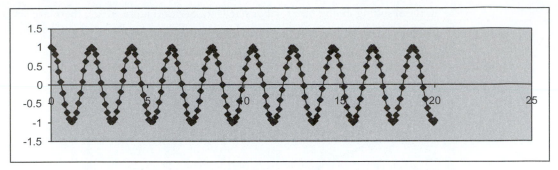

Figure A.79

The chart that is produced contains a gray background. To remove this we can right-click in the gray area and choose the option Format, Plot Area and then click in the appropriate boxes of the resulting dialog box to choose the option "none" for both border and area (Figure A.80).

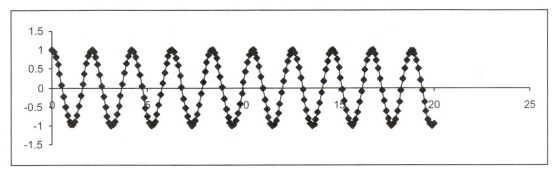

Figure A.80

To remove the markers, we right-click on the curve and select the option Format Data Series as shown in Figure A.81.

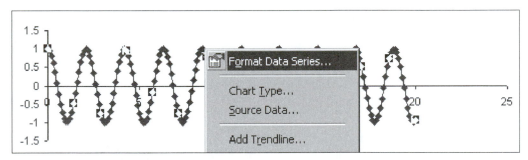

Figure A.81

In the resulting dialog box for Format Data Series (Figure A.82), there are several features of the chart that we can adjust. Here we click on the Patterns option and then click on the option None for the marker.

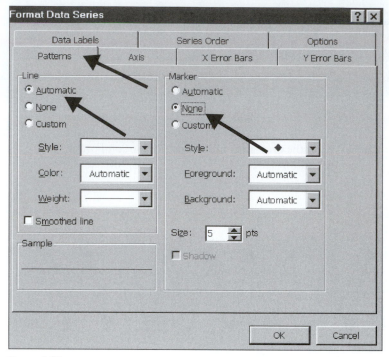

Figure A.82

Another modification that we will frequently need to make is to adjust the display of the axes. Usually the *xy*-graph produces longer axes than we may want. In Figure A.83, the *x*-axis extends to 25 although our values extend only to 20. To adjust it, we right-click on the *x*-axis, and from the resulting list we choose the option Format Axis. We must be careful to click on or very near the axis, for otherwise we may miss it.

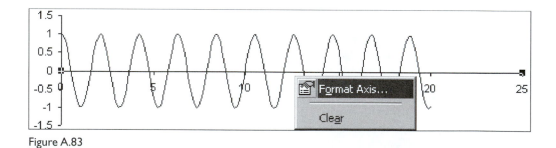

Figure A.83

Again the resulting dialog box provides us with many options, including the type of lines for the axis, the style and placement of the tick marks, their font and alignment, and the number of decimals. As shown in Figure A.84, we select the Scale tab. To remove the automatic scaling we can click in the respective slots provided and change the number to the one that we desire. Here we have set the x-range to cover the interval $0 \leqslant x \leqslant 20$ in steps of five units.

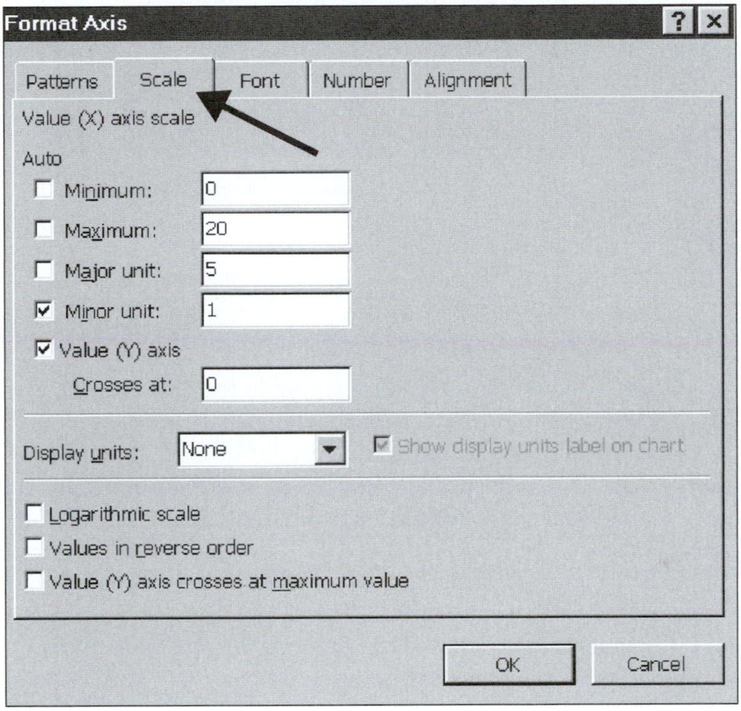

Figure A.84

We can use the same approach to adjusting the y-axis, so that the y-axis will be drawn over the interval $-1 \leqslant y \leqslant 1$. The resulting graph is shown in Figure A.85. There we have provided one more feature by right-clicking inside the graph's surrounding box and choosing the option Format, Chart Area and choosing the option None for the border.

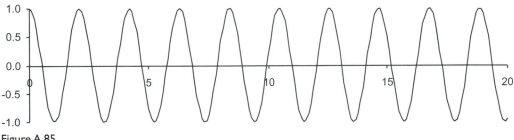

Figure A.85

Now let us look at a simple graph to illustrate one other feature that the spreadsheet provides us to help in our work (see Figure A.86). When we click in the chart, in the spreadsheet area the x- and y-series for the graphs are highlighted in color in the spreadsheet. This makes it easier for us to adjust the series if necessary.

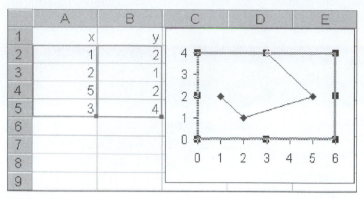

Figure A.86

In *Excel* we can adjust the chart region in the display in Figure A.86 by using the handles provided in the corners of the outlined areas.

Often we will want to include the graphs of more than one function in a spreadsheet graph. This is particularly easy to do if the functions are defined for the same points and their series can be listed in adjacent columns, as in the following example where we create the graphs of $f(x) = |x|$ and $g(x) = 4 - |2x|$ on the interval $-2 \leq x \leq 2$. The description of the model, which uses the absolute value function ABS, and its output are shown in Figure A.87 and Figure A.88. To create the graph we highlight the first three columns. The xy-graph uses the first column as the x-axis series and the next two columns as y-series of graphs.

x	f(x)	g(x)
-2	ABS(↑)	4-ABS(2 * ↑)
-1		
0		
1		
2		

Figure A.87

x	f(x)	g(x)
-2	2	0
-1	1	2
0	0	4
1	1	2
2	2	0

Figure A.88

The resulting *xy*-graph, created with both lines and markers, is shown in Figure A.89. Note that we change the format, color, and width of a line by right-clicking.

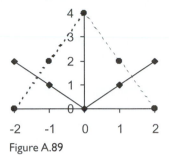

Figure A.89

A.4 Additional Spreadsheet Graphing Techniques

Sometimes we will have already created a graph of a function and then decide that we also want to plot the graph of another function over the same range of *x*-values. In the case shown in Figure A.90, we have already used the first two columns to create the graph of $f(x) = x^3 / 8$ over the interval $-2 \leq x \leq 2$. Now we decide to compute the values of $g(x) = 2 - x^2$ in the third column for the same values of *x*. This need not be adjacent to the previous *y*-series column, although in this case it is.

Using *Excel*, instead of going through the entire process of creating a new graph, we use the mouse to first highlight the new series. Then we put the mouse on the edge of the series. When we do this the mouse pointer changes from a white cross into a white arrow. We hold down on the left mouse button, drag the column until it is inside of the chart, and release the mouse button. We indicate part of the dragging process in Figure A.90.

	A	B	C
1	x	f(x)	g(x)
2	-2.00	-2.000	-2.00
3	-1.98	-1.941	-1.920
4	-1.96	-1.882	-1.842
5	-1.94	-1.825	-1.764
6	-1.92	-1.769	-1.686
7	-1.90	-1.715	-1.610
8	-1.88	-1.661	-1.534
9	-1.86	-1.609	-1.460
10	-1.84	-1.557	-1.386
11	-1.82	-1.507	-1.312
12	-1.80	-1.458	-1.240
13	-1.78	-1.410	-1.168
14	-1.76	-1.363	-1.098

Figure A.90

The spreadsheet interprets the data correctly as a new *y*-series and generates the graph that is shown in Figure A.91.

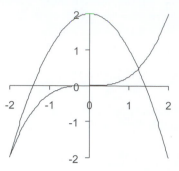

Figure A.91

Sometimes we wish to add a new set of (x, y)-coordinates for a new function that is defined for a different set of points from the first one. This is easy to do in *Excel*.

In the example illustrated in Figure A.92 and Figure A.93, we use the first two columns to create the graph of $f(x) - x^3 / 12$ at integer points ranging from -4 through 4. Next we want to include the graph of the inverse function formed by reversing the roles of the *x*- and the *y*-values. First we create the new series in the next two columns (Figure A.92 and Figure A.93).

x	f(x)		x	f⁻¹(x)
-4	-5.333			
-3	-2.250			
-2	-0.667			
-1	-0.083			
0	0.000			
1	0.083			
2	0.667			
3	2.250			
4	5.333			

Figure A.92

x	f(x)	x	f⁻¹(x)
-4	-5.333	-5.333	-4
-3	-2.250	-2.250	-3
-2	-0.667	-0.667	-2
-1	-0.083	-0.083	-1
0	0.000	0.000	0
1	0.083	0.083	1
2	0.667	0.667	2
3	2.250	2.250	3
4	5.333	5.333	4

Figure A.93

Now we use the mouse to select the two new columns and use the same process as before to drag them into the existing chart (Figure A.94).

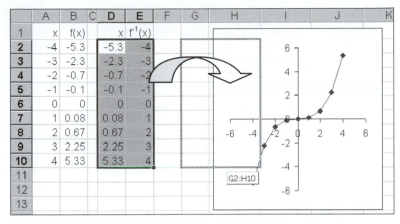

Figure A.94

This will generate a dialog box in which we indicate that the first column of the new block will represent points in an *x*-series (Figure A.95).

The resulting graph is shown in Figure A.96.

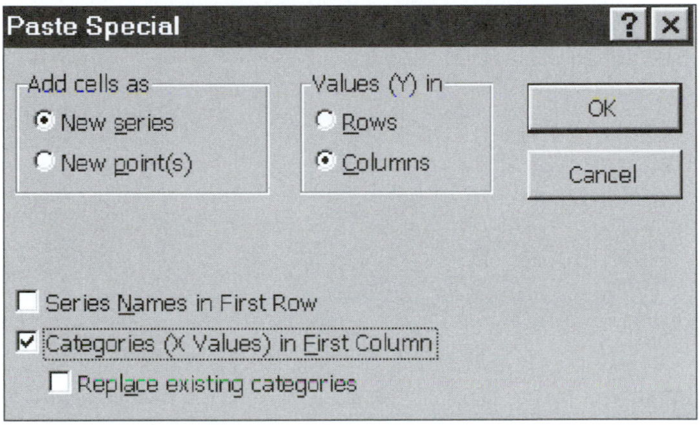

Paste Special [?] [X]

Add cells as
- ● New series
- ○ New point(s)

Values (Y) in
- ○ Rows
- ● Columns

[OK]

[Cancel]

☐ Series Names in First Row

☑ Categories (X Values) in First Column

   ☐ Replace existing categories

Figure A.95

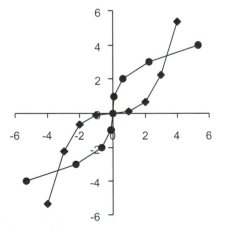

Figure A.96

Although in most of our applications we will use xy-graphs, there are times in which another graph type is appropriate. This typically is the case when we are representing a finite number of outcomes rather than a continuous function. The basic techniques for creating the graph are largely the same.

For example, Figure A.97 is a simple spreadsheet display showing the numbers of students who have received different possible scores on a four-question test. In the third column we compute the proportion of students receiving each possible score on the test (see Figure A.98). Either a column (vertical) or a bar (horizontal) is a good graph type to illustrate the proportions. One way to create it is to select the column of y-values, click on the Chart Wizard button, and then choose the column type. In the step of the Chart Wizard in which the y-series is listed, there is also a place for us to select the x-axis labels. Here we choose the first column for that purpose.

Score	Frequency	Proportion
0	9	/
1	25	
2	30	
3	35	
4	27	
Total	SUM( )	

Figure A.97

Score	Frequency	Proportion
0	9	0.071
1	25	0.198
2	30	0.238
3	35	0.278
4	27	0.214
Total	126	

Figure A.98

Depending on how we carry out the steps—if we eliminate the gridlines and legends as we create the graph and remove the gray background color afterward—we obtain the chart in Figure A.99. The remaining formatting is decided by the creator of the graph. You might compare the graphs in Figure A.99 and Figure A.100 and see some of the changes that could be made. We briefly list the changes that we have made in the following list. To carry out these changes, we generally right-click on the object that we want to change and select options from the list that is provided at that time.

- Format axes: Change font size, number of decimals.
- Format data series: Change fill patterns or color, column width, add data labels (values).
- Format chart area: Remove border.

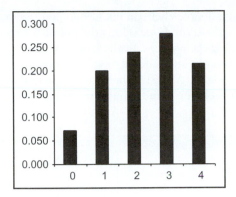

Figure A.99

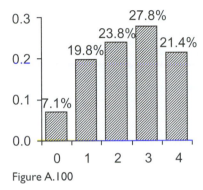

Figure A.100

We make one additional comment regarding the graphs in *Excel*. It is possible to vary the graph and the underlying model by clicking on a point of the curve in the chart and dragging it (either horizontally or vertically). This is a somewhat more advanced topic but one that allows us to develop some very interesting techniques.

First, let us generate a number of values of the function $f(x) = a /(1 + x)$ and then create an xy-graph in the usual way (see Figure A.101).

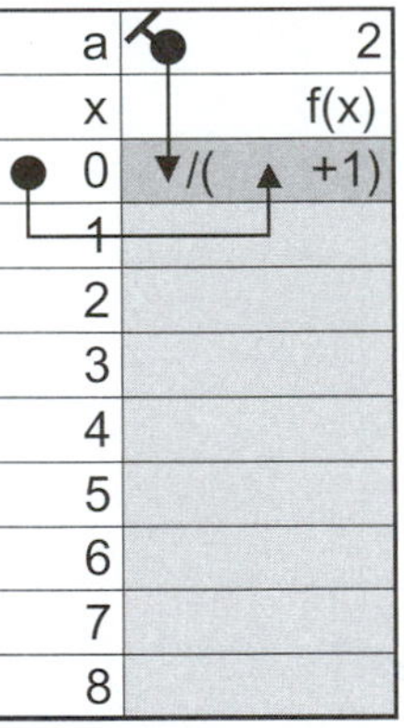

a	2
x	f(x)
0	/(+1)
1	
2	
3	
4	
5	
6	
7	
8	

Figure A.101

Next, we click twice (not a double-click) on the second point in the chart. The first click selects the curve, while the second one selects the point. We now hold down on the left mouse button and drag the point vertically from a y-value of 1.0 to a y-value of 0.5 (Figure A.102). We then obtain a small dialog box asking what cell we wish to change to move that point (Figure A.103). When we complete that by selecting the cell containing *a*, the Goal Seek command adjusts the value of *a* until the selected point has a y-coordinate of 0.5.

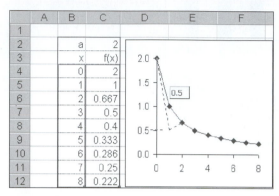

Figure A.102

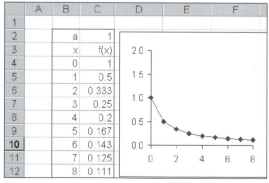

Figure A.103

The graph in Figure A.104 is produced by the process. Notice that the value of *a* and the entire set of values of the function have been changed.

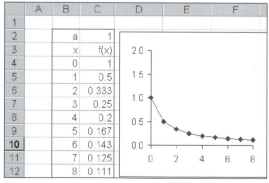

Figure A.104

If we drag a point of a graph whose coordinates are constant numbers rather than generated by a spreadsheet formula, a dialog box is not generated. Instead, the curve is adjusted as soon as we release the mouse button.

We can see this in the model displayed in Figure A.105 where we enter the (x, y)-coordinates for four points and create an xy-chart from these points. We then use *Excel*'s built-in INTERCEPT and SLOPE functions to compute the coefficients of the regression line that best fits the data. We then use the functions MIN and MAX to determine the largest and smallest values of x, and compute the y-coordinates of the endpoints of the regression line.

y-intercept	0.352		
slope	0.793		
x	y	X	Y
0.50	0.40	0.50	0.749
1.30	1.80	3.00	2.731
2.00	2.10		
3.00	2.50		

Figure A.105

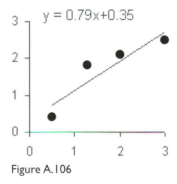

Figure A.106

We next highlight the last two columns and drag them into the existing chart to produce the graph shown in Figure A.106. At this point we click on one of the dots of the four data points to select the series, and then click a second time to select the specific point. At this time we hold down on the left mouse button and drag the point downward as shown in Figure A.107. Finally, when we release the mouse button we see that the point has been moved (Figure A.108).

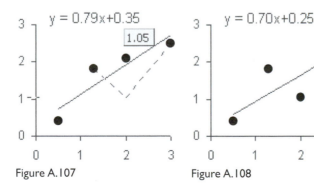

Figure A.107   Figure A.108

In addition, the corresponding cell in the model has been changed accordingly, and the expressions that reference are updated (see Figure A.109).

y-intercept	0.248		
slope	0.700		
x	y	X	Y
0.50	0.40	0.50	0.598
1.30	1.80	3.00	2.347
2.00	1.05		
3.00	2.50		

Figure A.109

1. Labels

It is often helpful to put labels on our graphs. *Excel* provides an easy way to display either the *x*- or the *y*-values as labels on each of the points. However, attaching other labels is a little more complex. We present one technique using the brief example shown in Figure A.110.

	A	B	C
2	Name	x	y
3	Tom	1	3
4	Mary	2	4
5	Sue	4	2

Figure A.110

We first select Column B and Column C and then use them to create an *xy*-graph as shown in Figure A.111. Suppose that ultimately we want to attach the respective names as labels of the points as in the graph in Figure A.112.

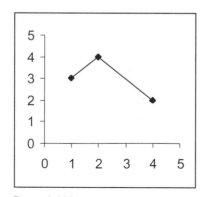

Figure A.111

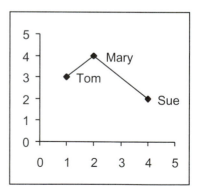

Figure A.112

Next, we right-click anywhere on the graph, and from the ensuing menu box we select the option Format Data Series, and then from the dialog box that is produced we select the tab Data Labels. Following this selection we then select the option Show Value as illustrated in Figure A.113. This will attach the y-values as labels for the points. We now right-click on one of these and choose the option Format Data Labels, as shown in Figure A.114.

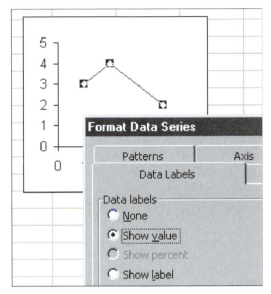

Figure A.113

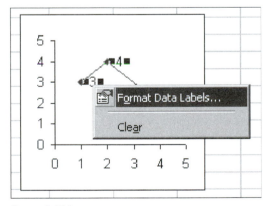

Figure A.114

Now we click twice (not double-click) on the first label, as indicated in Figure A.115 where we see the first label displayed in a box. We then click in the formula bar, type in =, and click in the cell containing the first label, Tom, as illustrated in Figure A.116. Finally, we press the Enter key and see that the name is now displayed as a label for the first point rather than the value.

We now repeat the process for the other points to obtain the desired labeling of the points of the graph.

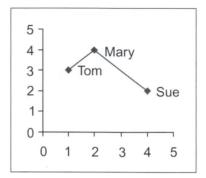

Figure A.115

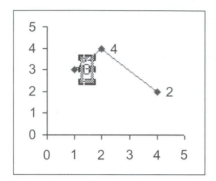

Figure A.116

2. Additional Formatting

We can also change the default settings for the markers and lines in an *xy*-graph. For example, suppose that we have created the graph shown in Figure A.117. To change the markers and lines that are used, we right-click anywhere on the series and then choose the option Format Data Series from the menu box that is generated.

Next, from the dialog box in Figure A.118, we choose the size, style, and color of the line and the marker that will be used. Here we have selected a circle marker, increased its size, and chosen a heavier line.

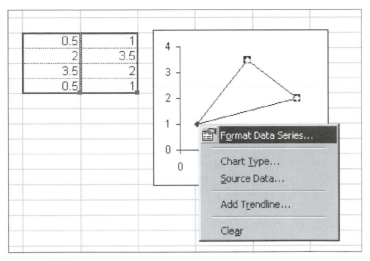

Figure A.117

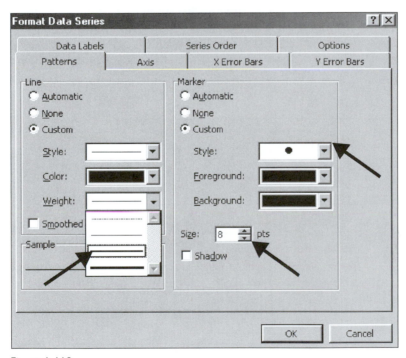

Figure A.118

The resulting graph is shown in Figure A.119. Notice that we have created a closed geometric figure by using the same (x,y)-coordinates for the first and last points.

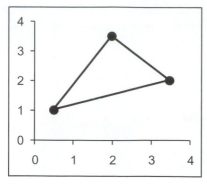

Figure A.119

3. Three-Dimensional Graphs

Although we primarily use two-dimensional graphs to illustrate our models, *Excel* can also produce three-dimensional, or surface, graphs. To illustrate the process of constructing such a graph, suppose that we wish to graph the function $f(x, y) = x^2 - y^2$. To do this we generate x-values down the first column and y-values across the top row. In creating the expression for the function, we need to fix the column reference for x and the row reference for y. To indicate doing this, we use the rail symbols as indicated in Figure A.120.

x\y	⊤-1.0	-0.8	-0.6	-0.4
◄▌ -1.0 ►	^2-▼ ^2			
-0.8				
-0.6				
-0.4				

Figure A.120

So that we can see what appears in *Excel*, we show the actual formulas (Figure A.121). Recall that we get formulas to appear on the screen by issuing the command Tools, Options, View, and then tick the option Formulas.

	A	B	C	D	E
1	x\y	-1	-0.8	-0.6	-0.4
2	-1	=$A2^2-B$1^2	=$A2^2-C$1^2	=$A2^2-D$1^2	=$A2^2-E$1^2
3	-0.8	=$A3^2-B$1^2	=$A3^2-C$1^2	=$A3^2-D$1^2	=$A3^2-E$1^2
4	-0.6	=$A4^2-B$1^2	=$A4^2-C$1^2	=$A4^2-D$1^2	=$A4^2-E$1^2
5	-0.4	=$A5^2-B$1^2	=$A5^2-C$1^2	=$A5^2-D$1^2	=$A5^2-E$1^2

Figure A.121

Over the range $-1 \le x \le 1$, $-1 \le y \le 1$, we obtain the output shown in Figure A.122. Note that we display the x- and y-values in bold. We now use the mouse to select the rest of the cells to define the graph.

x\y	-1.0	-0.8	-0.6	-0.4	-0.2	0.0	0.2	0.4	0.6	0.8	1.0
-1.0	0.0	0.4	0.6	0.8	1.0	1.0	1.0	0.8	0.6	0.4	0.0
-0.8	-0.4	0.0	0.3	0.5	0.6	0.6	0.6	0.5	0.3	0.0	-0.4
-0.6	-0.6	-0.3	0.0	0.2	0.3	0.4	0.3	0.2	0.0	-0.3	-0.6
-0.4	-0.8	-0.5	-0.2	0.0	0.1	0.2	0.1	0.0	-0.2	-0.5	-0.8
-0.2	-1.0	-0.6	-0.3	-0.1	0.0	0.0	0.0	-0.1	-0.3	-0.6	-1.0
0.0	-1.0	-0.6	-0.4	-0.2	0.0	0.0	0.0	-0.2	-0.4	-0.6	-1.0
0.2	-1.0	-0.6	-0.3	-0.1	0.0	0.0	0.0	-0.1	-0.3	-0.6	-1.0
0.4	-0.8	-0.5	-0.2	0.0	0.1	0.2	0.1	0.0	-0.2	-0.5	-0.8
0.6	-0.6	-0.3	0.0	0.2	0.3	0.4	0.3	0.2	0.0	-0.3	-0.6
0.8	-0.4	0.0	0.3	0.5	0.6	0.6	0.6	0.5	0.3	0.0	-0.4
1.0	0.0	0.4	0.6	0.8	1.0	1.0	1.0	0.8	0.6	0.4	0.0

Figure A.122

In the Chart Wizard, we select the Surface type and the upper left subtype (see Figure A.123).

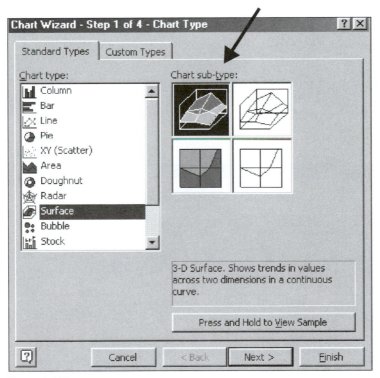

Figure A.123

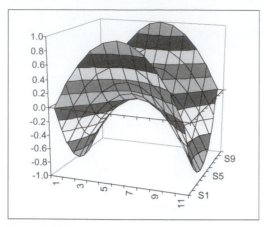

Figure A.124

After completing the wizard and removing the gray backgrounds as we did in previous graphs, we obtain the graph shown in Figure A.124. We can adjust the viewpoint in two ways. We can right-click in the chart area, choose the option 3-D View, and vary the perspective using the commands in the box shown in Figure A.126. Alternatively, we can right-click in the chart area, place the mouse pointer on one of the corner handles that appear, and then rotate the surface by dragging the corner to obtain the view shown in Figure A.125.

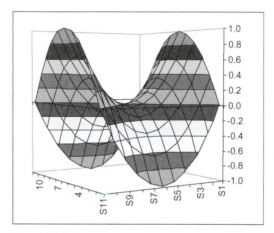

Figure A.125

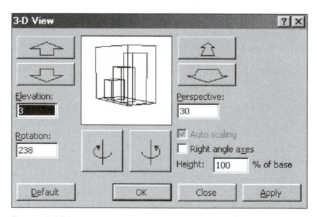

Figure A.126

A.5 Interactive Spreadsheet Tools

In this section we look at some powerful, interactive tools that are provided in *Excel*. We will examine scroll bars and buttons, the Goal Seek and Solver commands, and the implementation of simple macro programs.

1. The Scroll Bar

In our previous example for graphing the function $f(x) = \cos(ax)$, we can vary the value of the parameter a by typing new values into its cell. As we do this, the graph changes accordingly. We can improve on this technique by creating a scroll bar that is connected to the cell containing the parameter a. Then when we move the slider in the scroll bar the value of a will change, and vice versa. Since we can vary the value of the parameter in an essentially continuous manner, this provides us with a good way of adding animation effects to our dynamic models.

In fact there are two scroll bars provided in the PC version of *Excel* (the *Macintosh* version has only one). One is located on the Forms toolbar and the other is on the Control Toolbox toolbar. We first describe the one on the Forms toolbar, although the one on the Control Toolbox provides smoother animation effects.

To get the Forms toolbar to appear on the screen, we can enter the commands View, Toolbars, Forms. Alternatively, we can right-click in an empty space in the gray toolbar area at the top of the *Excel* display. In the dialog box that is produced (Figure A.127), we then click on the icon for the scroll bar.

Figure A.127

Next, we place the mouse pointer in the upper left corner of the location in the spreadsheet layout where we want to create the scroll bar. We then hold down on the left mouse button and drag to form a rectangle, and then release the mouse button. This process is illustrated in Figure A.128.

	A	B	C	D	E	F	G
1	a	0					
2	x	y					
3	0	1					
4	0.1	1	1.0				
5	0.2	1	0.5				
6	0.3	1	0.0				
7	0.4	1			5		10
8	0.5	1	-0.5				
9	0.6	1	-1.0				
10	0.7	1					

Figure A.128

After the scroll bar has been created, we right-click on it to edit its properties. From the resulting list of options that is provided, we choose the option Format Control (see Figure A.129).

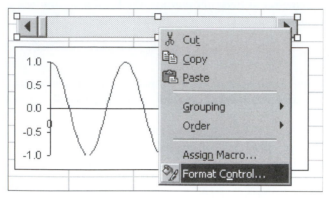

Figure A.129

This in turn gives us the dialog box that is shown in Figure A.130. The key step is for us to link the scroll bar to a cell. To do this we click inside of the bottom slot of the dialog box to enter the cell to which we link the scroll bar. To generate an entry in this cell, we then click on the cell that contains the value of the parameter a. From this point on, as we move the slider in the scroll bar the value of a will change, and

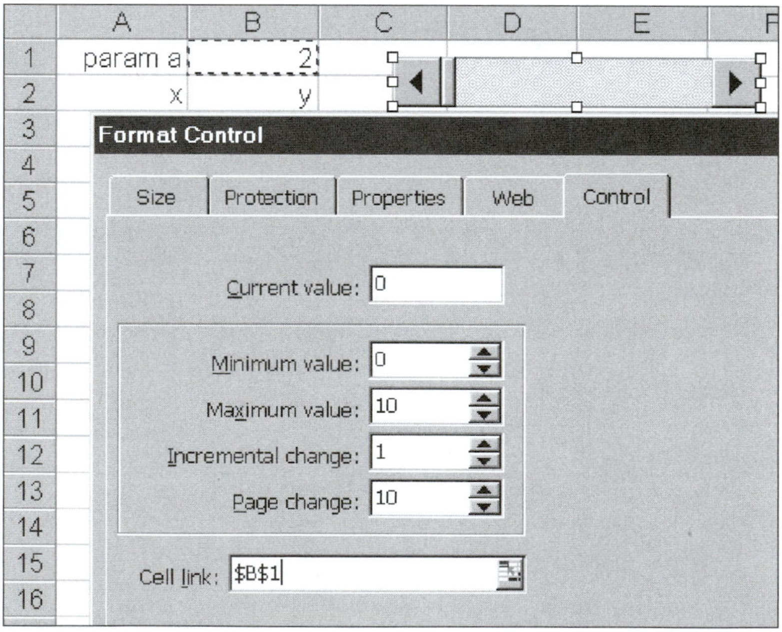

Figure A.130

vice versa. We can also establish the range over which *a* can vary. Here we choose to let *a* vary between 0 and 10. At this time we should point out that the scroll bar can only produce non-negative integer values.

If the slider on the scroll bar is moved completely to the left, then the value of *a* is set to 0. As we drag the slider to the right, the value of *a* increases through the values 0, 1, 2, up to 10. In addition, if we type a value in to the linked spreadsheet cell, then the slider on the scroll bar moves to the corresponding position on the scroll bar.

The resulting screen display with *a* = 3 is illustrated in Figure A.131.

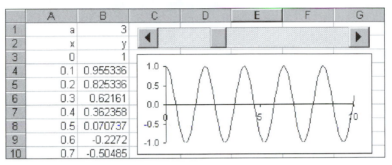

Figure A.131

However, we often will want a parameter such as *a* to take on noninteger values as well. To do this we must be somewhat clever. We use an auxiliary cell to contain an integer value that is connected to the scroll bar (see Figure A.132). For example, suppose that this integer is allowed to vary from 0 to 100 via the scroll bar. Then in the cell for the value of *a* we divide the value of the auxiliary cell by 10. Thus, as the scroll bar moves and sets the auxiliary cell to 15, the value of *a* becomes 1.5 (see Figure A.133). This way as the scroll bar varies from 0 to 100 in steps of size 1, the value of *a* varies from 0 to 10 in steps of size 0.1.

Figure A.132   Figure A.133

Similarly, if we want to have the value of *a* vary from −2.0 to 2.0 in increments of 0.01, then we could use the scroll bar to vary the value, *n*, of the auxiliary cell between 0 and 400, and generate the cell for *a* as $-2 + n/100$.

We need to observe that with these constructions we cannot enter a value for *a* directly in its cell. Doing so will replace the equation that links the value of the cell to the scroll bar.

Unfortunately, as we move the slider on the scroll bar created through the Forms toolbar, the spreadsheet is not fully recalculated until we stop moving the slider. This produces disappointing animation effects. Consequently, we will now look at the scroll bar on the Control Toolbox toolbar (see Figure A.134). We generate this as before, either selecting the commands View, Toolbars, Control Toolbox or by right-clicking in the gray toolbar area at the top of the *Excel* display.

Figure A.134

From the resulting toolbar, we must first ensure that the Design Mode button in the upper left is activated. We then click on the scroll bar button and drag it out just as before. The resulting screen display is shown in Figure A.135.

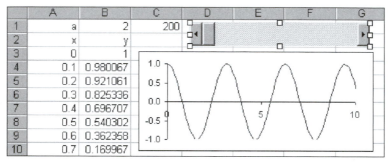

Figure A.135

Figure A.136

In this model we will generate an integer *n* in an auxiliary cell (C1) and generate the value of *a* in Cell B1 as *n*/100. We then link the scroll bar to the auxiliary cell C1, allowing it to vary over the interval between 0 and 500. In this way, *a* will vary between 0.00 and 5.00 in increments of size 0.01.

To make these settings, while we still have the Control Toolbox in Design Mode (the scroll bar will have open boxes as handles), we right-click in the scrollbar and select the option Properties in the menu that is generated (see Figure A.136)

In the Properties dialog box we must type in the location of the linked cell (C1) and set the maximum (500) and minimum (0) values that the scroll bar can attain (see Figure A.137). We may also want to set the small and large change settings (not shown) that determine the units of change in the scroll bar as we either click on its end arrows, or click in the slider track.

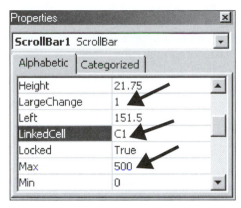

Figure A.137

After setting the values in the dialog box, we click on the Design Mode button to leave this mode. The scroll bar then becomes active, and as we move the slider the curve is redrawn in an essentially continuous fashion, producing effective animation. A view for the setting $a = 2.44$ is illustrated in Figure A.138.

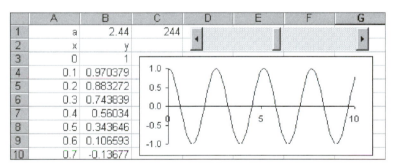

Figure A.138

2. Buttons and Macros

Another *Excel* tool that we use in different applications within the book is a button. Typically we will use this device to trigger a macro. A *macro* is a small program that we have written within the spreadsheet. We will do this by recording a series of steps that we perform in carrying out a set of spreadsheet operations. In several examples in this book we use macros to copy the values from one block of cells into another one. We will illustrate that approach here with an example designed in *Excel*.

In this example we build on earlier work in the Appendix to create another version of the compound interest computations. This time all of the work will be condensed into just a few rows that will be used repeatedly.

We begin with Figure A.139 by entering the size of a one-time deposit (1000) that earns interest at an annual rate (0.08 or 8.0%). We use the first column to count years, starting with Year 0 as the time of the initial deposit. We also enter a formula to copy the initial deposit as the balance of the account at the end of Year 0.

Next, to calculate the first year's interest we multiply the rate and the previous balance. We do not need to use an absolute reference for the rate because we are not going to copy the formula. Finally, in the right column we compute the new balance as the sum of the old balance and the new interest (Figure A.140).

deposit	rate	
1000	0.08	
year	interest	balance
0		
1+		

Figure A.139

deposit	rate	
1000	0.08	
year	interest	balance
0		1000
1		+

Figure A.140

The resulting output of our simple model is provided in Figure A.141.

deposit	rate	
1000	0.08	
year	interest	balance
0		1000
1	80	1080

Figure A.141

Next, we update the model by copying the values for Year 1 into the row above it. But before we do that, we will record our steps by issuing the command Tools, Macro, Record New Macro. When we do that, the macro dialog box displayed in Figure A.142 will appear on the screen. At this time we give the macro a name, *adjust*. When we click on the OK button all of the commands and spreadsheet operations that we perform will be recorded until we click on the Stop Record button that has been generated, or by issuing the Tools, Macro, Stop Recording command.

Figure A.142

We next want to copy the values, but not the formulas, from the Year 1 row up into the row above it, so that the Year 2 values will be generated in the bottom row. To do this, we first highlight the bottom row, as shown in Figure A.143. We then right-click in the highlighted block and select the Copy option.

Next, we highlight the row above and right-click on it, choosing the option Paste Special from the ensuing menu as shown in Figure A.144.

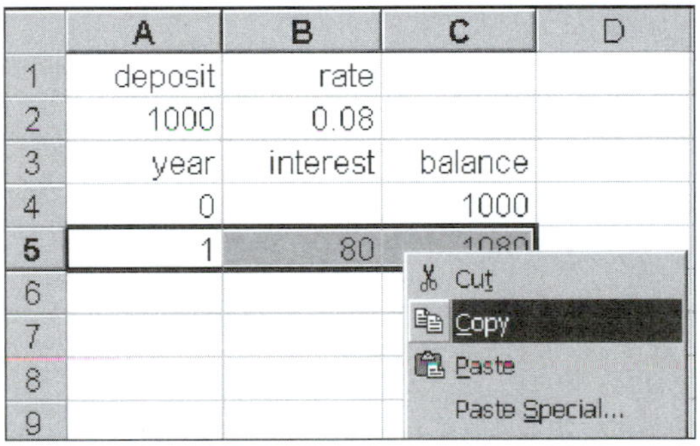

Figure A.143

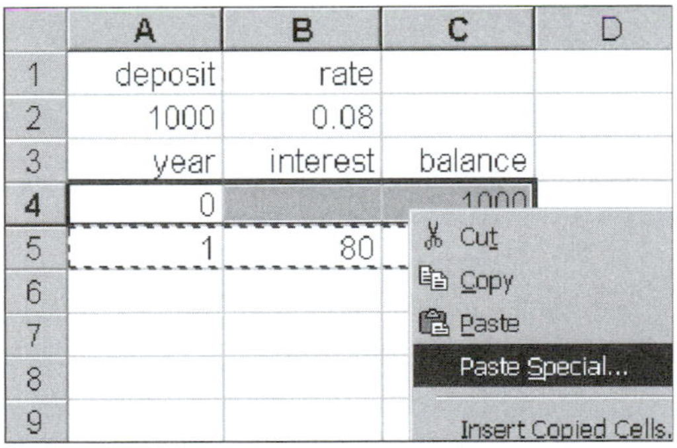

Figure A.144

Then we select the option Values (Figure A.145) to ensure that only values, rather than formulas, are copied into the top row.

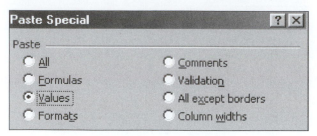

Figure A.145

We then click in another location to conclude this procedure. It is also convenient to press the [Esc] key at this time to remove any highlighting formats. We now turn the macro recorder off by the commands Tools, Macro, Stop Recording.

The resulting output is shown in Figure A.146. Notice that Year 1 now becomes the old year, and Year 2 becomes the new one.

deposit	rate	
1000	0.08	
year	interest	balance
1	80	1080
2	86.4	1166.4

Figure A.146

We can repeat this over and over again by using the same macro that we just recorded. One way to do this is to issue the commands Tools, Macro, Macros and then select the macro *adjust*. But to make this process easier we will create a button in our spreadsheet that we will use to trigger the macro.

To do this, we issue the command View, Toolbars and select the Forms toolbar. This time we click on the button icon and use the same technique that was used with the scroll bar to drag out a new button (Figure A.147).

	A	B	C	D	E
1	deposit	rate			
2	1000	0.08			
3	year	interest	balance		
4	1	80	1080		
5	2	86.4	1166.4		
6					
7		Button 1			
8					

Figure A.147

Either at a prompt provided by the program when we create the button or by right-clicking on it later, we choose the option of attaching the macro *adjust* to this button (Figure A.148).

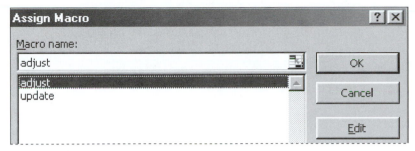

Figure A.148

We can also format the button so that it is labeled ADJUST. Now each time we click on the button, the updating process is repeated, and a new year's summary is generated. We illustrate this for the new year in the output shown in Figure A.149.

	A	B	C
1	deposit	rate	
2	1000	0.08	
3	year	interest	balance
4	2	86.4	1166.4
5	3	93.312	1259.712
6			
7		Adjust	

Figure A.149

We would probably want to create another macro with its own button to return the values of our model to their initial settings.

In using recorder macros we need to observe that they are dependent upon the specific cell locations that are used. Consequently, if we insert or delete rows and columns, or move the locations of cells referenced by a macro, after we have designed it, then the macro may not work correctly.

3. Goal Seek

Many spreadsheets have two built-in tools for solving equations that allow us to vary the value of one or more cells to reach a desired effect. For example, suppose that we consider the graph of the function $f(x) = x^3 - 2x + 0.5$ that is shown in Figure A.150.

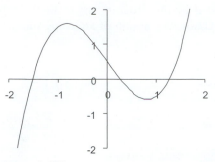

Figure A.150

One way for us to locate the zeroes of the function is to use the Goal Seek command. To do this we allocate a cell for the value of x and then enter the formula for $f(x)$ into an adjacent cell. Generally we try to find a reasonably close estimate for the zero to begin with. We show the spreadsheet construction in Figure A.151 where we start with an estimate of 0.2 for the zero. Clearly this is not the zero itself, since $f(0.2) = 0.108 \neq 0$ (see Figure A.152).

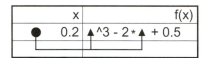

x	f(x)
0.2	▲^3 - 2*▲ + 0.5

Figure A.151

x	f(x)
0.2	0.108

Figure A.152

We next issue the Goal Seek command. When the dialog box displayed in Figure A.153 comes on the screen, we either type in constants or click in the appropriate cells on the spreadsheet. Here we say that we want to set the value of the cell containing $f(x)$ (that is, Cell B2) to the value 0 by changing the cell containing x (that is, Cell A2). We need to note that the value set must be a number; it cannot be a cell location.

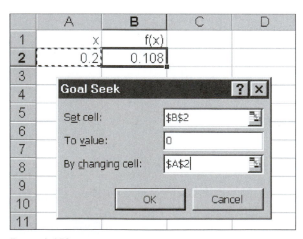

Figure A.153

After we click on the OK button, we see that the Goal Seek command has found a close approximation of the zero as $x = 0.258641$ with $f(x) = 0.0000203$. Note that a function may have several zeroes but that Goal Seek only locates one of them. The one that is located depends upon the initial estimate of the zero. To find more of the zeroes, we must adjust our initial estimate of where the zeroes are located and use the Goal Seek again (Figure A.154).

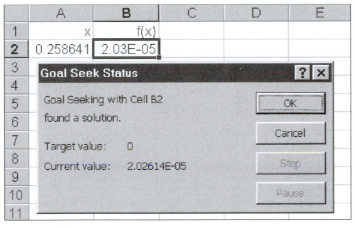

Figure A.154

4. Solver

The Solver is another *Excel* tool that provides us with even more versatility in adjusting the values of formulas. It is important to determine whether the Solver has been installed on your computer, since some standard installations do not provide it automatically. Also, because simply loading the solver uses a significant amount of memory, it may not be loaded automatically even if it is on your disk. If it is not listed under the Tools command menu option, then you can activate it through the commands Tools, Add-ins, Solver. You may need to use the program CD in the process.

With the same example and function, $f(x) = x^3 - 2x + 0.5$, as before, this time we seek to find the maximum value of the function for values of x that are in the interval $-2 \leq x \leq 0$. To do this we set as the target cell (see Figure A.155) the cell that contains $f(x)$ (that is, Cell B2). We then select the maximum option and click in the slot designated for the cells whose values change. Then we click on the cell that contains the value for x (that is, Cell A2). Although we have no need to use the facility here, the Solver also allows us to vary more than one cell in the process.

The maximum that the Solver tries to locate will be a relative maximum, that is, a point where the function is greater than nearby points. There may be several of these for a given function. Thus, we cannot be sure that we have found the overall maximum simply by using the Solver, and we must use intelligent initial estimates for the starting location.

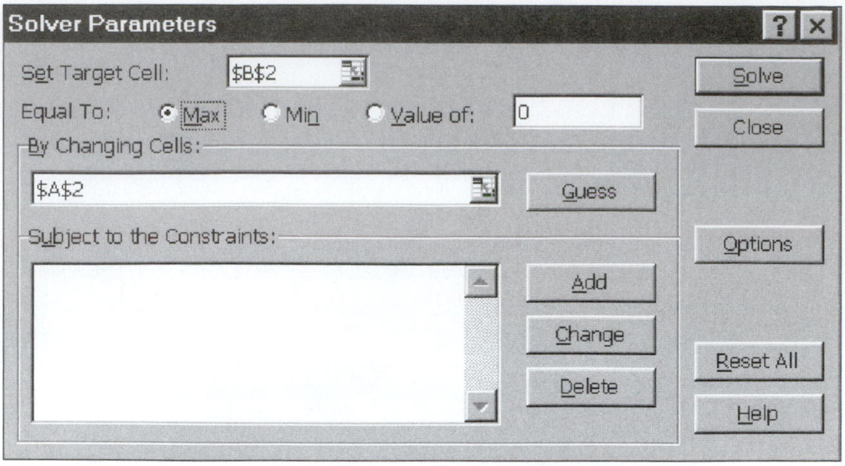

Figure A.155

Since we want to locate the largest value between the values –2 and 0, we set up two constraints by clicking on the Add button and then entering appropriate expressions and values in the subsequent boxes.

We show how to ensure that $x \leq 0$ in Figure A.156. We then construct the constraint for $x \geq -2$ similarly.

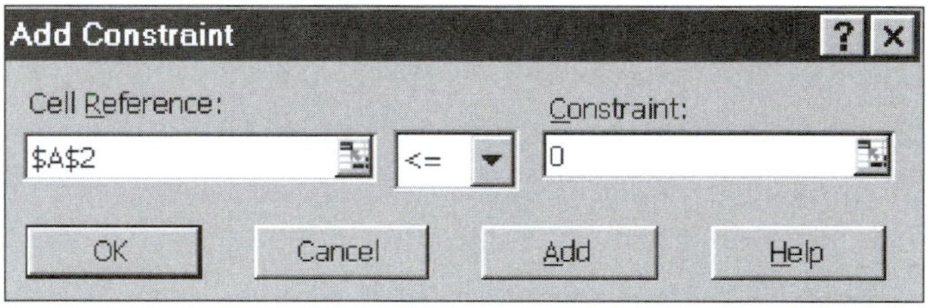

Figure A.156

We then click the OK button and see the whole setup, as shown in Figure A.157. If we confirm that the entries are correct, we then click on the Solve button.

This time we get a message indicating that the Solver has been successful (note: this will not always be the case). The resulting output is shown in Figure A.158. Note that the solver gives us additional options for the type of search techniques to employ; a choice of finding a maximum, a minimum, or the particular value for a function; and a variety of options reporting on different aspects of the solution. The Solver command also performs all of the tasks that the Goal Seek command does.

The solution that has been found is shown in Figure A.159. We can compare it to the graph of the function to see whether or not this is a reasonable answer.

The Solver and Goal Seek tools are often useful when we are investigating the graphs of functions. Illustrations of this approach are contained in the sections on graphs.

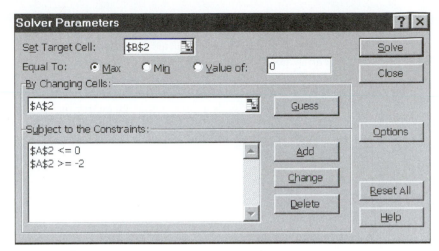

Figure A.157

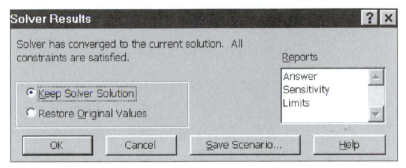

Figure A.158

x	f(x)
-0.8165	1.588662

Figure A.159

A.6   Reformatting Models for User Interaction

Our approach to creating spreadsheet models is to build them in a way that is as natural as possible. Thus, we are not overly concerned with some formatting aspects of the final layouts of our models. However, when a model is subsequently completed and will be used repeatedly, there are a number of ways in which we can improve upon its format and layout. In this section we will look at some ways to do this with one of our book's models. On the CD that accompanies this book, we have done this with amplified versions of each of our basic models.

I. General Layout

After a model has been designed, adjusted, and works correctly, it is generally a good idea to incorporate some good basic design principles to reformat the layout. In this section we will show some of the fundamental process in reformatting our logistic growth model. First, we need to thwart users from redesigning or modifying our model and provide them with a format that facilitates their ability to interpret the results as they interact with the model's values. Our initial modifications are shown in Figure A.160.

Logistic Growth			
Parameters			
	Base Growth Rate		0.05
	Population Capacity		5000
Initial Conditions			
	Initial Population		1000
Output			
	Year		6
	Population at Start		1211.8415
	Percent Saturation		0.2423683
	Growth in Population		45.906478
Computations			
Period	Animals	Saturation	Change
1	1000.00	0.200	40.0
2	1040.00	0.208	41.2
3	1081.18	0.216	42.4
4	1123.55	0.225	43.6
5	1167.11	0.233	44.7
6	1211.84	0.242	45.9
7	1257.75	0.252	47.1

Figure A.160

A model should be designed so that the parameters, and perhaps the initial conditions, are not entered within the model's computation section. Thus, as a first step we set aside an area at the top of the layout for users to input the values of the model's parameters and its initial conditions. In this example the base growth rate and the area's population capacity constitute the parameters, and the starting population provides the initial condition.

Next we set aside a section for the model's output. In some cases this section may provide locations for a user to enter numbers that indicate the particular items of output that are desired. In this case we allow a user to input a given year and read the corresponding results for that year. These come from the use of table lookup functions. For example, in the formula of Cell E13 (Figure A.162) the vertical lookup function VLOOKUP looks for the content of Cell E11 (which is 6) and returns the value from the second column of that row in the Block B20:E119. This is the population in Year 6, or 1211.84. The other functions return similar data from the other columns.

Finally, we provide an area for the computations that constitute our model. This section should be designed so that users cannot modify it easily. Indeed, in the next section we discuss how to protect anything in this area from casual or unwanted changes.

2. Formatting and Protection

As a second step, we can modify the format of our display and thereby improve its interaction efficiency for users. One approach for doing this is shown in the display in Figure A.161. In practice we make extensive use of color in the outlining and shading of various cells and blocks of cells, though here we have primarily used simple outlining. Among the other features that we have used are bolding and underlining the section headings, aligning text within cells or across blocks of cells, and inserting empty rows to separate the components of the model.

In a completed model that will be used by many individuals, it is important that the users not change the model's formulas either by accident or in attempts to modify the model's design. *Excel* and other spreadsheets have means for doing just this. In *Excel*'s default setting, each cell is "locked" while the spreadsheet's protection feature is turned off. Before turning the protection feature on, we first must unlock those cells that we want users to be able to change. Typically these are the cells that are used for input. In our model such cells are double-outlined. To carry out the protection process, we select the cells that are permitted to be changed and issue the command Format, Cells, Protection. In the resulting dialog box we remove the tick in the Locked box and click on the OK button. All other cells remain locked. We now invoke *Excel*'s protection feature by issuing the command Tools, Protection, Protect Sheet. We can even install a password if that is desired. We finish the process by clicking on the OK button. Now if anyone attempts to make an entry in a locked cell, then an error message will be generated.

Logistic Growth

Parameters

	Base Growth Rate	5.0%
	Population Capacity	5000

Initial Conditions

	Initial Population	1000

Output

	Year	6
	Population at Start	1211.8
	Percent Saturation	24.24%
	Growth in Population	45.9

Computations

Period	Animals	Saturation	Change
1	1000.00	0.200	40.0
2	1040.00	0.208	41.2
3	1081.18	0.216	42.4
4	1123.55	0.225	43.6
5	1167.11	0.233	44.7

Figure A.161

	C	D	E
11	Year		6
12			
13	Population at Start		=VLOOKUP($E$11,$B$20:$E$119,2)
14	Percent Saturation		=VLOOKUP($E$11,$B$20:$E$119,3)
15	Growth in Population		=VLOOKUP($E$11,$B$20:$E$119,4)

Figure A.162

There is yet another formatting technique that we can use to make our model more usable. While the spreadsheet's gridlines and row and column labels are generally quite useful in the process of designing the model, at this stage we can remove them so that users see only the format that we provide for them. If this is well done, our format will direct users to the input and output areas. This is illustrated in the layout in Figure A.163. Such a layout is improved further by the effective use of color.

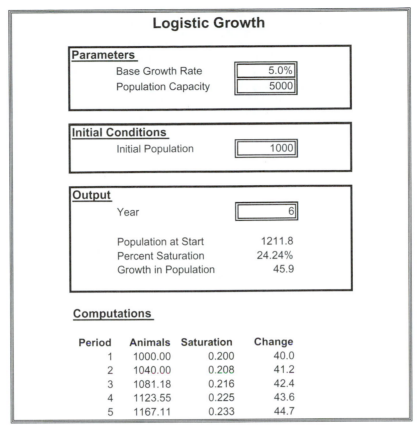

Figure A.163

3. Comments

While generally our layout itself will carry the essential ideas of the model, it is also desirable to include comments for a user in an inobtrusive way. To do this, we select the cell into which we wish to insert a comment, and right-click on it. In the resulting menu we select Insert Comment (Figure A.164). When we have done this, we then type a comment into the text box that is generated.

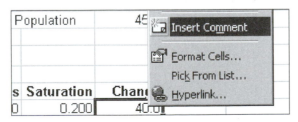

Figure A.164

This leaves a small colored triangle in the upper right corner of the cell. If we place the mouse cursor over it, the comment appears, as is illustrated in Figure A.165.

Saturation	Change	
0.200	40.0	Designer: Multiply the base growth rate by (1-staturation) obtaining the current growth rate, and multiply that by the population to get the population change.
0.208	41.2	
0.216	42.4	
0.225	43.6	
0.233	44.7	

Figure A.165

A.7 Adding Interactive Tools to Microsoft Excel Models

In this section we will illustrate the use of some of the spreadsheet tools discussed earlier in the appendix to embellish two of the introductory models of this book.

1. Logistic Growth

We will start with a slightly modified version of the logistic growth model that we developed earlier (see Figure A.166). In this version we have generated the growth projections through year 100. In addition, we have created an xy-chart from the output in Column A and Column B. Initially we opt to plot both lines and markers. After the chart has been created we reformat this series to remove the markers. Later we will add another series in which we want to use a marker, and this is easier to do if the initial construction utilizes markers.

Because we want to vary the base growth rate by using a scroll bar, we enter an integer n that lies between 0 and 1000 in Cell D3 and the formula for $n/10000$ in Cell D2. Then as n varies between 0 and 1000, Cell D2 will vary between 0 and 0.1 in increments of 0.001 (that is, between 0.0% and 10.0% in increments of 0.1%).

In Cell D6 we enter a year of interest and use a lookup function to find the population for that year from the first two columns of output.

In Column F and Column G we enter the coordinates for the endpoints of the horizontal line that shows the limit of possible population growth. The y-values are generated by a formula that reproduces the value of Cell D4 (that is, =D4). In Column I and Column J we enter formulas to reproduce the year and population from Cell D8 and Cell D9.

We enhance the graph by selecting the Block F12:G13 and dragging it into the existing chart as a new series, with the series in columns and the first column giving the x-values. We then right-click on the line and reformat it so that it is dashed and it does not include markers. Next, we repeat the process with the Block I12:J12. This

	A	B	C	D	E	F	G	H	I	J	K
1	**Parameters**										
2		Base Growth Rate		5.0%							
3		Auxiliary		500							
4		Population Capacity		5000							
5	**Initial Conditions**										
6		Initial Population		1000							
7	**Output**										
8		Year		6							
9		Population at Start		1211.8							
10	**Computations**					**Extra Graphics**					
11	**Period**	**Animals**	**Sat. Pct.**	**Change**		**Limit x**	**Limit y**		**Point x**	**Point y**	
12	1	1000.00	0.200	40.0		0	5000		6	1211.8	
13	2	1040.00	0.208	41.2		100	5000				
14	3	1081.18	0.216	42.4							
15	4	1123.55	0.225	43.6							
16	5	1167.11	0.233	44.7							
17	6	1211.84	0.242	45.9							
18	7	1257.75	0.252	47.1							
19	8	1304.82	0.261	48.2							
20	9	1353.03	0.271	49.3							
21	10	1402.38	0.280	50.5							
22	11	1452.83	0.291	51.5							
23	12	1504.36	0.301	52.6							
24	13	1556.95	0.311	53.6							
25	14	1610.56	0.322	54.6							
26	15	1665.15	0.333	55.5							
27	16	1720.68	0.344	56.4							

Figure A.166

time we show only the marker. We may also want to alter the default colors and marker styles to enhance the graph further. In the graph in Figure A.167 we have also included a value label for the point.

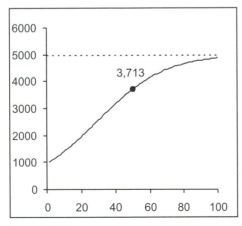

Figure A.167

Creative graphics can improve the quality of our models greatly, and the use of scroll bars allows us to include effective animation in the graphics. We will include three scroll bars in this model. To identify them we set aside a convenient area near the graph and parameter values and label them with text. We then create the scroll bars as described earlier through the Control Toolbox. To review the process, we right-click in the scroll bar area and choose the Control Toolbox option (see Figure A.168). Next, we click on the Design Mode button followed by the scroll bar button, and drag out a space for the scroll bar. We then repeat that process for the other two scroll bars. The respective scroll bars will be used to vary the base rate, the initial population, and a point on the curve.

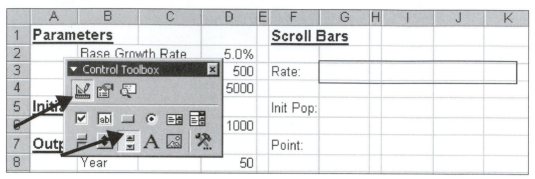

Figure A.168

When the rate scroll bar is created and we are still in the design mode, we right-click on the scroll bar and select the Properties menu option. There we set D3 as the linked cell, 0 as the minimum value, and 1000 as the maximum value. In addition, we set the large change to 50. This allows us to scroll at a faster rate by clicking in the slider track. These settings are displayed in the dialog box shown in Figure A.169.

We then repeat similar steps, so that the scroll bar for the initial population varies between 0 and 2000, while that of the selected point varies from 1 to 100.

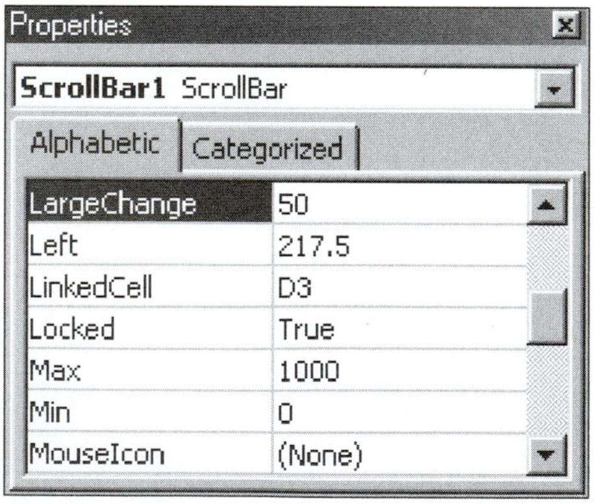

Figure A.169

The resulting screen display is shown in Figure A.170. The model allows us to vary parameters in an essentially continuous model and visualize the results through the animated graph that has been produced.

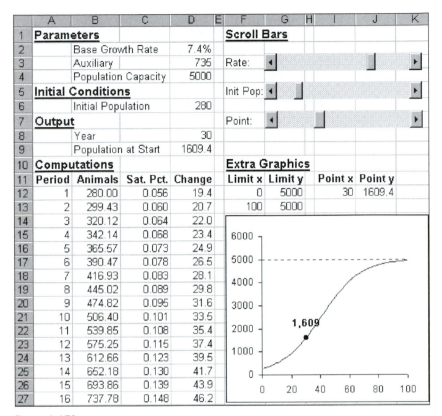

Figure A.170

2. Harvesting

In this example we will look at some of the ways in which we can incorporate formatting and spreadsheet tools into our constant harvesting model. In the screen shot in Figure A.171, we observe that we have arranged the model in the manner discussed earlier in this appendix. The model also incorporates three scroll bars. The first scroll bar is linked to an auxiliary cell for the rate (D4) and is set to vary over the range $0 \leq r \leq 1000$. The formula in Cell D3 references this cell to compute the base growth rate as $n/1000$. Thus, as the slider moves on the scroll bar, the rate varies from 0.0% to 100.0% in increments of 0.1%.

The second scroll bar is linked to Cell D5 giving the amount of the yearly harvest. This scroll bar is set to range from 0 to 200. The final scroll bar is linked to Cell D8 and varies the initial population from 0 to 500.

Recall that each year we start with a certain population. During the year a fixed number of the population that are harvested (harvest), leaving the rest (cont) to continue into the next year. In addition, new members (birth) of the population are born during the year.

	A	B	C	D	E	F	G
1							
2	**Parameters**						
3		Base growth rate		0.6	◄	▮	►
4		Auxiliary (rate)		600			
5		Yearly harvest		25	◄ ▮		►
6		Pop limit		1000			
7	**Initial Conditions**						
8		Initial population		200	◄	▮	►
9							
10	**Model Computations**						
11	**Year**	**StartPop**	**GrRate**	**Birth**	**Harvest**	**Cont**	
12	1	200.0	0.48	96.0	25.0	175.0	
13	2	271.0	0.44	118.5	25.0	246.0	
14	3	364.5	0.38	139.0	25.0	339.5	
15	4	478.5	0.31	149.7	25.0	453.5	
16	5	603.2	0.24	143.6	25.0	578.2	

Figure A.171

We have already presented a description of the construction of this model earlier in the text by using our standard notation. For comparison, Figure A.172 and Figure A.173 provide an alternate view using traditional formulas and cell references.

	A	B	C	D	E	F
11	Year	StartPop	GrRate	Birth	Harvest	Cont
12	1	=D8	=(1-B12/$D$6)*$D$3	=C12*B12	=$D$5	=B12-E12
13	=1+A12	=D12+F12	=(1-B13/$D$6)*$D$3	=C13*B13	=$D$5	=B13-E13
14	=1+A13	=D13+F13	=(1-B14/$D$6)*$D$3	=C14*B14	=$D$5	=B14-E14
15	=1+A14	=D14+F14	=(1-B15/$D$6)*$D$3	=C15*B15	=$D$5	=B15-E15
16	=1+A15	=D15+F15	=(1-B16/$D$6)*$D$3	=C16*B16	=$D$5	=B16-E16

Figure A.172

	D
3	=D4/1000

Figure A.173

Our model will be enhanced significantly by including graphs. A column chart can be used with Column A and Columns D:F to show yearly the three components of the population (new births, units harvested, and units continuing from the previous year).

The cobweb model is created in Columns H:N. First, the defining curves are computed in Columns H:J (see Figure A.174).

	H	I	J	K	L	M	N
10	**Supplementary Graphs**						
11	x	$g_1(x)$	$g_2(x)$		pt	x	y
12	0	0	-25		0	200.0	0.0
13	10	10	-9.06		1	200.0	271.0
14	20	20	6.76		0	271.0	271.0
15	30	30	22.46		1	271.0	364.5
16	40	40	38.04		0	364.5	364.5
17	50	50	53.5		1	364.5	478.5
18	60	60	68.84		0	478.5	478.5
19	70	70	84.06		1	478.5	603.2

Figure A.174

Column H generates x-values in increments of 10 (see Figure A.175), while Column I computes y-values on the line $g_1(x) - x$, and Column J produces y-values on the curve

$$g_2(x) - (1 + \text{rate})\, x - (\text{rate/capacity})\, x^2 - \text{harvest}.$$

	H	I	J
11	x	$g_1(x)$	$g_2(x)$
12	0	=H12	=(1+$D$3)*H12-($D$3/$D$6)*H12^2-$D$5
13	=10+H12	=H13	=(1+$D$3)*H13-($D$3/$D$6)*H13^2-$D$5
14	=10+H13	=H14	=(1+$D$3)*H14-($D$3/$D$6)*H14^2-$D$5
15	=10+H14	=H15	=(1+$D$3)*H15-($D$3/$D$6)*H15^2-$D$5
16	=10+H15	=H16	=(1+$D$3)*H16-($D$3/$D$6)*H16^2-$D$5

Figure A.175

The cobweb is generated in Columns L:N (see Figure A.176). Column L serves as a toggle switch, alternately taking on the values 0 and 1. When the value in this column is 1, the x-coordinate remains unchanged and the y-coordinate moves to the curve. This produces a vertical-line segment. When the value is 0 the value of y remains unchanged and x takes on that value. This produces a horizontal-line segment going from the curve to the line $y = x$.

	L	M	N
11	pt	x	y
12	0	=D8	0
13	=MOD(1+L12,2)	=IF(L13=1,M12,N13)	=IF(L13=1,(1+$D$3)*M13-($D$3/$D$6)*M13^2-$D$5,N12)
14	=MOD(1+L13,2)	=IF(L14=1,M13,N14)	=IF(L14=1,(1+$D$3)*M14-($D$3/$D$6)*M14^2-$D$5,N13)
15	=MOD(1+L14,2)	=IF(L15=1,M14,N15)	=IF(L15=1,(1+$D$3)*M15-($D$3/$D$6)*M15^2-$D$5,N14)
16	=MOD(1+L15,2)	=IF(L16=1,M15,N16)	=IF(L16=1,(1+$D$3)*M16-($D$3/$D$6)*M16^2-$D$5,N15)

Figure A.176

A view of the resulting screen display is shown in Figure A.177.

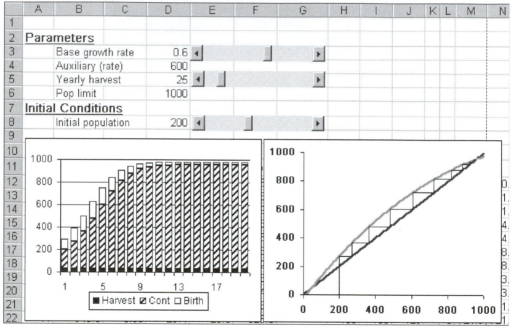

Figure A.177

References

Adair, Robert K. The Physics of Baseball, 3rd ed. New York: Perennial: Harper-Collins, 2002.

Arganbright, Deane E. *Mathematical Applications of Electronic Spreadsheets*. New York: McGraw-Hill, 1985.

Arganbright, Deane E. *Practical Handbook of Spreadsheet Curves and Geometric Constructions*. Boca Raton, FL: CRC Press, 1993.

Arganbright, Deane E. "Mathematical Modeling with Spreadsheets." *A Computer Science Reader: Selections from ABACUS*. New York: Springer-Verlag, 1988, pp. 167–179.

Balinsky, Michel L. and H. Peyton, Young. *Fair Representation: Meeting the Ideal of One Man, One Vote*. 2nd ed. Washington, DC: Brookings Institution Press, 2001.

Barnsley, Michael. *Fractals Everywhere*. San Diego, CA: Academic Press, 1988.

Beltrami, Edward. *Mathematics for Dynamic Modeling*. Boston: Academic Press, 1987.

Boyer, Carl B. and Uta C. Merzbach. *A History of Mathematics*, 2nd ed. New York: John Wylie and Sons, 1989.

Burden, Richard and J. Douglas Faires. *Numerical Analysis*, 7th ed. Pacific Grove, CA: Thomson Brooks/Cole, 1993.

Crow, James F. *Basic Concepts in Population, Quantitative, and Evolutionary Genetics*. New York: W. H. Freeman and Company, 1986.

Daley, Daryl and Joe Gani. *Epidemic Modelling*. Cambridge: Cambridge University Press, 1999.

De Mestre, Neville. *The Mathematics of Projectiles in Sport*. Cambridge: Cambridge University Press, 1990.

De Villiers, Michael and Leslie J. Nielsen. *Is Democracy Fair?: The Mathematics of Voting and Apportionment*. Emeryville, CA: Key Curriculum Press, 1997.

Dudley, Underwood. *Elementary Number Theory*, 2nd Ed. New York: W. H. Freeman and Company, 1978.

Edelstein-Keshet, Leah. *Mathematical Models in Biology*. Boston: McGraw-Hill, 1988.

Eigen, Manfred and Ruthild Winkler. *Laws of the Game*. Princeton, NJ: Princeton University Press, 1993.

Eves, Howard. *An Introduction to the History of Mathematics*, 6th ed. Fort Worth, TX: Saunders College Publishing, 1990.

Filby, Gordon, ed. *Spreadsheets in Science and Engineering*. Berlin: Springer-Verlag, 1998.

Fusaro, B. A. and P. C. Kenscraft, eds. *Environmental Mathematics in the Classroom*. Washington, DC: Mathematical Association of America, 2003.

Garfunkel, Solomon. *For All Practical Purposes,* 5th ed. New York: W. H. Freeman and Company, 2000.

Giordano, Frank R., Maurice D. Weir, and William P. Fox. *A First Course in Mathematical Modeling,* 3rd ed. Pacific Grove, CA: Thomson Brooks/Cole, 2003.

Hadlock, Charles. *Mathematical Modeling in the Environment*. Washington, DC: Mathematical Association of America, 1998.

Hardisty, J., D. M. Taylor, S. E. Metcalfe. *Computerised Environmental Modelling*. Chichester, UK: John Wiley & Sons, 1993.

Hoppensteadt, F. C. and C. S. Peskin. *Mathematics in Medicine and the Life Sciences*. New York: Springer-Verlag, 1992.

Hoppensteadt, F. C. *Mathematical Methods of Population Biology*. Cambridge: Cambridge University Press, 1982.

Horelick, Brindell and Sinan Koont. *Prescribing Safe and Effective Dosage*. UMAP Unit 72. COMAP, Lexington, MA.

Kalman, Dan. *Elementary Mathematical Models*. Washington, DC: Mathematical Association of America, 1997.

Kaplan, Daniel and Leon Glass. *Understanding Nonlinear Dynamics*. New York: Springer-Verlag, 1995.

Kreith, Kurt and Don Chakerian. *Iterative Algebra and Dynamic Modeling*. New York: Springer-Verlag, 1999.

Meerscharet, Mark M. *Mathematical Modeling*. Boston: Academic Press, Inc., 1993.

Mooney, Douglas and Randall Swift. *A Course in Mathematical Modeling*. Washington, DC: Mathematical Association of America, 1999.

Morrison, Foster. *The Art of Modeling Dynamic Systems*. New York: John Wiley & Sons, 1991.

Nielsen, Leslie Johnson and Michael de Villiers. *Is Democracy Fair? The Mathematics of Voting and Apportionment*. Emeryville, CA: Key Curriculum Press, 1997.

Additional Entries for References: Neuwirth and Arganbright

Neuwirth, Erich. "Spreadsheets as Tools in Mathematical Modeling and Numerical Mathematics," *Spreadsheets in Science and Engineering*, Gordon Filby, ed., Berlin: Springer Verlag, 1998.

Neuwirth, Erich. "Visualizing Structural and Formal Relationships with Spreadsheets," DiSessa et al, *The Design of Computational Media to Support Exploratory Learning*, Berlin: Springer-Verlag, 1995.

Neuwirth, Erich. "Visualizing Correlation with Spreadsheet," *Teaching Statistics,* Vol. 12, 1990.

Pielou, E. C. *Mathematical Ecology*. New York: John Wiley & Sons, 1977.

Reinhardt, H. E. "Some Statistical Paradoxes, Teaching Statistics and Probability." *NCTM 1981 Yearbook,* Reston, VA: National Council of Teachers of Mathematics, 1981.

Renshaw, Eric. *Modelling Biological Populations in Space and Time*. Cambridge: Cambridge University Press, 1991.

Saaty, Thomas L. and Joyce Alexander. *Thinking with Models*. Oxford: Pergamon Press, 1981.

Sandefur, James T. *Discrete Dynamical Systems: Theory and Applications*. Oxford: Clarendon Press, 1990.

———. *Discrete Dynamical Modeling*. New York: Oxford University Press, 1993.

———. *Elementary Mathematical Modeling*. Pacific Grove, CA: Thomson Brooks/Cole, 2003.

Sigmund, Karl. *Games of Life: Explorations in Ecology, Evolution, and Behaviour*. Oxford, UK: Oxford University Press, 1993.

Smith, J. Maynard. *Mathematical Models in Ecology*. Cambridge: Cambridge University Press, 1974.

Stewart, B. M. *Theory of Numbers*. New York: The Macmillan Company, 1952.

Uspensky, J. V. *Introduction to Mathematical Probability*. New York: McGraw-Hill Book Company, 1937.

Von Neumann, John and Oskar Morgenstern. *Theory of Games and Economic Behavior*. Princeton, NJ: Princeton University Press, 1953.

Winston, Wayne and S. Christian Albright. *Practical Management Science*, 2nd ed. Pacific Grove, CA: Thomson Brooks/Cole, 2001.

Zill, Dennis. *A First Course in Differential Equations with Modeling Applications*. 7th ed. Pacific Grove, CA: Thomson Brooks/Cole, 2001.

Web Sites

http://sunsite.univie.ac.at/Spreadsite/ Erich Neuwirth's wide-ranging site with many spreadsheet references and links.

http://www.j-walk.com/ss/ John Walkenbach's spreadsheet site, providing many tips and utilities.

http://www.xl-logic.com/ Exploring advanced concepts in spreadsheet formulas and logic.

http://office.microsoft.corn/ Microsoft's Office Web Site, with information about *Excel*.

http://www.sie.bond.edu.au Web Site of the journal Spreadsheets in Education.

Index